IET ENERGY ENGINEERING 117

Structural Control and Fault Detection of Wind Turbine Systems

Other volumes in this series:

Volume 1 **Power Circuit Breaker Theory and Design** C.H. Flurscheim (Editor)
Volume 4 **Industrial Microwave Heating** A.C. Metaxas and R.J. Meredith
Volume 7 **Insulators for High Voltages** J.S.T. Looms
Volume 8 **Variable Frequency AC Motor Drive Systems** D. Finney
Volume 10 **SF_6 Switchgear** H.M. Ryan and G.R. Jones
Volume 11 **Conduction and Induction Heating** E.J. Davies
Volume 13 **Statistical Techniques for High Voltage Engineering** W. Hauschild and W. Mosch
Volume 14 **Uninterruptible Power Supplies** J. Platts and J.D. St Aubyn (Editors)
Volume 15 **Digital Protection for Power Systems** A.T. Johns and S.K. Salman
Volume 16 **Electricity Economics and Planning** T.W. Berrie
Volume 18 **Vacuum Switchgear** A. Greenwood
Volume 19 **Electrical Safety: A guide to causes and prevention of hazards** J. Maxwell Adams
Volume 21 **Electricity Distribution Network Design, 2nd Edition** E. Lakervi and E.J. Holmes
Volume 22 **Artificial Intelligence Techniques in Power Systems** K. Warwick, A.O. Ekwue and R. Aggarwal (Editors)
Volume 24 **Power System Commissioning and Maintenance Practice** K. Harker
Volume 25 **Engineers' Handbook of Industrial Microwave Heating** R.J. Meredith
Volume 26 **Small Electric Motors** H. Moczala *et al.*
Volume 27 **AC-DC Power System Analysis** J. Arrillaga and B.C. Smith
Volume 29 **High Voltage Direct Current Transmission, 2nd Edition** J. Arrillaga
Volume 30 **Flexible AC Transmission Systems (FACTS)** Y.-H. Song (Editor)
Volume 31 **Embedded generation** N. Jenkins *et al.*
Volume 32 **High Voltage Engineering and Testing, 2nd Edition** H.M. Ryan (Editor)
Volume 33 **Overvoltage Protection of Low-Voltage Systems, Revised Edition** P. Hasse
Volume 36 **Voltage Quality in Electrical Power Systems** J. Schlabbach *et al.*
Volume 37 **Electrical Steels for Rotating Machines** P. Beckley
Volume 38 **The Electric Car: Development and future of battery, hybrid and fuel-cell cars** M. Westbrook
Volume 39 **Power Systems Electromagnetic Transients Simulation** J. Arrillaga and N. Watson
Volume 40 **Advances in High Voltage Engineering** M. Haddad and D. Warne
Volume 41 **Electrical Operation of Electrostatic Precipitators** K. Parker
Volume 43 **Thermal Power Plant Simulation and Control** D. Flynn
Volume 44 **Economic Evaluation of Projects in the Electricity Supply Industry** H. Khatib
Volume 45 **Propulsion Systems for Hybrid Vehicles** J. Miller
Volume 46 **Distribution Switchgear** S. Stewart
Volume 47 **Protection of Electricity Distribution Networks, 2nd Edition** J. Gers and E. Holmes
Volume 48 **Wood Pole Overhead Lines** B. Wareing
Volume 49 **Electric Fuses, 3rd Edition** A. Wright and G. Newbery
Volume 50 **Wind Power Integration: Connection and system operational aspects** B. Fox *et al.*
Volume 51 **Short Circuit Currents** J. Schlabbach
Volume 52 **Nuclear Power** J. Wood
Volume 53 **Condition Assessment of High Voltage Insulation in Power System Equipment** R.E. James and Q. Su
Volume 55 **Local Energy: Distributed generation of heat and power** J. Wood
Volume 56 **Condition Monitoring of Rotating Electrical Machines** P. Tavner, L. Ran, J. Penman and H. Sedding
Volume 57 **The Control Techniques Drives and Controls Handbook, 2nd Edition** B. Drury
Volume 58 **Lightning Protection** V. Cooray (Editor)
Volume 59 **Ultracapacitor Applications** J.M. Miller
Volume 62 **Lightning Electromagnetics** V. Cooray
Volume 63 **Energy Storage for Power Systems, 2nd Edition** A. Ter-Gazarian
Volume 65 **Protection of Electricity Distribution Networks, 3rd Edition** J. Gers
Volume 66 **High Voltage Engineering Testing, 3rd Edition** H. Ryan (Editor)
Volume 67 **Multicore Simulation of Power System Transients** F.M. Uriate
Volume 68 **Distribution System Analysis and Automation** J. Gers
Volume 69 **The Lightening Flash, 2nd Edition** V. Cooray (Editor)
Volume 70 **Economic Evaluation of Projects in the Electricity Supply Industry, 3rd Edition** H. Khatib
Volume 72 **Control Circuits in Power Electronics: Practical issues in design and implementation** M. Castilla (Editor)
Volume 73 **Wide Area Monitoring, Protection and Control Systems: The enabler for smarter grids** A. Vaccaro and A. Zobaa (Editors)
Volume 74 **Power Electronic Converters and Systems: Frontiers and applications** A.M. Trzynadlowski (Editor)
Volume 75 **Power Distribution Automation** B. Das (Editor)
Volume 76 **Power System Stability: Modelling, analysis and control** B. Om P. Malik
Volume 78 **Numerical Analysis of Power System Transients and Dynamics** A. Ametani (Editor)
Volume 79 **Vehicle-to-Grid: Linking electric vehicles to the smart grid** J. Lu and J. Hossain (Editors)
Volume 81 **Cyber-Physical-Social Systems and Constructs in Electric Power Engineering** S. Suryanarayanan, R. Roche and T.M. Hansen (Editors)
Volume 82 **Periodic Control of Power Electronic Converters** F. Blaabjerg, K.Zhou, D. Wang and Y. Yang
Volume 86 **Advances in Power System Modelling, Control and Stability Analysis** F. Milano (Editor)
Volume 87 **Cogeneration: Technologies, optimisation and implementation** C.A. Frangopoulos (Editor)
Volume 88 **Smarter Energy: From smart metering to the smart grid** H. Sun, N. Hatziargyriou, H.V. Poor, L. Carpanini and M.A. Sánchez Fornié (Editors)
Volume 89 **Hydrogen Production, Separation and Purification for Energy** A. Basile, F. Dalena, J. Tong and T.N. Veziroðlu (Editors)
Volume 90 **Clean Energy Microgrids** S. Obara and J. Morel (Editors)
Volume 91 **Fuzzy Logic Control in Energy Systems with Design Applications in MATLAB®/Simulink®** İ.H. Altaş
Volume 92 **Power Quality in Future Electrical Power Systems** A.F. Zobaa and S.H.E.A. Aleem (Editors)
Volume 93 **Cogeneration and District Energy Systems: Modelling, analysis and & optimization** M.A. Rosen and S. Koohi-Fayegh
Volume 94 **Introduction to the Smart Grid: Concepts, technologies and evolution** S.K. Salman
Volume 95 **Communication, Control and Security Challenges for the Smart Grid** S.M. Muyeen and S. Rahman (Editors)
Volume 97 **Synchronized Phasor Measurements for Smart Grids** M.J.B. Reddy and D.K. Mohanta (Editors)
Volume 98 **Large Scale Grid Integration of Renewable Energy Sources** A. Moreno-Munoz (Editor)
Volume 100 **Modeling and Dynamic Behaviour of Hydropower Plants** N. Kishor and J. Fraile-Ardanuy (Editors)
Volume 101 **Methane and Hydrogen for Energy Storage** R. Carriveau and D.S.-K. Ting
Volume 104 **Power Transformer Condition Monitoring and Diagnosis** A. Abu-Siada (Editor)
Volume 108 **Fault Diagnosis of Induction Motors** J. Faiz, V. Ghorbanian and G. Joksimoviæ
Volume 110 **High Voltage Power Network Construction** K. Harker
Volume 111 **Energy Storage at Different Voltage Levels** Technology, integration, and market aspects Ed. by Zobaa, Ribeiro, Aleem and Afifi
Volume 112 **Wireless Power Transfer: Theory, technology and application** N. Shinohara
Volume 117 **Structural Control and Fault Detection of Wind Turbine Systems ed. by** Prof Hamid Reza Karimi
Volume 119 **Thermal & Power Plant Control and Instrumentation: The control of boilers and HRSGs, 2nd Edition** D. Lindsley, J. Grist and D. Parker
Volume 123 **Power Systems Electromagnetic Transients Simulation, 2nd Edition** N. Watson and J. Arrillaga
Volume 124 **Power Market Transformation** B. Murray
Volume 130 **Wind and Solar Based Energy Systems for Communities** R. Carriveau and D.S.-K Ting (Editors)
Volume 131 **Metaheuristic Optimization in Power Engineering** J. Radosavljević
Volume 905 **Power System Protection, 4 volumes**

Structural Control and Fault Detection of Wind Turbine Systems

Edited by
Hamid Reza Karimi

The Institution of Engineering and Technology

Published by The Institution of Engineering and Technology, London, United Kingdom

The Institution of Engineering and Technology is registered as a Charity in England & Wales (no. 211014) and Scotland (no. SC038698).

First published 2018

The Institution of Engineering and Technology
Michael Faraday House
Six Hills Way, Stevenage
Herts, SG1 2AY, United Kingdom

www.theiet.org

British Library Cataloguing in Publication Data
A catalogue record for this product is available from the British Library

ISBN 978-1-78561-394-4 (hardback)
ISBN 978-1-78561-395-1 (PDF)

Typeset in India by MPS Limited

To my wife Baharak

Contents

About the editor

Hamid Reza Karimi, born in 1976 in Saveh, Iran, received the B.Sc. (First Hons.) degree in power systems from the Sharif University of Technology, Tehran, Iran, in 1998, and the M.Sc. and Ph.D. (First Hons.) degrees in control systems engineering from the University of Tehran, Tehran, Iran, in 2001 and 2005, respectively. From 2009 to 2016, he was a full professor of Mechatronics and Control Systems at University of Agder in Grimstad, Norway. Since 2016, he has been a professor of Applied Mechanics with the Department of Mechanical Engineering, Politecnico di Milano, Milan, Italy. His current research interests include control systems and mechatronics with applications to automotive control systems and wind energy.

Dr. Karimi is a member of the Agder Academy of Science and Letters and also a member of the IEEE Technical Committee on Systems with Uncertainty, the IEEE Technical Committee on Industrial Cyber-Physical Systems, the IFAC Technical Committee on Mechatronic Systems, the IFAC Technical Committee on Robust Control, and the IFAC Technical Committee on Automotive Control. He is a Senior Member of IEEE and awarded the 2016 and 2017 Web of Science Highly Cited Researcher in Engineering, Alexander-von-Humboldt Research Fellowship (in Germany), JSPS Research Award (in Japan), DAAD Research Fellowship (in Germany) and August-Wilhelm-Scheer Award (in Germany). Dr Karimi has been invited as visiting professor at a number of universities including University of Bremen (in Germany), Harbin Institute of Technology (in China), Technical University of Munich (in Germany), University of Picardie Jules Verne (in France), and University of Lorraine (in France). He is currently the *Editor-in-Chief* of the Journal of *Cyber-Physical Systems* (Taylor & Francis), the *Editor-in-Chief* of the Journal of *Machines* (MDPI Switzerland), *Editor-in-Chief* of the Journal of *Designs* (MDPI Switzerland), and an Editorial Board Member for some international journals, such as the *IEEE Transactions on Industrial Electronics*, the *IEEE Transactions on Circuit And Systems – I: Regular Papers*, the *IEEE/ASME Transactions on Mechatronics*, *IEEE Transactions on Systems, Man, and Cybernetics: Systems*, *Information Sciences*, *IFAC-Mechatronics*, *Neurocomputing*, the *Asian Journal of Control*, the *Journal of The Franklin Institute*, the *International Journal of Control, Automation, and Systems*, the *International Journal of Fuzzy Systems*, the *International Journal of e-Navigation and Maritime Economy*, and the *Journal of Systems and Control Engineering*. Dr. Karimi has published more than 500 articles in peer-reviewed international scientific journals including *IEEE Transactions and Automatica*.

Chapter 1
Introduction

Hamid Reza Karimi[1]

Electricity plays a very important role in the human life, and how to generate it by effective methods is an interested case for the scientists all over the world. Wind-energy deployment through wind turbines has attracted global interest in recent years; for instance, according to The World Wind Energy Association the worldwide wind energy reached a capacity of 254,000 MW in June 2012 [1]. And, this capacity follows an increasing trend due to the high environmental and political impacts of the wind-energy production.

Currently, most large wind turbines around the world are installed on land with sparse population and vast land. However, in many countries, most inhabitants are concentrated in places along coastlines where land is scarce and power is in high demand. Therefore, utilizing offshore wind resources is more beneficial which will both reduce electricity transmission loss and reserve more land space for people, animals and plants. Besides, offshore wind quality has been evaluated to be much better than that onshore. According to [2], a wind farm located offshore could experience wind speeds that are, on average, 90% greater than that over land. Therefore, it is widely considered that global wind-energy exploitation and development will gradually move to offshore [3]. On this line, cost analysis and environmental potentials make a major influence on feasibility and applicability of the wind-energy resources. On top of that, integration of social and technical developments of this energy has been highly considered in both academic and industrial researches. For instance, this integration should be implemented in planning, designing and construction of wind turbines or wind farms.

Wind turbine as a large-scale system is an aeroelastic mechanical system that converts kinetic wind energy into electrical power. Performance limitation of wind turbines in dealing with some problems such as vibration, corrosion and temperature changes, e.g. in the motors, sensors, blades and gearbox could affect the production capability and may cause remarkable downtime of the entire system. Therefore, considering the high cost of wind-turbine maintenance, increasing the reliability of wind-turbine operation becomes a key point for wind-farm operators. In other words, the maintenance strategy and reliability of wind turbine are depending on each

[1]Department of Mechanical Engineering, Politecnico di Milano, Italy

other, which leads us to the reliability-centred maintenance (RCM) and implement an optimized maintenance program for wind turbine.

Over the last few years, many researches were carried out by academicians or practitioners on implementation of RCM for wind turbines and the literature of this work is condensed by developing algorithms or methodologies in theory or practical aspects for dealing with structural load analysis or health monitoring of wind turbines for diagnostics or prognostics of various faults or possible failures in the system. To this aim, the current book is devoting to present some recent results in the field of structural control and fault detection of wind turbines. In the following, a review on the organization of the book and main contributions of each chapter is presented in detail.

Strong potentials for offshore wind energy have been found in the deep-sea areas. Although offshore wind farms (OWFs) are advantageous in many ways, their availability is still not comparable to their onshore counterparts. It was found that while the onshore availability is reaching 99%, the OWFs for the United Kingdom at Barrow North, Hoyle, Kentish Flats have availability figures between 67% and 85% [4]. The downtimes are longer because of the finite weather windows in which personnel can perform the required maintenance. Besides, the maintenance costs are also comparatively higher at the range of 18%–23% of the life-cycle costs owing to their location and associated difficulty of maintenance [5]. According to extensive experiences in offshore industry, floating foundations for wind turbines are considered as an economical and applicable solution. So far, plenty of numerical investigations in [6,7] have been conducted by world-wide research institutions, and different kinds of prototype programs have also been launched, including OC3-Hywind, MIT/NREL TLP, ITI Barge and Principle Power WindFloat, etc.

The authors in the chapter entitled 'Wave loads on monopile-supported offshore wind turbines: current methods and future challenges' provide an overview of the methods for the prediction of the wave loads on fixed-bottom wind turbines. More specifically, the role of nonlinear wave contributions in the assessment of the dynamic response of the support structure is considered in this study. Also, the principles of wave motions will be presented to numerically solve the fully nonlinear problem and the main commonly used analytical wave theories are reviewed. Then, the authors discuss on the available hydrodynamic loading models and an overview of the main literature findings and open issues in modelling of the nonlinear resonant effects is provided. Some key examples of the effects of highly nonlinear waves on the response of a 5-MW offshore wind turbine (OWT) will be presented before concluding the chapter with the current situation and future trends of the study.

It is well known that the design process of wind turbines is a multidisciplinary problem; the chapter entitled 'Numerical and experimental tools for small wind turbine load analysis' is mainly devoted on the design of a small diffuser-augmented wind turbine, a kind of turbines being of great interest on the prosumer market, specifically for modest wind speeds prevailing in Poland. Specifically, multiple methods were applied to design and verify the construction, ranging from an application of simple analytical methods through complex numerical simulations, to finish with the experimental tests of models in different scales. Moreover, an influence of the turbine duct (diffuser) made it impossible to use standard methods of turbine design and forced one to combine analytical methods with computational fluid dynamics (CFD) predictions of the flow. The authors have used a set of different numerical models, ranging from the 2D CFD ADM (actuator disk model) up to 3D FRM (fully resolved rotor model), which took the most important elements of the final construction into account. Those simulations provided data on aerodynamic loads, further used to verify the design with analytical methods. Some trials of the load analysis employing the fluid–structure interaction were carried out. They proved to be useful for the design load analysis. However, it is shown that significant requirements of computer resources for simulations of the complex turbine model as well as high deformations of polyamide blades limited the practical application of that method within this study so far. Nevertheless, continuous progress in the computer performance and numerical methods, combining the fluid flow and structural solvers, should provide tools for detailed design analysis in the near future. At the same time, the Fourier transform analysis can be performed on the rotor power signal to provide the fault tolerance and to increase the turbine load factor.

So far, most of the large wind turbines in the world have only been installed on land or shallow offshore sites. However, stronger potentials have been found in the deep-sea areas. Main advantages of installing large wind turbines in the deep-sea area include strong and consistent wind with less turbulence, lack of constraints, no visual or noise annoyance, near to high-populated coastal cities. Consequently, large attention in the wind industry has been diverted to the promising deep offshore wind resource.

According to the extensive experience of marine industry, floating foundations for wind turbines are considered to be an economical and applicable choice in deep-sea areas. Several traditional floating platforms have been proposed to support large wind turbines in deep offshore regions, including spar-buoy, tension leg, barge, semi-submersible, etc. Different with fixed-bottom wind turbines, the big challenge for floating wind turbines is the platform motion, which will heavily increase the load on the nacelle and tower due to the high inertial and gravitational forces. Large platform motion will cause severe fatigue and ultimate loading on the tower base and nacelle-tower bearing and even cause the failure of pitch control strategy. Therefore, an effective control strategy for platform motion reduction is necessary for the design of floating wind turbines.

It is well known in the literature that a more direct approach for wind-turbine load reduction is to adopt structural control devices, such as tuned mass dampers (TMDs). These resonance absorbers have been widely used in large civil structures, e.g. skyscrapers and bridges, thus are also seen promising to deal with wind-turbine load problem, especially for offshore types. For high-fidelity simulation, the state-of-the-art numerical wind turbine simulator FAST was modified by Lackner [8], in order to characterize the dynamic influence of passive and active structural control. With this so-called code FAST-SC, different parameter-optimization methods were used to find the optimal parameters of passive structural control devices, see [9]–[10]. However, passive TMD has to bring a huge mass into the system, usually made of concrete or steel, which is unfavourable to the system cost. It was also shown that more load reduction could be achieved when introducing active structural control [11], but the improvement is at the expenses of more power consumption and longer mass stroke. Besides, active mass damper highly depends on the actuation control strategy, which might bring instability to the system. Regarding these issues, semi-active structural control device, such as tuned-liquid column damper (TLCD), shows a compromise due to its low cost and decent performance. In [12], Colwell *et al.* explored the structural responses of a fixed-bottom OWT with a TLCD installed in nacelle.

The authors in the chapter entitled 'Structural control concept for load reduction of offshore wind turbines' emphasized on the structural control concept for load reduction of OWTs. Theories for both passive and active structural control are introduced. Particularly, a spar-type floating OWT (FOWT) is used as the study case to demonstrate the load mitigation effectiveness of the proposed structural control methods. For the passive case, design optimization process is performed on a TMD installed in the spar platform, and the obtained numerical simulation results have indicated both their effectiveness and limits regarding different system parameters and installations. Regarding the active case, a gain scheduling H_2/H_∞ active structural full-state feedback controller is designed for a hybrid mass damper installed at the tower top of a spar-type FOWT, aiming at both reducing tower bottom load and mitigating the aerodynamic disturbance. Results demonstrate that more load reduction could be achieved at the expense of more energy consumption. On the other hand, the full-state feedback controller is not very practical from a technical point of view due to the lack of sensors and measurement inaccuracy. Therefore, the authors have proposed a number of research directions to further investigate different structural control designs for load reduction of OWTs.

The purpose of the chapter entitled 'Advanced control of wind turbine system' is to present information about design, implementing and testing of advanced control algorithm for wind turbines to engineers. First, the state of-the-art wind-turbine controller and the available control design methods are provided. Then, an advanced wind-turbine pitch controller for loads reduction by using available control methods is explained including some simulation cases for showing the performance of the presented control algorithm. Moreover, a generator torque control for the purpose of drivetrain vibration suppression is described. Finally, the sensorless generator control techniques are presented for the drivetrain control.

The main objective in the chapter entitled 'Magnetic bearing for wind turbine power generator shaft: an emulator prototype design for vibration control' is to present a practical demonstration of the relevant role that can be played by micro-sized wind turbines (mSWTs) in research and education. To this end, the authors have designed an experimental platform for active magnetic bearing vibration control of a rotary shaft, which provides an excellent illustration of the mechatronical multidomain approach and, at the same time, makes it possible to perform advanced full-scale experimental studies on fundamental aspects of mSWTs, such as vibration control, fault detection or health monitoring. The platform prototype has been designed following the principles of cost-efficiency and limited usage of resources, which can be critical factors for its practical application. To illustrate the effectiveness of the proposed approach, a simple vibration control strategy has been implemented using a low-cost digital microcontroller.

In chapter entitled 'Toward farm-level health management of wind turbine systems: status and scope for improvements', the authors investigate a farm-level health management approach for increasing availability of wind turbines and reducing maintenance costs of OWFs simultaneously. While the condition monitoring (CM) or health monitoring is the knowledge of whether the system is healthy or faulty, the health management is the capability to make intelligent, informed, appropriate decisions about maintenance and logistics actions based on diagnostics/prognostics information, available resources and operational demand [13]. It is improving from current CM of specific zones of the WT to a solution that handles maintenance of the whole wind farm, intelligently. The CM systems should feed into maintenance planning rather than operate as end-products to assist in reactive maintenance tasks. In order to achieve such farm-level health management, an architecture based on RCM and condition-based maintenance methodologies is presented in this chapter. The architecture describes the necessary elements of a health management scheme that plans maintenance tasks and correspondingly the resources, schedules and logistics based on the asset health and failure predictions.

In literature, several fault detection systems have been successfully developed for wind turbines. Most of the proposed approaches rely on the physical model of wind turbines and based on which, the well-established model-based fault diagnosis techniques can be directly applied, see [14]–[23]. However, the authors in the chapter entitled 'Health monitoring of wind turbine: data-based approaches' study the health-monitoring issue of wind turbine. Specifically, a robust data-based fault detection approach is proposed with application to a wind-turbine benchmark. The main challenges of the wind-turbine fault detection lie in its nonlinearity, unknown disturbances as well as significant measurement noise. To overcome these difficulties, a subspace-aided fault detection scheme is designed with robust residual generators directly constructed from available process data. A performance index and an optimization criterion are given to achieve the robustness of the residual signals related to the disturbances. For the residual evaluation, a proper evaluation approach as well as a suitable decision logic is provided to make a correct final decision. The effectiveness of the proposed approach is finally illustrated by simulations on the wind-turbine benchmark model.

In the chapter entitled 'Fault diagnostics for electrically operated pitch systems in offshore wind turbines', the authors investigate the feasibility and applicability of current signature analysis for pitch motors in typical operating profiles. In order to determine pitch system operation profiles, the 5-MW reference wind turbine is simulated in FAST analysis tools developed by National Renewable Energy Laboratory (NREL) [24]. The pitch systems however pose significant challenge in terms of intermittent, start-stop operating profiles and low-speed operations. The main contribution of this chapter is therefore twofold: (1) to develop a detailed physical modelling of various motor faults and study their effect on motor currents in pitch system operating profiles; and (2) to determine the feasibility of current signature analysis in such operating profiles.

The authors in the chapter entitled 'Condition monitoring and diagnostics of wind turbine power train' focus on CM, fault identification and failure analysis of wind turbines via signal-processing tools. The main objective is to study maintenance policies and diagnostics tools in order to minimize the maintenance costs and operation failures. Furthermore, in the chapter entitled 'Robust fuzzy fault tolerant control wind energy system subject to actuator and sensor faults', the authors' focus is on the development of robust fuzzy fault tolerant controllers for wind energy systems subject to time-varying actuator faults, sensor faults and parameter uncertainties and maximizing the output power from the wind turbine.

Finally, it is understood that the selected topics and chapters are not a comprehensive representation of the area of this book on structural control and fault diagnostics methods for wind-turbine systems, they represent the rich and many-faceted knowledge that we have the pleasure of sharing with the readers.

References

[1] Wilkes J, Pineda I, Corbetta G. Wind energy scenarios for 2020. *A Report by European Wind Energy Association (EWEA).* 2014.

[2] Archer C, Jacobson M. Evaluation of global wind power. Journal of Geophysical Research. 2005, 110, D12110, doi:10.1029/2004JD005462.

[3] Kaldellis JK, Kapsali M. Shifting towards offshore wind energy – recent activity and future development. Energy Policy. 2013, 53, 136–148.

[4] Tavner PJ, Long H, Feng Y. Early experiences with UK round 1 offshore wind farms. Proceedings of the ICE – Energy, Nov. 2010, 163, 167–181.

[5] van Bussel G, Henderson AR. State of the art and technology trends for offshore wind energy: operation and maintenance issues, in [Online] Delf University of Technology, European Commission, 2001.

[6] Jonkman J, Matha D. A quantitative comparison of the responses of three floating platform concepts. In: European Offshore Wind 2009 Conference & Exhibition. 2014, p. 14–6, 1985.

[7] Jonkman J. Dynamics modeling and loads analysis of an offshore floating wind turbines. *Ph.D. thesis. Department of Aerospace Engineering Sciences, University of Colorado.* 2007.

[8] Lackner MA, Rotea MA. Passive structural control of offshore wind turbines. Wind Energy. 2011, 14(3), 373–388.

[9] Si Y, Karimi HR, Gao, H. Modelling and optimization of a passive structural control design for a spar-type floating wind turbine. Engineering Structures. 2014, 69, 168–182.

[10] Stewart G, Lackner M. Offshore wind turbine load reduction employing optimal passive tuned mass damping systems. IEEE Transactions on Control Systems Technology. 2013, 21(4), 1090–1104.

[11] Lackner MA, Rotea MA. Structural control of floating wind turbines. Mechatronics. 2011, 21(4), 704–719.

[12] Colwell S, Basu B. Tuned liquid column dampers in offshore wind turbines for structural control. Engineering Structures. 2009, 31(2), 358–368.

[13] Integrated Product Support Element Guidebook. *Defense Acquisition University, Department of Defense (US).* p. 265–301, Dec. 2011.

[14] Lu B, Li Y, Wu X, *et al.* A review of recent advances in wind turbine condition monitoring and fault diagnosis. In: PEMWA 2009. IEEE. IEEE; 2009. p. 1–7.

[15] Odgaard PF, Stoustrup J. Unknown input observer based scheme for detecting faults in a wind turbine converter. In: Proceedings of the 7th IFAC Symposium on Fault Detection, Supervision and Safety of Technical processes. vol. 1; 2009. p. 161–166.

[16] Polinder H, Lendenmann H, Chin R, *et al.* Fault tolerant generator systems for wind turbines. In: Electric Machines and Drives Conference, 2009. IEMDC09. IEEE International. IEEE; 2009. p. 675–681.

[17] Johnson KE, Fleming PA. Development, implementation, and testing of fault detection strategies on the National Wind Technology Centers controls advanced research turbines. Mechatronics. 2011, 21(4), 728–736.

[18] Hameed Z, Hong Y, Cho Y, *et al.* Condition monitoring and fault detection of wind turbines and related algorithms: a review. Renewable and Sustainable Energy Reviews. 2009, 13(1), 1–39.

[19] Hatch C. Improved wind turbine condition monitoring using acceleration enveloping. Orbit. 2004, 61, 58–61.

[20] Yang W, Tavner P, Wilkinson M. Condition monitoring and fault diagnosis of a wind turbine synchronous generator drive train. IET Renewable Power Generation. 2009, 3(1), 1–11.

[21] Yang W, Tavner PJ, Crabtree CJ, *et al.* Cost-effective condition monitoring for wind turbines. IEEE Transactions on Industrial Electronics. 2010, 57(1), 263–271.

[22] Caselitz P, Giebhardt J, Mevenkamp M, *et al.* Application of condition monitoring systems in wind energy converters. In: EWEC-Conference. Bookshop for Scientific Publications; 1997. p. 579–582.

[23] Schlechtingen M, Santos IF. Comparative analysis of neural network and regression based condition monitoring approaches for wind turbine fault detection. Mechanical Systems and Signal Processing. 2011, 25(5), 1849–1875.

[24] Jonkman J, Butterfield S, Musial W, Scott G. Definition of a 5-mw reference wind turbine for offshore system development. *Technical Report NREL/TP-500-38060, National Renewable Energy Laboratory.* 2009.

Chapter 2

Wave loads on monopile-supported offshore wind turbines: current methods and future challenges

Enzo Marino[1], Agota Mockutė[1], Claudio Borri[1], and Claudio Lugni[2,3]

2.1 Introduction

Wind is an important source of energy accounting for 3.7% of the total energy consumption (in 2015) and 10.4% of the European energy market on an average year [1]. Moreover, 2016 was the unique year when, after the record breaking investment in offshore wind turbines (OWTs), the price of offshore wind energy fell below the cost of onshore wind [2]. Indeed, anticipating the target for 2020, already in 2016 the price of MW h fell well below 100 euros [2]. However, the 8–12-MW wind-turbine requirements of larger rotors and taller towers make the whole system increasingly sensitive to dynamic effects induced by complex environmental conditions. In this context, more advanced simulation tools become fundamental to design safer and more cost-efficient structures with longer lifetime.

In the OWT technology, the turbine can be placed on a variety of support structures, the two main categories being fixed and floating. Most common types are illustrated in Figure 2.1: fixed monopile, jacket and tripod, and floating semisubmersible, TLP (tension leg platform) and spar buoy. The optimal choice depends on a number of factors, among which water depth and soil conditions are the most important ones. In shallow-water condition, i.e. for water depth up to 30 m, the monopile is the most common solution because of easier design and low costs of installation and maintenance. In the common engineering practice, under the assumption of long waves, an extended Morison equation is used to evaluate the wave-related loads for the monopile OWT, with the wave kinematics providing the velocity and acceleration fields around the structure. In mild environmental conditions, the linear wave is used. On the contrary, when highly nonlinear waves with breaking and near-breaking events occur, possibly with slamming phenomena, the use of nonlinear models able to properly predict the wave kinematics as well as the local wave–structure interaction

[1]Department of Civil and Environmental Engineering, University of Florence, Italy
[2]CNR-INSEAN, Marine Technology Research Institute – Italian Research Council, Italy
[3]NTNU-AMOS, Center for Autonomous Marine Operation Systems, Norway

Figure 2.1 Different offshore wind turbine support structures. Illustration by Joshua Bauer, National Renewable Energy Laboratory

become unavoidable. In such cases, the wave modelling must account for the nonlinear contributions and requires a larger computational effort.

The aim of this contribution is to provide an overview of the methods recently proposed for the prediction of the wave loads on fixed-bottom wind turbines, with particular emphasis on the role of nonlinear wave contributions in the assessment of the dynamic response of the support structure. After a brief presentation of the common numerical approaches to the simulation of OWTs given in Sections 2.2 and 2.3, we review the fundamental equations governing the wave motions, briefly present a numerical method for the numerical solution of the fully nonlinear (FNL) problem and finally summarize the main commonly used analytical wave theories. In Section 2.4, the available hydrodynamic loading models are discussed, and in Section 2.5 an overview of the main literature findings and open issues in modelling the nonlinear resonant effects follows. Section 2.6 presents some key examples of the effects of highly nonlinear waves on the response of a 5-MW OWT. Finally, the conclusive Section 2.7 summarizes the current situation and future trends.

2.2 Hydro–aero–servo–elastic coupled simulation models

The design of OWTs requires predictions of the coupled system dynamic response as well as of the extreme and fatigue loads associated with given environmental conditions [3,4]. Fixed-bottom OWT is a complex system whose dynamical behaviour involves the coupling between deformable structures, i.e. the monopile and the wind turbine, forced by time-varying aero- and hydrodynamic loads, with a proper control strategy for the pitch angle of the rotor blades. The monopile is not rigidly clamped to the sea bottom, which means that soil–structure interaction matters. A comprehensive mathematical and numerical model should account for two phase flows, air and water, each described by means of the solution of Navier–Stokes equations, interacting through the air–water interface and acting on local and global deformable structures, possibly with large deformations (typical for wind-turbine blades). Such a fully integrated model is highly complex and unfeasible to be used on the time scales characteristic for the environmental conditions where the fixed-bottom OWTs typically operate, that is short- or long-crested sea with turbulent wind. Indeed, this requires long simulation time (usually some hours in full scale) to allow for a proper statistical analysis of the results.

A remarkable effort is therefore made to increase the accuracy of design-oriented tools which are based on simplified and semi-empirical models. Computer-aided engineering tools such as Bladed (DNV-GL), FAST (NREL), ADAMS (MSC), HAWC2 (DTU WE) and Flex5 (DTU Lingby) are typically used by the offshore wind community for the fixed-bottom OWT design [5]. Blade Element Momentum theory and linear wave model with Morison equation are the most commonly used approaches to calculate the aerodynamic forces on the blades and the hydrodynamic loads on the monopile, respectively. Bladed, Flex5 and FAST are modal-based solvers, for which the structural dynamics are approximated by mode shapes with modal masses, damping and stiffness. On the contrary, HAWC2 and ADAMS are based on multi-body assumptions, i.e. the flexible structure is described as a finite number of rigid bodies linked with inelastic joints. In this way, the PDE system giving the structural dynamics is approximated by a finite number of ordinary differential equations. More details are available in [5].

Because of the open-source strategy and the possibility to include additional modules, our studies use FAST [6] as the reference solver. In particular, our main contribution, which is also the focus of the present chapter, addresses the issues related to the modelling of the hydrodynamic loads acting on the monopile, possibly causing hydro-structural as well as aero–hydro-structural coupling. Generally, FAST uses the Morison equation to predict the nonlinear hydrodynamic loads on a slender cylinder. Several authors [7–10] stressed the importance of the higher order wave loads, i.e. up to third order, at small and medium wave steepness, relating to the needs of satisfying the nonlinear free-surface conditions. However, looking at large-wave steepness, recently Kristiansen and Faltinsen [10] highlighted an important role of the flow separation around the cylinder [larger Keulegan–Carpenter (*KC*) numbers at higher steepness] and the subsequent interaction with the free-surface flow. This causes a local run-up and steep local wave, the associated load of which is believed to be the origin of significant higher order loads (higher than third order).

2.3 Wave kinematics

2.3.1 Governing equations

An inertial coordinate system with origin in o is set with the xy-plane along the still-water level and the z-axis vertical and upwardly oriented.

From Euler's equations valid for an incompressible and inviscid fluid (see e.g. [11]) the additional hypothesis of irrotational flow allows one to describe FNL water waves by means of a potential flow model, i.e. Laplace equation for the potential function $\phi(t,p)$ holds in the whole hydrodynamic field $\Omega(t)$:

$$\nabla^2\phi(t,p) = 0 \quad \forall p \in \Omega(t). \tag{2.1}$$

The initial boundary value problem (BVP) is completed by the boundary conditions and the solution at the initial time instant. In particular, for a water domain with depth h, the impermeability condition at the bottom reads

$$\nabla\phi \cdot \bar{n} = 0, \tag{2.2}$$

where $\cdot$ indicates the standard scalar product and $\bar{n}$ is the unit outward normal vector.

On the free-surface Γ_f, the unknown potential function and deformation z_f, often also denoted by η, require that both dynamic

$$\frac{D\phi}{Dt}(t,p) = -\frac{p_a}{\rho} - gz_f + \frac{1}{2}\nabla\phi(t,p)\cdot\nabla\phi(t,p) \quad \forall p \in \Gamma_f, \tag{2.3}$$

and kinematic

$$\frac{D\bar{r}}{Dt}(t,p) = \bar{v}(t,p) = \nabla\phi(t,p) \quad \forall p \in \Gamma_f, \tag{2.4}$$

boundary conditions are satisfied, with D/Dt indicating the Lagrangian (total) derivative. Here, p_a is the atmospheric pressure and ρ the water density. By rewriting the Lagrangian derivative as $D/Dt = \partial/\partial t + \nabla\phi\cdot\nabla$, it is immediate to recast the Eulerian formulation. Radiation conditions on the free-surface far-field and a prescribed solution at $t = 0$ complete the mathematical formulation of the water-wave problem.

2.3.2 Method of solution

One of the most efficient approaches to the numerical solution of the FNL potential flow initial BVP is based on the mixed Eulerian–Lagrangian (MEL) scheme which consists of a repeated two-step procedure. This method of solution was first proposed by Longuet–Higgins and Cokelet in [12] and then further developed by Dold and Peregrine [13] who introduced a new efficient procedure to integrate in time the dynamic and kinematic boundary conditions on the free-surface with the advantage of solving the Laplace equations involved at each time step by making use of the same system matrix. After these pioneering works, the method continued to be developed over the years, significantly contributed by [14–18]. A wider literature survey is available in [19,20].

The two steps of the MEL scheme are as follows:

1. *Eulerian step:* At a fixed time t, the free-surface Γ_f and the velocity potential $\phi_{|p\in\Gamma_f}$ on it are known. A Laplace BVP for the potential ϕ can then be solved to determine the solution in the whole field.
2. *Lagrangian step:* Velocity potential and free-surface profile are updated in time providing boundary conditions for the next Eulerian step by means of common time integration schemes, e.g. Runge–Kutta fourth order.

2.3.2.1 The HOBEM solver

Several methods can be used to solve the BVP for the Laplace equation at the Eulerian step. Among them the most widely used are the finite-element (e.g. [21]) and the boundary-element methods (e.g. [22]). Recently, a novel efficient and accurate solver based on the use of the harmonic polynomial cell has been proposed in connection with the water wave problem [23]. The related numerical challenges for steep wave have been detailed in [24]. In our studies, we numerically solved the BVP at the Eulerian step by using the direct boundary-element method [22]. Numerical solutions presented in the next sections are obtained by implementing a high-order boundary-element method (HOBEM) in which the domain boundary is discretized into quadratic elements, consisting of three nodes each, therefore providing enhanced performances in the case of severe deformation of the free-surface, such as the case of breaking waves.

Laplace equation is turned into the following boundary integral equation (BIE)

$$c(p_c)\phi(p_c) + \int_\Gamma \phi(p)q^*(p,p_c)\mathrm{d}\Gamma - \int_\Gamma \phi^*(p,p_c)q(p)\mathrm{d}\Gamma = 0, \tag{2.5}$$

where $\phi^*(p,p_c)$ is the Green function, p_c is the collocation point, $q = \nabla\phi \cdot \bar{n}$ and $q^* = \nabla\phi^* \cdot \bar{n}$ [22]. The BIE can be discretized by using constant, linear or higher order boundary elements. For example, in the case of isoparametric quadratic elements, the discretized version of (2.5) becomes

$$c(p_c)\phi(p_c) + \sum_{j=1}^{n_e}\int_{\Gamma_j}\sum_{k=1}^{3}\varphi_k(s)\phi_k^{(j)}q_{p_c}^{*(j)}\mathrm{d}\Gamma - \sum_{j=1}^{n_e}\int_{\Gamma_j}\sum_{k=1}^{3}\varphi_k(s)q_k^{(j)}\phi_{p_c}^{*(j)}\mathrm{d}\Gamma, \tag{2.6}$$

where n_e is the number of boundary elements, and φ_k with $k = 1,2,3$ being the quadratic shape functions such that on the jth element we have the following approximations $\phi^{(j)} = \sum_{k=1}^{3}\varphi_k(s)\phi_k^{(j)}$ and $q^{(j)} = \sum_{k=1}^{3}\varphi_k(s)q_k^{(j)}$. If the collocation point p_c is applied to all n_n boundary nodes in sequence, (2.6) becomes

$$c_i\phi_i + \sum_{j=1}^{n_e}\sum_{k=1}^{3} h_{ik}^{(j)}\phi_k^{(j)} - \sum_{j=1}^{n_e}\sum_{k=1}^{3} g_{ik}^{(j)}q_k^{(j)} = 0 \quad \text{with} \quad i = 1,2,\ldots,n_n, \tag{2.7}$$

where the following definitions have been set

$$h_{ik}^{(j)} = \int_{-1}^{1} q_i^{*(j)}(s)\varphi_k(s)j^{(j)}(s)\mathrm{d}s, \tag{2.8}$$

$$g_{ik}^{(j)} = \int_{-1}^{1} \phi_i^{*(j)}(s)\varphi_k(s)j^{(j)}(s)\mathrm{d}s, \tag{2.9}$$

with $j^{(j)}$ being the Jacobian necessary to transform the integral over the element Γ_j into an integral over the normalized domain $[-1, 1]$ of the shape functions.

After an assembling process [25], (2.7) can be written as $H_{ij}\phi_j = G_{ij}q_j$ where ϕ_j and q_j are the nodal velocity potential and flux, respectively. After collecting the unknowns in a vector $\bar{x}$ made of ϕ_j on the boundaries where Neumann conditions are assigned and q_j where the Dirichlet conditions are assigned, a standard algebraic system is obtained in the form $A_{ij}x_j = b_i$, where A_{ij} is the system matrix and b_i the known boundary conditions. Corner potential continuity [15,26] and smoothing and regridding techniques [12,27,28] are also normally introduced to ensure the numerical stability. A more exhaustive presentation of the HOBEM solver can be found in [25].

We conclude this section by emphasizing that the HOBEM solver briefly summarized above and the consequent time integration do not introduce any approximation neither on the velocity potential nor on the dimension of typical wave parameters such as wave height, wave period, water depth and wavelength.

2.3.3 Analytical wave theories

Exact solution of the FNL initial boundary potential flow problem discussed above does not exist in general. The simplest and most commonly used approach to find analytical solutions to the problem consists of linearizing the problem assuming that the wave height H is much smaller than the wavelength L and the water depth h. There is a wide selection of excellent textbooks which present wave theories in a systematic and detailed way; therefore, in the following we simply provide a brief summary of the main analytical theories typically used, while for full details the reader may refer to [19,20,29–31] among others.

Linear (or Airy) wave theory: Widely used theory at the design stage. Also known as small amplitude wave theory, it is based on the linearization of the water wave BVP around the undisturbed free-surface configuration. A basic solution for the linear wave model is represented by a progressive sinusoidal (single-harmonic or regular) wave with a certain amplitude, period-wavelength (related through the dispersion relation) and phase. Because of the linearization, it is strictly valid for waves of small amplitude in deep to intermediate waters (refer to Figure 2.2).

Stokes theory: Also known as finite amplitude wave theory. Introduced by Stokes in 1847, it is based on a series perturbation expansion of the nonlinear water wave problem around the undisturbed free-surface configuration. The perturbation parameter is the wave steepness $\varepsilon = ka$, with k the wavenumber and a the wave amplitude; larger wave steepness means higher nonlinear contributions. Referring to Figure 2.2, Stokes theory is normally used until the third, fifth order in infinite and intermediate water depth, enabling a better prediction of higher wave steepness. The considered high-order effects become more significant causing steeper wave crests and shallower troughs, thus losing the wave elevation symmetry along horizontal axis and, because of the Eulerian formulation of the free-surface boundary conditions, still keeping the symmetry along the vertical axis.

Cnoidal theory: Stokes theory has some restrictions on the applicability in shallow waters. The cnoidal theory supplies a proper description for finite amplitude long waves in shallow waters. It is limited in wave steepness by Airy theory on one

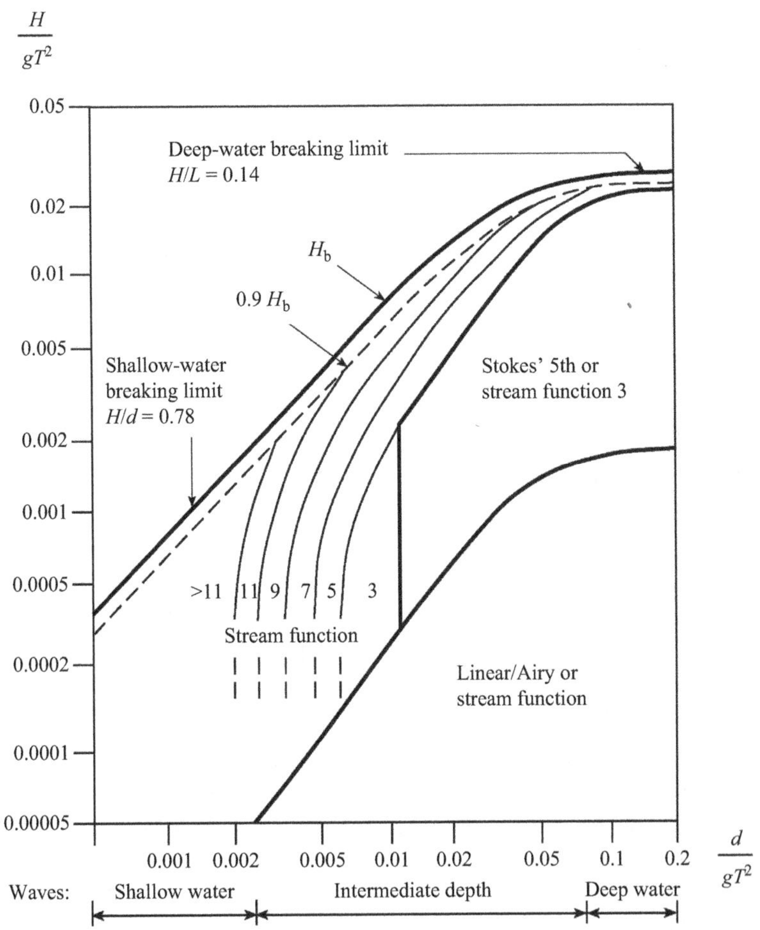

Figure 2.2 Applicability of wave theories. H, T, d, g denote the wave height, wave period, water depth, gravity acceleration, respectively. Figure from [4][1]

end and a solitary wave on the other end. A cnoidal wave has a typical shape consisting of sharper crests separated by wide troughs, but still it maintains its symmetry along vertical axis.

Stream function: Similarly to Stokes theory, the stream function satisfies the Laplace equation. It is derived from the stream function representation of the

[1]The authors thank the International Electrotechnical Commission (IEC) for permission to reproduce Information from its International Standards. All such extracts are copyright of IEC, Geneva, Switzerland. All rights reserved. Further information on the IEC is available from www.iec.ch. IEC has no responsibility for the placement and context in which the extracts and contents are reproduced by the author, nor is IEC in any way responsible for the other content or accuracy therein.

flow rather than potential in Stokes theory. The choice of a reference system moving with the wave allows for the solution of a steady free-surface problem. As a consequence, it provides a better fit of the free-surface boundary conditions than the Stokes theory, with a perfect fit of the kinematic boundary condition. It can then be extended to higher order terms depending on the accuracy required without any stability problem, e.g. it is suited for the shallow-water depths where Stokes theory fails, and, depending on its order, can be appropriate for higher wave heights than Stokes in the deep water, as evident from Figure 2.2.

Fully nonlinear (Rienecker–Fenton): FNL wave kinematic model can be achieved by implementing Rienecker–Fenton wave theory [32]. It is a perturbation theory with no analytical approximation except for the truncation of Fourier series; therefore, the model is valid for deep and shallow waters and any wave steepness, with exceptions for the solitary wave limit and re-entering of breaking wave. It allows for full nonlinearity of the wave profile with sharper, nearly vertical crests and much shallower troughs, which lead to more extreme associated wave kinematics.

The use of an inappropriate wave theory may lead to major inaccuracies in the wave kinematics and in the assessment of associated loads. The limits of each of the above theories are shown in Figure 2.2 on a normalized wave height to normalized water depth graph.

2.3.4 Nondeterministic representation of waves

If we measure the instantaneous free-surface elevation at any point in the open sea, we immediately realize that deterministic models, as those described above, fail in reproducing the actual nature of the waves which, instead, is intrinsically random. A unidirectional random wave can be represented by linear superposition of infinite regular waves with random phase angle. The linear wave theory is straightforwardly used to obtain the wave elevation of a long-crested irregular sea propagating along the direction x as follows:

$$\eta(x,t) = \sum_{m=1}^{N} A_m \cos(k_m x - \omega_m t - \varepsilon_m), \tag{2.10}$$

where A_m, ω_m, k_m and ε_m are the wave amplitude, circular frequency, wavenumber and random phase angle of the mth wave component, respectively. The amplitude A_m of each single-harmonic wave is related to an energy density spectrum $S(\omega)$ usually expressed in terms of wave frequencies (or wave periods) as follows $A_m = \sqrt{2S(\omega)\Delta\omega}$ where $\Delta\omega$ is the constant step size used to discretize the spectrum $S(\omega)$. Pierson–Moskowitz and JONSWAP are two of the most commonly used spectra for design applications [4,20,33–36].

For the second-order wave theory, a formulation was proposed by Sharma and Dean in 1981 [37] based on a perturbation approach. At a fixed location, the free-surface elevation $\eta(t)$ is described as a sum of a first-order term $\eta_1(t)$ [linear term given in (2.10)] and a second-order term $\eta_2(t)$. The second-order contributions to the wave elevation are obtained as summations of differences and sums of frequencies as follows:

$$\eta_2(t) = \eta_2^-(t) + \eta_2^+(t), \tag{2.11}$$

where

$$\eta_2^{\pm}(t) = \sum_{m=1}^{N}\sum_{n=1}^{N} [A_m A_n \{B_{mn}^{\pm} \cos((\Psi_m \pm \Psi_n)\}], \tag{2.12}$$

in which $\Psi_m = \omega_m t - \varepsilon_m$ and $\Psi_n = \omega_n t - \varepsilon_n$. A_m, A_n, ω_m and ω_n are, respectively, the amplitude and the circular frequency of the *m*th and *n*th wave components. B_{mn}^- and B_{mn}^+ are the second-order transfer functions [37]. Similarly, the second-order velocity potential, not reported here, is obtained as a sum of a first- and a second-order terms $\Phi(t) = \Phi_1(t) + \Phi_2(t)$. Details on the implementation can be found in [38].

In the case of short-crested and multidirectional sea, the extension of the linear and second-order model is quite straightforward, requiring a two-dimensional spectrum $S(\omega, \theta) = S(\omega)f(\theta)$, with $f(\theta)$ giving the spreading function related to the angle θ measuring the wave propagation direction of the single-wave component.

We emphasize that the higher the order of the wave model, the higher the complexity of the methods to reproduce irregular waves. Typically for nondeterministic models larger than second order, numerical models are required. The most efficient is definitely the higher order spectral (HOS) method [39,40]. It solves the nonlinear wave propagation problem modelled by the Euler equations. In particular, the associated initial and BVP for Laplace equation with nonlinear boundary conditions on the free-surface is solved by means of a perturbation expansion of the potential velocity field up to a prescribed order of nonlinearities. The HOS method solves for nonlinear wave–wave interactions (up to an arbitrary specified order *M*) of a large number $N = O(1{,}000)$ of free waves (i.e. Fourier modes) on the undisturbed free-surface plane. A pseudo-spectral technique is used to solve the associated BVP. This ensures a computational cost which scales linearly with N and M, guaranteeing high computational efficiency for large spatial domains. Recently, it has been successfully used to predict the occurrence of rogue waves during the 1-h sea state of Hurricane Joaquin [41], demonstrating its ability on large-scale simulation of real ocean waves.

2.3.5 A domain decomposition approach to account for fully nonlinear random waves

To account for the FNL effects, the numerical solver presented in Sections 2.3.2 and 2.3.2.1 should be used on the entire spatial domain and for the whole duration of the time-domain simulation. Moreover, multiple simulations are needed to get a statistically representative response, resulting in an unfeasible approach at the design stage. To circumvent this problem, in [42] a domain decomposition strategy was proposed based on coupling the irregular linear analytic wave model (2.10) with the FNL HOBEM solver. The domain decomposition algorithm is summarized as follows: statistics of met-ocean data, or simplified sea state classification, provide values for the mean wind speed, significant wave height and peak spectral period

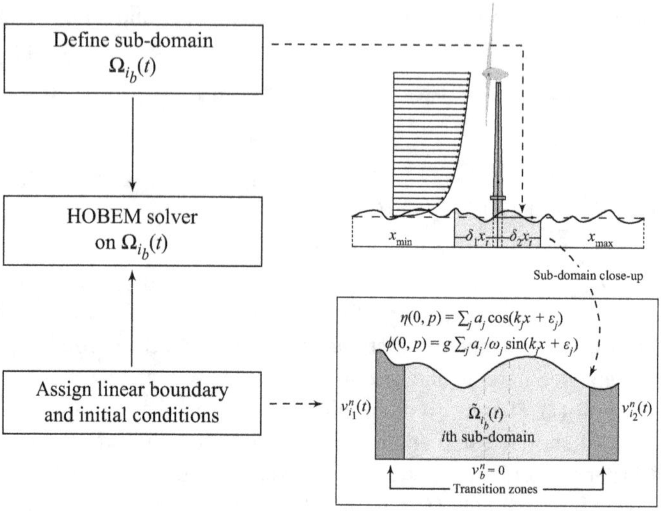

Figure 2.3 Schematic representation of the domain decomposition method for a generic sub-domain Ω_{i_b}. Figure from [43]

to be used as input to the model. A suitable wave spectrum is chosen and the linear theory is used to simulate a long-crested irregular sea on the global space–time domain. A zero-crossing analysis of the wave elevation time history in a small area around the monopile location identifies temporal sub-domains over which nonlinear waves are expected. On these sub-domains, the linear wave theory is only used to provide initial boundary conditions to the HOBEM solver whose results in terms of velocity and acceleration are used to assess the actual hydrodynamic forces acting on the monopile. A schematic representation of the domain decomposition approach is shown in Figure 2.3 and full details on the method can be found in [42].

2.4 Hydrodynamic loading models

2.4.1 Morison equation

Morison equation [44] is the most commonly used method to account for hydrodynamic loading on slender body. It is given by

$$F = \int_{-h}^{\eta} C_m \rho \pi R^2 \frac{\partial u}{\partial t} dz + \int_{-h}^{\eta} C_d \rho R \, |u| \, u dz, \tag{2.13}$$

where η is instantaneous wave elevation, h the water depth, ρ the water density, C_m the inertia coefficient, C_d the drag coefficient, R the cylinder radius and u the horizontal velocity.

Equation (2.13) is a semi-empirical formula consisting of linear inertia component and quadratic drag which are integrated along the length of the pile. Depending on the regime in which the slender structure is placed, i.e. inertia or drag dominated, one of these components is usually neglected. The inertia and drag coefficients, needed in the formula, must be estimated by the help of experiments and/or theoretical best practice guidelines. This loading model is commonly used due to its simplicity and fast implementation and has been deemed suitable for most cases. However, when it comes to more complex cases such as third-order effects of steep waves or shallow waters, the suitability of Morison equation has been doubted [9,45–47].

2.4.2 Slender body theory

Free-surface area is not only where forces and arm for the overturning moment are the largest, but also where nonlinear effects gather both in kinematics and in loading; therefore, it requires special attention in the studies. To account for the free-surface and other nonlinear effects, in 1989 Rainey [7] proposed corrections to Morison equation and further developed the model later in 1995 [48].

The theory separates the inertia term from Morison equation and adds an axial divergence force A, (2.14), integrated along the length of the pile to the instantaneous free water surface, and a contribution due to the surface intersection I, (2.15), applied at the instantaneous free-surface, which together correct the linear Morison inertia term to second order in Stokes expansion. Finally a third-order term D, (2.16), is also applied at the instantaneous free-surface, which accounts for the surface distortion due to the piercing cylinder.

$$A = \int_{-h}^{\eta} \rho \pi R^2 \frac{\partial w}{\partial z} u dz, \tag{2.14}$$

$$I = -\frac{\rho \pi R^2}{2} u^2 \frac{\partial \eta}{\partial x}, \tag{2.15}$$

$$D = \frac{7 \rho \pi R^2}{2g} u^2 \frac{\partial u}{\partial t}. \tag{2.16}$$

In (2.14)–(2.16), ρ is the water density, R the cylinder radius, g the gravitational acceleration, u, w the horizontal and vertical water particle velocity, respectively, η the instantaneous wave elevation and h the water depth.

2.4.3 Perturbation theories

Perturbation theories can be expanded to take any order of hydrodynamic loading components into account, at the cost of increasing complexity and computational effort. Both of the best established theories, FNV [8] and Malenica and Molin [45], have been derived with interest to study phenomena associated with higher order loading components (see detailed discussion in Section 2.5).

Malenica and Molin [45] is a perturbation theory, formulated in diffraction regime for finite depth, following a classical Stokes perturbation technique to the third order.

The main difference from FNV, according to [45], is that FNV assumes the radius of the structure to be of the same order as the wave amplitude.

FNV theory, named after its originators Faltinsen, Newman and Vinje [8], considers the local disturbance of the wave field due to the presence of the cylinder and takes into account third-order hydrodynamic loading components. The main limitations include neglection of viscosity and exclusion of breaking wave. It was formulated in the long-wave regime, originally for deep-water regular waves in 1995 [8], but was quickly extended to irregular waves by Newman in 1996 [49] and generalized for cylinders of non-circular cross-sections by Faltinsen in 1999 [50]. The theory has only been derived for a fixed cylinder but including the effect of body motion is possible [50].

The newest development was the generalization of FNV theory for finite depth [10]. It now consists of a horizontal force integrated along the cylinder up to the instantaneous free water surface and a point load applied at the still-water level in accordance to the original wave theory given by

$$F_x = \int_{-h}^{\eta} F'(z,t)dz + F^{\psi}. \tag{2.17}$$

The integrated force F', which includes the nonlinear loading due to wave kinematics (first term) as well as the effect of hydrodynamic added mass (second term), is given as follows:

$$F' = \rho\pi R^2\left(\frac{\partial u}{\partial t} + u\frac{\partial u}{\partial x} + w\frac{\partial u}{\partial z}\right) + a_{11}\left(\frac{\partial u}{\partial t} + w\frac{\partial u}{\partial z}\right), \tag{2.18}$$

where a_{11} is the two-dimensional added coefficient. The point load F^{ψ}, which is a force due to nonlinear free-surface conditions, is kept as derived in the original theory [8] and given by

$$F^{\psi} = \frac{\rho\pi R^2 4}{g}u^2\frac{\partial u}{\partial t}. \tag{2.19}$$

This theory is suitable for regular waves, irregular long-crested waves and numerical wave tanks [10].

2.5 Nonlinear resonant effects

2.5.1 Ringing and springing

Ringing is defined as a nonlinear motion with burst-like characteristics. It is a resonant response that builds up in a time interval of the order of one wave period [9,51,52]. In contrast to springing, which is a more steady resonant phenomenon, ringing has a non-Gaussian nature, generates very high stress levels within a burst of few oscillations, and occurs only during the passage of very steep waves. First attention was paid to ringing and springing phenomena in offshore oil and gas industry in the 1990s during the Heidrun concrete hull model tank tests [53]. Ringing and springing are illustrated

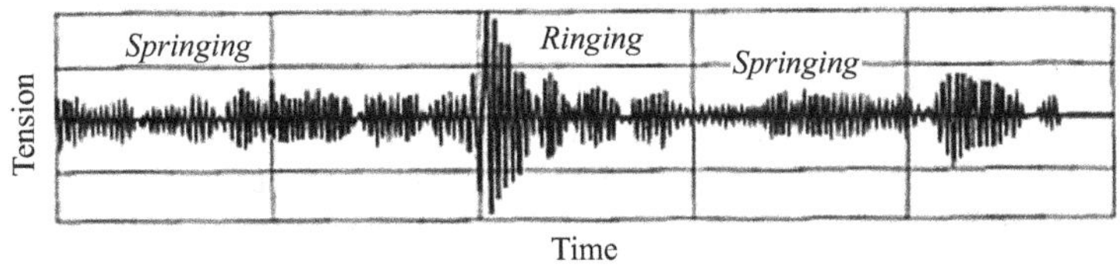

Figure 2.4 Illustration of ringing and springing responses in tension over time. Adapted from [53]

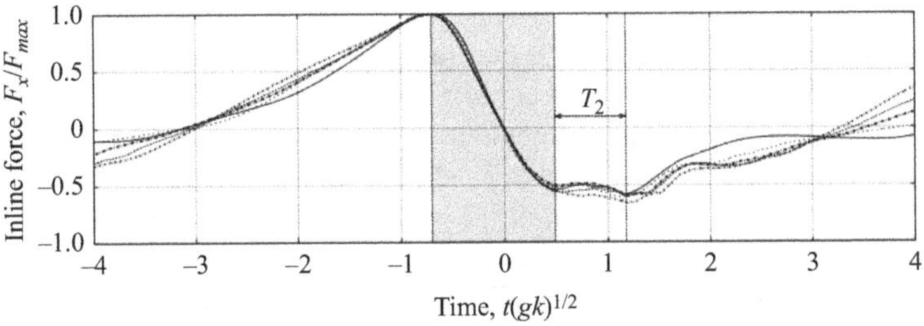

Figure 2.5 Illustration of secondary loading cycle (T_2). Multiple lines represent different values of kh, where k is the wavenumber and h the flume depth. Figure from [47]

in Figure 2.4 and have been deemed likely to occur for structures with relatively low natural periods, therefore including surface-piercing gravity base and TLP structures. Springing had already been easily modelled with the analytic methods of that time, i.e. second-order theory, whereas that was insufficient to predict ringing.

2.5.2 Ringing and secondary loading cycle

Over the last decades, a number of studies have focused on ringing in order to provide an adequate understanding of the underlying physical processes. From the first experiments, ringing has been associated with the secondary loading cycle (*slc*) [9,52,54]. Secondary loading cycle (also called 'secondary force oscillation', 'secondary force cycle', 'secondary load cycle') appears as a second smaller peak in loading, taking place around one quarter of period after the main peak, as illustrated in Figure 2.5, but what it is governed by has been a topic of discussion.

Grue *et al.* [54] described *slc* as an additional loading due to a suction force caused by free-surface effects, taking place below the free water surface. They did not observe flow separation and suggested that this may have induced the secondary loading. They concluded that it was closely related to Froude number (*Fr*) – appearing when *Fr* exceeds 0.35 and becoming pronounced from $Fr = 0.4$.

The *slc* was also observed in the experiments by Stansberg *et al.* [55], where the results of two sets of experiments, one with a single cylinder and the other with an array of cylinders, were reported. In both cases, the cylinders were considered rigid (their frequency was 9 Hz, while the wave frequency ranged from 0.3 to 1.5 Hz). The tests were performed in regular and irregular waves. The Keulegan-Carpenter number KC (defined as $\pi A/R$, A being the wave amplitude and R the cylinder radius) and the slenderness ratio kR (k being the wavenumber) were 2.6 and 0.285, respectively. Measured forces were compared with those numerically predicted using Morison equation (inertia term only) employing a second-order wave kinematic model. The study highlighted the inadequacy of Morison equation in capturing second-order effects and that measured forces were largely influenced by the third-order components.

A wider comparative study was presented by Stansberg [56] where, in addition to the case of [55], data of two experimental tests with different pile diameters were presented. Regular wave tests with increasing wave steepness were performed comparing measured forces with different slenderness rations and KC numbers. It was shown that by increasing the wave steepness kA approximately from 0.13 to 0.30, the total signal of the measured force exhibited a rather low sensitivity to the variations of wave steepness, slenderness ratios and KC numbers. On the contrary, more evident differences appeared in the high-frequency part of the measured force, obtained by filtering out the basic harmonic from total signal. These differences became apparently more pronounced when passing from the pair $(KC, kR) = (1.4, 0.28)$ to $(KC, kR) = (2.3, 0.17)$, but still a low sensitivity to the variation of wave steepness was found (see Figures 4 and 5 of [56]). Moreover, first-, second- and third-order components of the measured forces were compared with those predicted by the FNV theory [8] with linear wave kinematics as input. The theoretical model gave good results on the first-order component, while it apparently over-predicted second- and third-order components when the steepness increased. Larger kR values seemed to have contrasting effects: pejorative on the second-order and beneficial on the third-order components.

Chaplin *et al.* [9] reported the results of seven focused waves with increasing steepness (from 0.208 to 0.345) and the same period of approximately 1.05 s. The water depth was 0.525 m. Three cylinders with diameters 0.127, 0.1 and 0.07 m were tested resulting in three slenderness rations kR: 0.22, 0.18, 0.12, respectively. The KC number ranged from 3 to 10. With respect to a (kA, kR)-plane (refer to Figure 2.6), experiments covered a test grid of approximately $(0.2 - 0.35) \times (0.12 - 0.23)$, resulting in 21 test cases. Good agreement was shown between the (normalized) measured bending moment peak and the one predicted by the Morison equation (inertia term only) augmented by the 'axial divergence force' (A) and 'surface intersection force' (I), (2.14) and (2.15), respectively, in the case of focusing waves.

Note that the majority of the experimental conditions being in the KC number range 3–10 implies that the tested cylinders are in the inertia-dominated regime instead of drag or diffraction, which is also confirmed by the Stokes first-order diffraction effects starting at $kR \approx 0.5$, outside the scope of their experiment. On the other hand, the threshold of Stokes third-order diffraction effects, determined as $kR = 0.05$ by Malenica and Molin [45], would be applicable to the region where secondary loading cycle occurs but is dismissed by the authors due to the local breaking of free water surface around the cylinder, which cancels the validity of all perturbation schemes and the asymptotic forms derived from them. The data points falling to the right of $kA = 0.1$ denote that substantial phase speed nonlinearities are probable. The significance and relevance of ringing to the full-scale offshore structures are shown by the

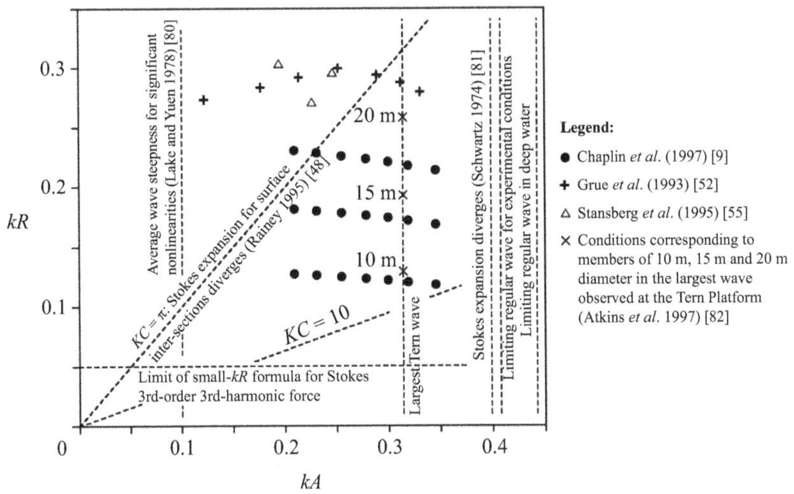

Figure 2.6 Experimental and other conditions on the kA, kR-plane, where k is the wavenumber, A the wave amplitude and R the cylinder radius. Adapted from [9]

authors by augmenting the figure with the conditions corresponding to 10-, 15- and 20-m diameter cylinders being placed in the North Sea (at the location of Tern Platform) during the highest extreme wave recorded during the 1990–92 (crosses on the vertical dashed line at $kA = 0.314$). For our following discussion of OWTs, it is worth pointing out that in the experiments of [9,52,54], *slcs* were observed in the region $(0.2 - 0.35) \times (0.1 - 0.2)$ of the (kA, kR)-plane at Froude (Fr) number $Fr \geq 0.4$.

Even though in the previous studies, relating to ringing flow viscosity had been considered negligible, ringing was shown to be induced in viscous flow for the structures with large wavelength to diameter ratios and viscous models for ringing were developed by [57–59].

Krokstad and Solaas [60] described the secondary loading cycle phenomenon as ‘hydraulic jump’ at the front of the cylinder, caused by local velocities, which then propagates around the cylinder and collides at the back of it causing an upward jet. They observed a direct correlation between the amplitude of the jump and both wave steepness and wavelength, and that it was occurring even in the conditions where diffraction effects are not expected. This evidence appears to be consistent with the assumptions of Chaplin *et al.* [9]. However, they found no correlation between the hydraulic jump and ringing as the ringing force occurs later than the collision of the jumps.

Huseby and Grue [51] studied the higher harmonics in the wave forcing and found that the first harmonic is always the dominant one. However, as the wave amplitude increases to be of the same order as the radius of the cylinder, the total higher harmonic part becomes as large as 10% of the first harmonic. Even though the fifth to seventh harmonics become somewhat smaller, the measured second, third and fourth harmonic forces are no longer negligible and, with larger kA values, start deviating from the analytical predictions by the perturbation theories. It is of significance because these are the harmonics with which ringing is normally associated. Continuing their work on higher harmonics, Grue and Huseby [52] confirmed that ringing occurs for the same wave parameters as the secondary loading cycle, in particular when the Froude number exceeds 0.4, just as defined by Grue *et al.* [54]. They experimentally observed the *slc* for the kR values in the range between 0.1 and 0.33, but the kA range differed in the small and moderate scale experiments: in moderate scale, it was seen from $kA > 0.2$, while in small scale only from $kA > 0.3$ (note: instead of kA Grue and Huseby [52] use $k\eta_m$, where η_m is the maximum wave elevation). It was wondered whether the reason was the more evident flow separation in the small-scale experiments rather than moderate scale, which would lead to conclusion that the occurrence of secondary loading cycle is diminished by flow separation, contradicting the work on viscous models for ringing [57–59]. For reference, Huseby and Grue [51] observed no flow separation at all. Grue and Huseby [52] concluded that the secondary loading cycle contributes to larger higher harmonic forces and implied that ringing relates to the fourth-order harmonic loads. Grue [61] confirmed that ringing is induced by the higher frequency components in the wave loading and, in addition to underlining the importance of flow separation, speculated about whether the *slc* is caused by a resonating effect between the structure and the local flow created due to the presence of the structure itself.

Rainey [62] suggested that ringing may be associated with higher than third-order loading components based on the observation that ringing was seen even when the natural frequency of the cylinder was raised to higher multiples of the wave frequency. He discussed the differences between weakly and strongly nonlinear phenomena and claimed that ringing is strongly nonlinear; therefore, the weakly nonlinear third-order perturbation theories of FNV [8] and Malenica and Molin [45] are insufficient to explain it. Rainey also confirmed the connection between ringing and the secondary loading cycle and explained the latter by local wave breaking and the collapsing of the cavity left behind the cylinder. Such a mechanism sounds similar to Krokstad and Solaas [60]; however, Krokstad and Solaas emphasized no correlation between the collapse of water flows behind the cylinder and ringing.

Similarly to Rainey [62], Paulsen *et al.* [47] also suggested that the analytical theories FNV [8] and Malenica and Molin [45] are insufficient to portray secondary loading cycle, because after an analysis of the harmonic structure of the inline force the *slc* was not captured even by the first six harmonics. However, on the contrary to the previous research, based on these findings, they suggested that ringing and secondary loading cycles are not directly connected. Instead, the *slc* would be acting as an indicator of highly nonlinear flow, where the harmonic force content is strong and therefore ringing may be likely, but without directly inducing ringing. Moreover, their numerical study explains the mechanism of secondary loading cycle: the passing wave crest creates a cavity at the back of the cylinder, which is filled by the diffracted wave and local vortex is created. The interaction between this vortex and the surrounding flow continuing in the downstream direction temporarily reduces the pressure at the downstream side of the cylinder, which is observed as the *slc*. This would lead to the conclusion that three-dimensional effects need to be taken into account for successful modelling of *slc*. The relation of diffraction to the occurrence of secondary load cycle supplements the observations of Chaplin *et al.* [9], Krokstad and Solaas [60] and Grue and Huseby [52], who observed the *slc* in the *kR* range before the diffraction effects were expected.

Recently in [63,64], the secondary load cycle has been tried to be numerically recreated using FNL wave kinematics from HOBEM solver as input to three hydrodynamic loading models: Morison equation, slender-body theory and finite-depth FNV theory. However, none of the models successfully captured the secondary load cycle on a fixed rigid cylinder in this two-dimensional study. Such results are in accordance with the findings of Paulsen *et al.* [47] and Kristiansen and Faltinsen [10], namely that to model the *slc* numerically, consideration of either three-dimensional effects, or inclusion of higher order loading than third (to which the perturbation theories are derived), are needed.

2.5.3 Role of wave kinematics and hydrodynamic model

Even though it is agreed that ringing is a nonlinear effect, some uncertainties still exist on which nonlinearity source dominates and mostly contributes to its cause. Wave nonlinearities have an important effect as demonstrated in a number of works [42,43,58,65,66] where first-order wave kinematics have repeatedly failed to predict ringing and similar responses. Swan *et al.* [67] believe that ringing is caused by the nonlinearities in the wave particle acceleration rather than the hydrodynamic model, and for this they recommend Rienecker–Fenton theory [32], discussed in Section 2.3.3, as it was the only one to reproduce the accelerations accurately. On the other hand, Paulsen *et al.* [47] noted that secondary loading cycles imply strong nonlinearities in wave forcing rather than in wave kinematics. But it is worth noting that they also claim that ringing and *slc* may not be directly related.

In the next section, we examine and review the main results recently obtained focusing on the role of nonlinear wave kinematics only. In all discussed cases, hydrodynamic forces are modelled with Morison equation in its standard form.

2.6 Dynamic response of an offshore wind turbine

OWTs mounted on fixed-bottom monopiles fall in the category of offshore structures sensitive to potentially dangerous resonant phenomena induced by exciting forces associated with severe environmental conditions. These types of structures are typically characterized by low natural periods (in the range of 3–5 s) such that, when exposed to harsh wave conditions, they may incur high-frequency loads and vibrations, such as ringing and springing.

In the following, we examine some of the most recent literature studies which analyse the role of FNL wave kinematics as well as other environmental parameters, such as the mean wind speed and the turbulence intensity, in the assessment of the dynamic response of an OWT. All presented cases refer to the NREL 5-MW baseline

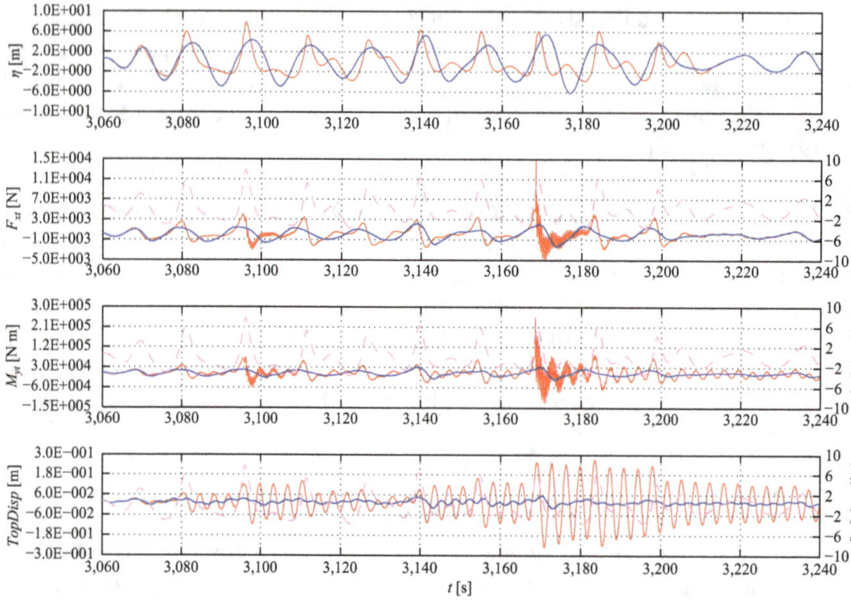

Figure 2.7 Comparison of linear (blue) and fully nonlinear (red) free-surface elevation (top panel), tower-base shear force F_{xt}, tower-base overturning bending moment M_{yt} and tower-top displacement TopDisp. Figure from [42]

wind turbine model [68]: a variable-speed, pitch-controlled wind turbine with a rated rotor speed of 12.1 rpm. The rated wind speed is 11.4 m/s, the hub height is 90 m above the mean sea level and the rotor diameter is 126 m. For all cases, a water depth of 20 m is considered. The monopile, with a diameter of 6 m, is assumed to be rigidly connected to the seabed.

2.6.1 *Effects of fully nonlinear waves*

The effects of FNL wave kinematics, including breaking wave cases, were addressed in [42]. In that study, the novel numerical approach to assess the influence of FNL waves on the dynamic response of OWT developed in [69] was adopted. The global strategy was based on the domain decomposition method recalled in Section 2.3.5 consisting of incorporating the FNL wave kinematics into the hydro–aero–servo–elastic code FAST [6]. Hydrodynamic forces were computed with the standard Morison equation for non-breaking wave cases, whereas in the case of wave breaking, the impact model proposed by Wienke and Oumeraci [70] was used (the reader can refer to [25,69,71,72] for the details about the impact model).

Figure 2.7 shows a comparison of free-surface elevation and the structural response associated with FNL and linear (L) wave kinematics during a sea state given by $H_s = 7.5$ and $T_p = 15$ (JONSWAP spectrum).

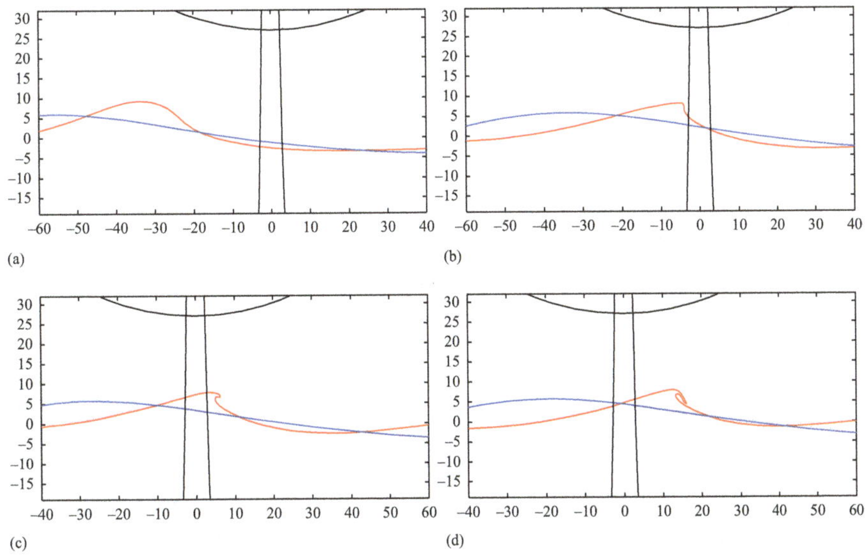

Figure 2.8 Evolution of the plunging breaker. Blue solid line: linear prediction, red solid line: fully nonlinear numerical solution. (a) t = 3166.50 s, (b) t = 3168.25 s, (c) t = 3168.88 s and (d) t = 3169.44 s. Figure from [42]

Around 3,080 s, a steep and asymmetric wave triggers a resonant vibration. At about 3,095 s a steeper wave amplifies the oscillation amplitudes for approximately one T_p. At the subsequent smaller wave peak, a smaller resonant phenomenon occurs and so on until the strong wave impact. Then the impact event (at 3,168.5 s) induces two different high-frequency scales that are superimposed one on another in the force and moment: (1) a high-frequency vibration (2.3 Hz) that persists for about one period ($T_p = 15$ s) of the sea state; (2) the resonant vibration of the structure at its first natural frequency of 0.28 Hz. In the same time interval, an amplification of the tower oscillation amplitudes of about 80% occurs (see bottom panel of the same figure). It is noted that the impulsive load excites the first in-plane and out-of-plane natural frequencies of the blades causing their oscillation in a coupled mode [73]. The first natural edgewise bending frequency of the blades is 2.3 Hz, while the flapwise frequency is around 1.05 Hz. The former, at 2.3 Hz, excites the tower in the fore–aft motion affecting the TwrBsFxt and TwrBsMyt. It is pointed out that a similar pattern, even though less pronounced, is observed around 3,095 s due to the passage of a non-breaking wave.

Figure 2.8 shows four snapshots of the wave breaking against the monopile structure at around 3,168.2 s.

2.6.1.1 Response in parked and power production states

The dynamic response of OWT structures, in contrast to the static structures in the oil and gas industry, are also affected by the aerodynamics of the rotor blades, leading to rather different responses depending on the state of the turbine. The wave nonlinearities cause significant effects in the case of parked configuration [42,43,66,74], which is more similar to the static oil and gas structures because of the rotor idling and the three blades fully pitched to minimize aerodynamic loads. Whereas, if the turbine is in power production, effects of higher order excitation are largely damped out by the aerodynamic damping [42,66,74].

In a sea state given by a mean wind speed $U = 18$ m/s, a spectral peak period $T_p = 12.3$ s and a significant wave height $H_s = 7.5$ m [38], in the case of parked condition, Figure 2.9(a) shows that the passage of a steep wave at approximately 1,977 s triggers a resonant vibration of the tower that persists for about six wave periods. This vibration is strongly amplified when a second steep non-breaking wave occurs approximately at 2,020 s. At 2,034 s, an impact takes place, giving rise to a very large peak in the shear force and bending moment. Very similarly to what is observed in Figure 2.7, a high-frequency oscillation (superimposed to the first natural frequency still excited by the previous steep waves) follows. Note that also in this case the linear prediction of the structural response significantly underestimates the actual response, completely omitting the resonance of the system.

When the turbine is in power production [see Figure 2.9(b)], no resonant vibration of the structure is observed for the top displacement of the monopile (see bottom panel). However, the steep nonlinear waves preceding the impact event induce an increase of the peaks of the internal forces TwrBsFxt and TwrBsMyt.

2.6.2 Effects of different wind conditions

In this section, we discuss the effects that different wind conditions, represented by different mean wind speeds and turbulence intensity levels, have on the system response when linear or nonlinear wave models are employed. We summarize some key results from the literature; the complete study can be found in [66]. The wave conditions are characterized by a significant wave height, H_s, equal to 5.60 m and a wave spectral peak period, T_p, equal to 10.82 s. These are the conditional mean values of H_s and T_p, for a hub-height wind speed, U, of 33.43 m/s, which is above the cut-out wind speed. Wave conditions are kept the same, while the wind speed and turbulence intensity levels are changed. Variation of mean wind speeds from 12 to 32 m/s in increments of 4 m/s is considered. For each mean wind speed, three IEC [3] turbulence categories – A, B and C – are considered.

2.6.2.1 Parked state

An example of a significant nonlinear wave is observed at approximately 222.5 s, see Figure 2.10. The passage of this steep wave generates higher order loading components absent in the linear wave modelling, which trigger resonant vibrations of the tower. It is noted that, with the passing of this wave, the tower oscillates at its fundamental frequency of 0.28 Hz (see the TTDspFA time series in the bottom panel of Figure 2.10). The excitation due to the nonlinear waves is persistent – the response damps out and returns to similar levels as that predicted by the linear wave model only around $t = 400$ s. The enhanced response of the tower triggered by the nonlinear

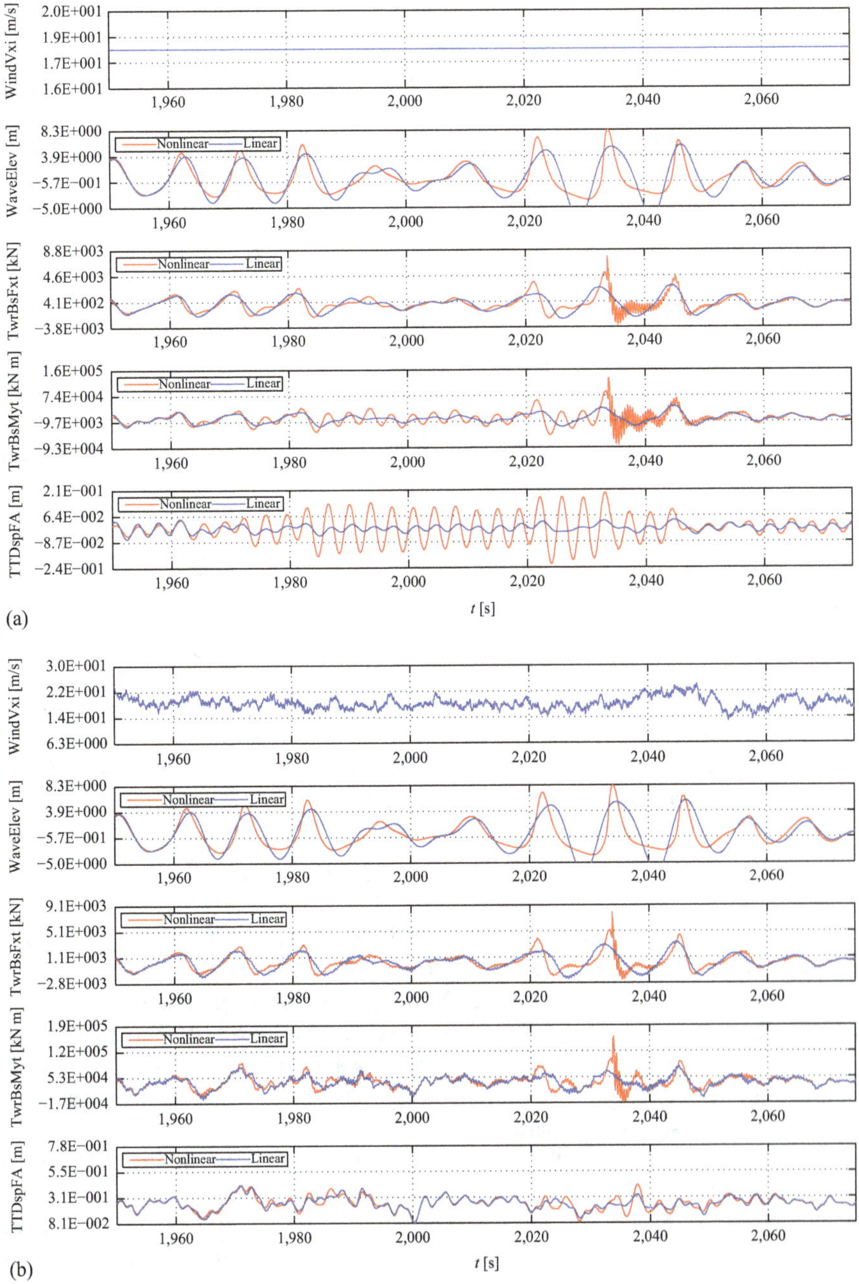

Figure 2.9 Time series of (from top to bottom) horizontal wind speed (WindVxi), wave elevation (WaveElev), tower-base shear force (TwrBsFxt), tower-base bending moment (TwrBsMyt) and tower-top fore–aft displacement (TTDispFA) associated with linear (blue) and fully nonlinear (red) hydrodynamics: (a) parked and (b) power production. Figure from [43]

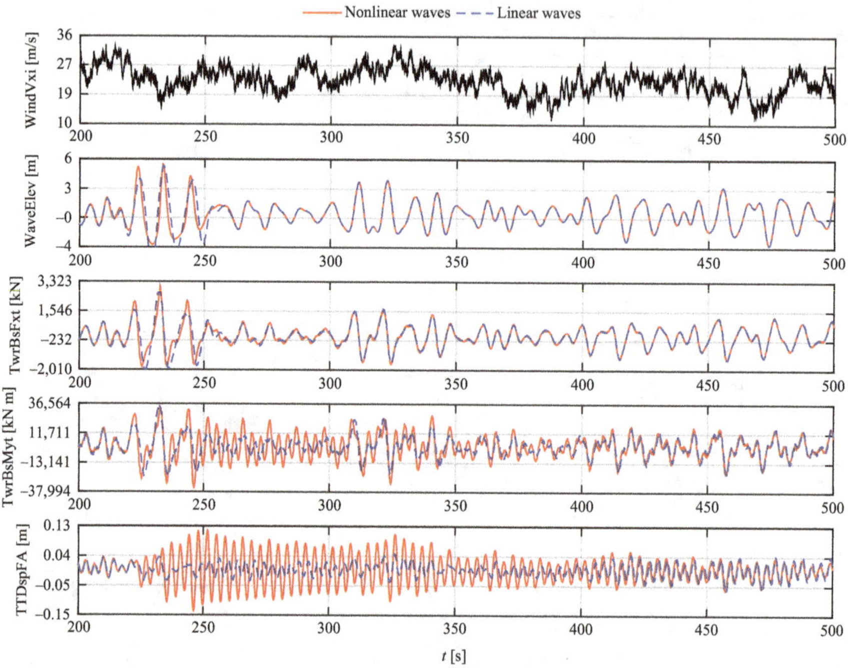

Figure 2.10 Time series of the hub-height longitudinal wind velocity (for a mean wind speed of 24 m/s), wave elevation, tower-base fore–aft shear force, tower-base fore–aft bending moment and tower-top fore–aft deflection. Linear (dashed line) and nonlinear (solid line) responses are compared. The turbine is in a parked state. Figure from [66]

wave contributions is also reflected in the TwrBsMyt time series, where significantly higher response peaks compared to the linear wave case are observed.

Statistics of the tower-base fore–aft bending moment (TwrBsMyt) are presented in Figure 2.11. As expected, the mean response (top panel) grows monotonically with mean wind speed. No significant differences in mean response are observed with the different turbulence categories or between the linear and nonlinear wave kinematics. In contrast, the response standard deviation (s.d.), middle panel of Figure 2.11, is about 10% larger with the FNL wave kinematics compared to the linear (L) wave kinematics. In both cases, the s.d. values increase slightly with increasing mean wind speed. In general, it is observed that the response s.d. is sensitive both to wind speed and to wave modelling choice, with nonlinear waves resulting in significantly higher s.d. response levels than linear waves.

The 1-h maximum loads (refer to bottom panel of Figure 2.11), computed as the mean value of the peaks over a selected threshold (equal to 1.4 s.d.), are up to 15% larger in the case with FNL wave kinematics compared to the L wave kinematics case. Turbulence has a minor effect on the maximum values.

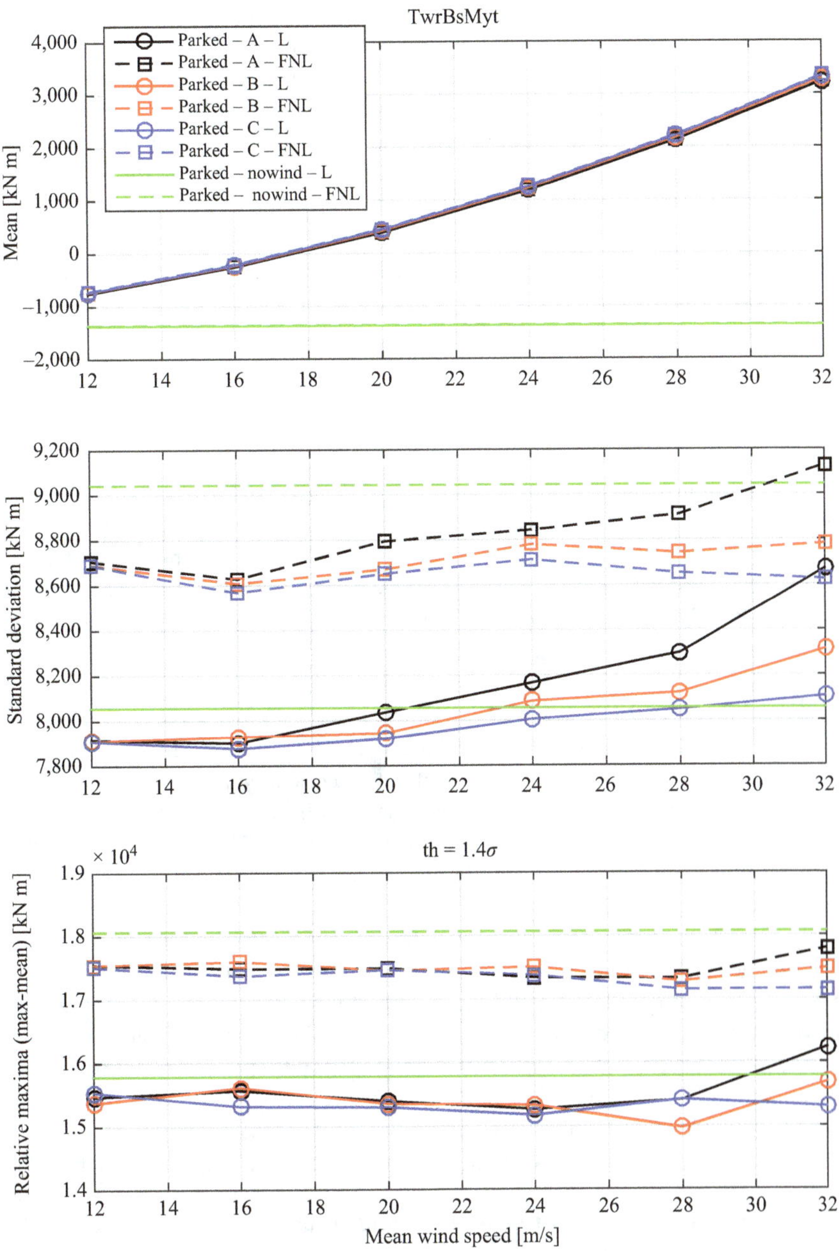

Figure 2.11 TwrBsMyt: statistics in the case of nonlinear (dashed lines with squares) and linear (solid lines with circles) wave kinematics for the three turbulence categories, A (black), B (red), C (blue). The turbine is in a parked state. Figure from [66]

The PSDs of the tower-base fore–aft bending moment are shown in Figure 2.12. The two dominant frequencies, namely the wave frequency of about 0.09 Hz (wave peak spectral period, T_p = 10.82 s) and the first tower fore–aft natural frequency of 0.28 Hz are clearly visible (see upper panel). A close-up of the first spectral peak (middle panel of Figure 2.12) shows that the modelling choice of nonlinear wave kinematics leads to a slightly lower energy at low frequencies. On the contrary, the response of the structure at the tower fundamental frequency (see bottom panel of Figure 2.12) is rather sensitive to the wave modelling choice as well as to the mean wind speed. The maximum energy occurs around 0.28 Hz at an intermediate mean wind speed (20 m/s), while at the highest wind speed (32 m/s), the frequency content of the response is spread over a wider range of frequencies and reaches lower peaks (see orange dashed curve in Figure 2.12). For wind speeds up to 24 m/s, wave action dominates; for higher winds, turbulence leads to excitation of more modes and a greater spread of the energy around 0.28 Hz is seen. For a more detailed discussion of these results, we refer to [66].

A key influence of aeroelastic interaction is easily understood by studying the PSDs (see the green dashed curves in the bottom panel of Figure 2.12) associated with the admittedly unrealistic case consisting of waves alone. In the absence of wind, the aeroelastic interactions are minimized and the entire system experiences minimum damping. It is for this reason that the wave excitation produces its maximum effect on the overall structural response.

2.6.2.2 Power-production state

Figure 2.13 shows statistics of the tower-base fore–aft bending moment for the turbine in a power-production state. Due to the effect of collective pitch control, the mean TwrBsMyt (top panel of Figure 2.13) decreases with increasing mean wind speed. The mean value of the tower bending moment is mainly influenced by wind loading, while negligible effects are induced by the wave models and turbulence intensities (see top panel of Figure 2.13).

The response standard deviation (middle panel of Figure 2.13) suggests virtually no difference between the L and FNL wave kinematics models. This is due to the aeroelastic damping when the turbine is operating and the higher order terms in the wave kinematics are dissipated. While the blade pitch control is able to reduce the mean response at all wind speeds, for the standard deviation and the extreme response (see middle and bottom panels of Figure 2.13, respectively), we observe that the pitch control is effective for mean wind speeds up to about 16–18 m/s; above this level, no reduction of the response results in increasing the pitch angle beyond about 13 deg.

2.6.2.3 A remark on wind–waves misalignment

From the results discussed above one would conclude that in the case of a power-production state, nonlinear wave contributions are not as dangerous as in the case of a parked state and they may possibly be neglected. Before drawing this conclusion, attention should be paid to the wind–wave directionality. We remark that all the studies discussed in the previous sections are conducted with co-directional wind and waves. A study looking at the effect of wind–wave misalignment on the fatigue

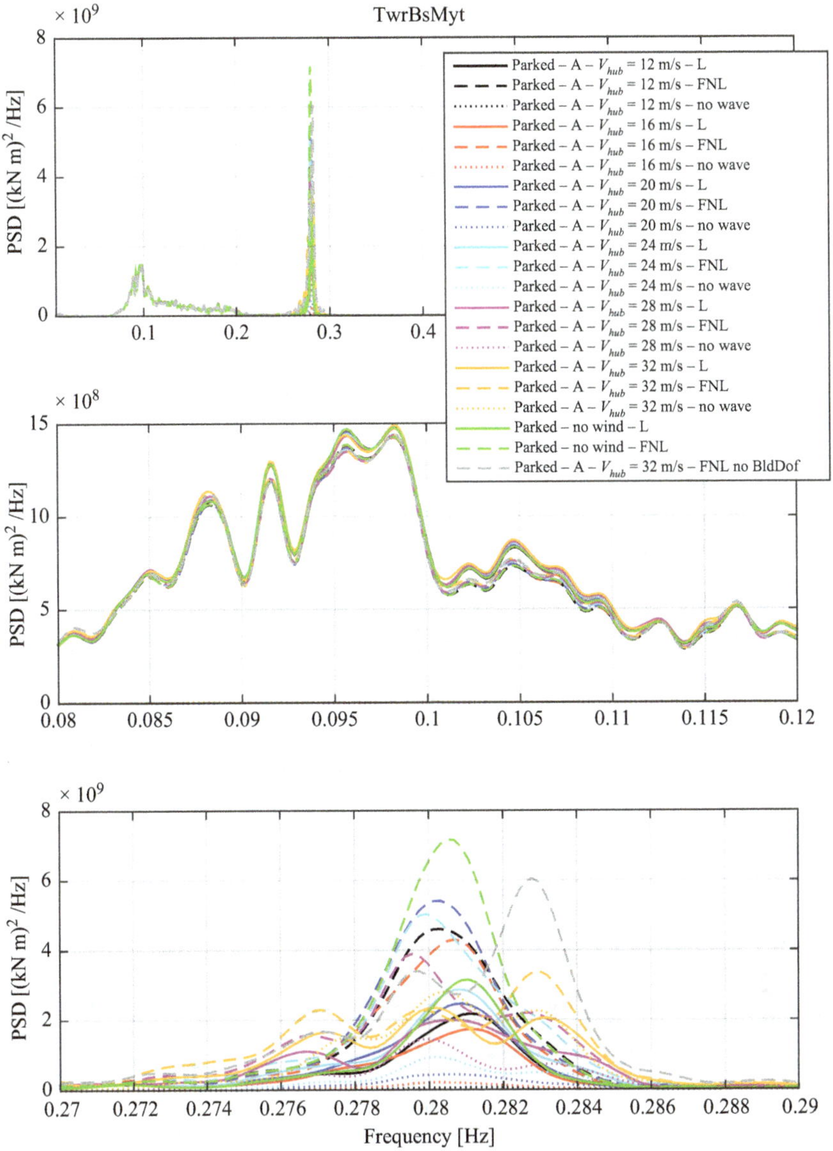

Figure 2.12 Power spectral density functions of TwrBsMyt associated with several wind speeds, turbulence categories A, for nonlinear (dashed lines) and linear (solid lines) wave kinematics. The turbine is in a parked state. Figure from [66]

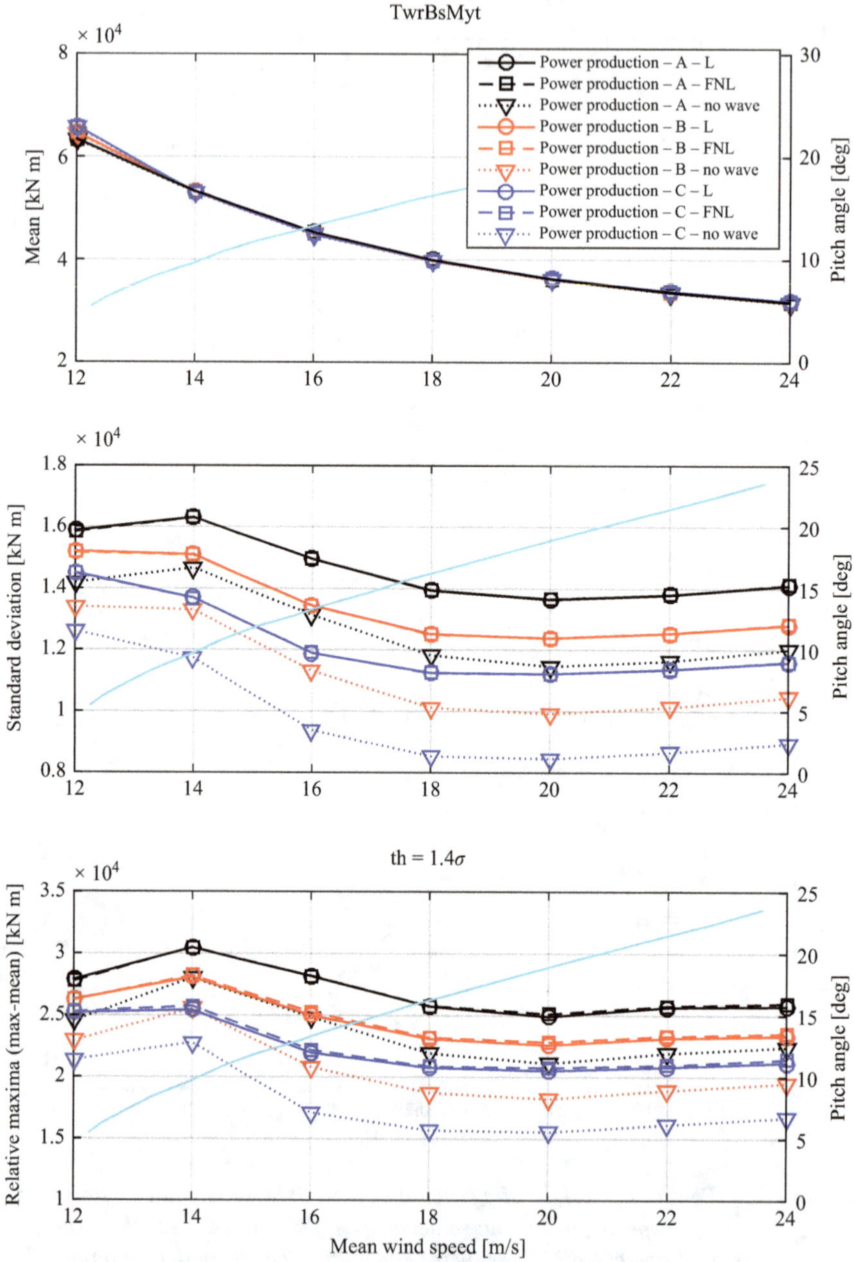

Figure 2.13 TwrBsMyt: statistics in the case of nonlinear (dashed lines with squares) and linear (solid lines with circles) wave kinematics for the three turbulence categories A (black), B (red), C (blue). The turbine is in a power-production state. The cyan curves with ordinate on the right-hand side of each panel indicate the pitch angle variation with mean wind speed. Figure from [66]

loads of OWTs (even though with only second-order wave model) found highest accumulation of fatigue when the wind and waves were misaligned, stating reduced aerodynamic damping as the cause [75]. A study with FNL wave kinematics simulated with HOBEM model and seven wind–wave misalignment angles was recently conducted [76] and found that in the power production case the impact of wave nonlinearities on the fatigue damage equivalent loads in the tower-base bending moments indeed increased with increasing angles of misalignment – it resulted 1.5% larger with 90 deg misalignment, in comparison to 0.2% with co-directional case in power production configuration. The Danish Wave Loads project has also looked at the lack of aerodynamic damping in the case of misalignment, and how it is substituted by other types of damping, e.g. hydrodynamic radiation, soil or structural damping [77].

2.6.3 Limitation of the second-order wave model

Effects of second-order wave model, see Section 2.3.4, on the response of OWTs are significant in terms of both fatigue and extreme loads [38,75]. However, in this paragraph we emphasize that second-order models are not able to capture actual resonant effects which are ascribed to higher order wave components. To show this we summarize the main results obtained in [78,79].

A regular wave case is considered with circular frequency $\omega = 0.6283$ rad/s ($T = 10$ s) and increasing steepness ka from 0.05 to 0.3 with increment of 0.05. Both parked and power production turbine states are considered. A total simulation time of $30T$ is considered for each condition. An example of the system response when $ka = 0.2$ is given in Figure 2.14, which shows the time series of the wave elevation (WaveElev), tower-base fore–aft bending moment (TwrBsMyt) and tower-top fore–aft deflection (TTDspFA) corresponding to linear, second-order and FNL wave kinematics when the turbine is parked and in power production.

The maximum wave elevations obtained with the linear, second-order and FNL models are 3.86, 4.80 and 5.7 m, respectively. In the parked case, linear and second-order wave kinematics provide similar results for the tower-base shear force (not shown in figure) and bending moment. In contrast, the FNL produces an increase of about 20% on TwrBsFxt and 50% on TwrBsMyt. The maximum tower motion increases dramatically from approximately 2 cm (linear and second order) to 10 cm (FNL). A clear low-frequency component of 0.02 Hz is observed. This contribution is much more significant in the FNL case because it is associated with the difference frequency $3f - f_n$, where $f = 0.1$ Hz is the wave frequency and $f_n = 0.28$ Hz is the first natural bending frequency of the tower.

When the blades rotate with a rotor speed of 12.1 rpm, with a pitch angle of 17.5° and a hub-height constant wind of 20 m/s, TwrBsFxt (not reported here) and TwrBsMyt maximum values increase both by about 15%. Due to the rotor thrust caused by the wind action, the tower top oscillates around a mean value of 22.7 cm. The FNL minimum peak is 15.5 cm while the linear and second-order minimum peaks are 19 cm. We observe that the low-frequency oscillation, characterizing the tower motion in parked condition, almost disappears in the present case.

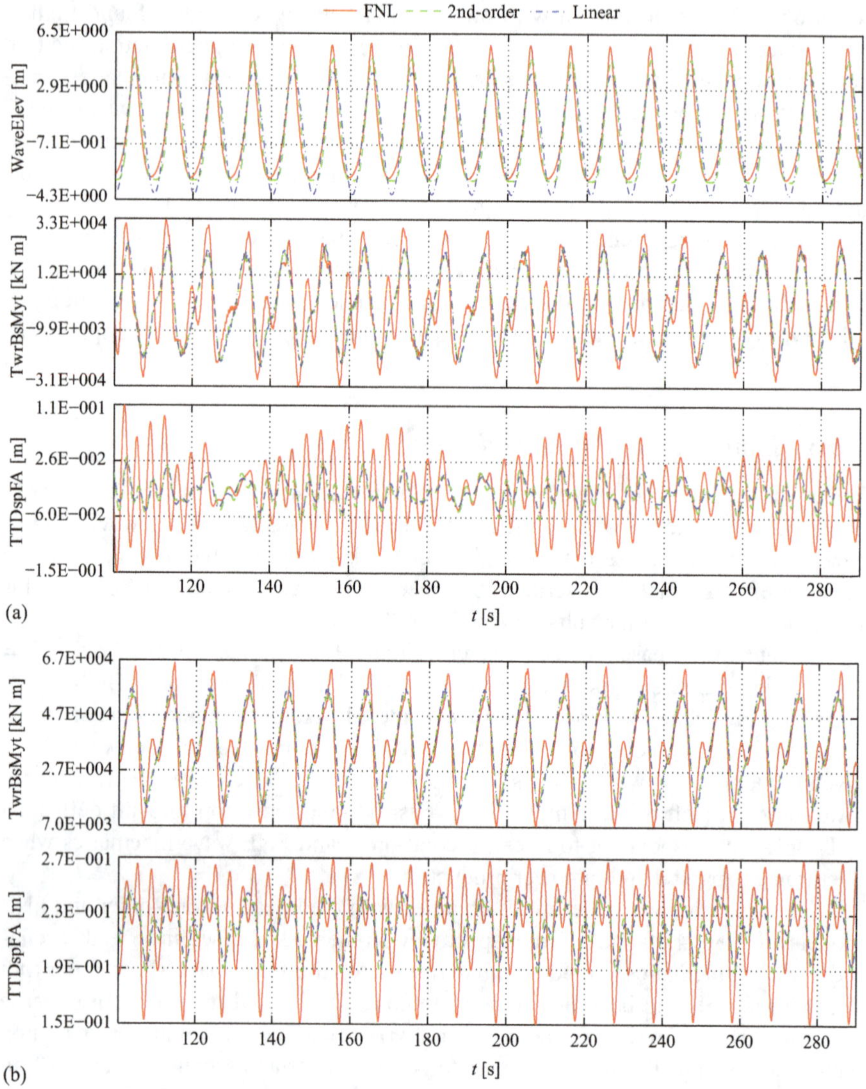

Figure 2.14 Time series of the wave elevation (first panel from top), tower-base fore–aft bending moment (TwrBsMyt) and tower-top fore–aft deflection (TTDspFA) for the parked (second and third panels from top) and operation (fourth and fifth panels from top), corresponding to the linear (blue dotted line), second-order (green dashed line), FNL with (red solid line) wave kinematic models with ka 0.20. (a) Parked state and (b) power-production state. Figure from [79]

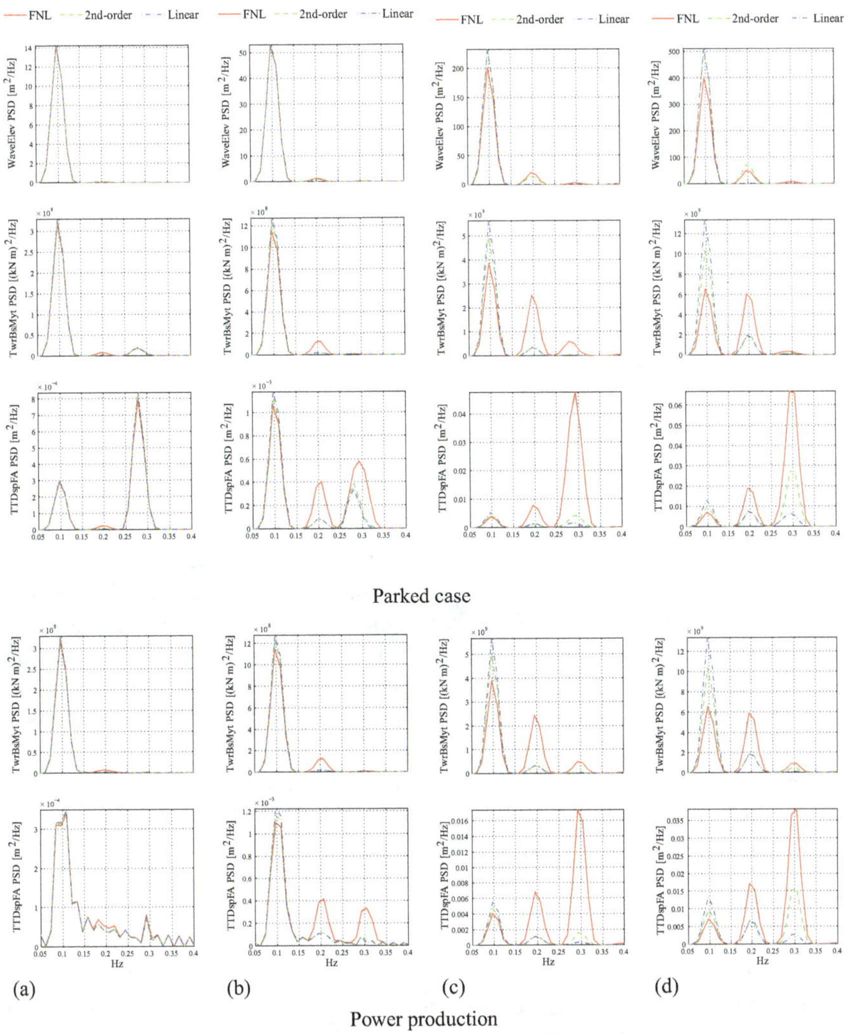

Figure 2.15 Spectra of the wave elevation (first row), tower-base fore–aft bending moment and tower-top fore–aft deflection for the parked (second and third rows) and operation (fourth and fifth rows) conditions corresponding to the linear (blue dashed dotted), second-order (green dashed), FNL with (red solid) wave kinematic models with increasing ka. (a) ka = 0.05, (b) ka = 0.10, (c) ka = 0.20 and (d) ka = 0.30. Figure from [79]

2.6.3.1 Power spectral densities

Power spectral densities (PSDs) of the wave elevation, tower-base fore–aft bending moment and tower-top fore-aft deflection corresponding to linear, second-order and

FNL wave kinematics when the turbine is parked (upper group) and in power production (lower group) are shown in Figure 2.15. The wave steepness increases from 0.05 to 0.30. For $ka = 0.05$, no differences exist in the PSDs of the structural loads by using the three different models for the wave kinematic as well as by changing the operating conditions.

In the PSD of the free-surface elevation process, the dominant peak occurs at 0.1 Hz (since $T = 10$ s). As the steepness increases, peaks gradually appear at $2f$ and $3f$, that is at 0.2 and 0.3 Hz. With $ka = 0.05$ [Figure 2.15(a)], the nonlinear load contributions are negligible as a consequence of the almost linear forcing. As the steepness increases, more and more energy is provided at $3f$; therefore, as confirmed in Figures 2.15(b)–2.15(d) (see PSDs of TTDspFA), the peak moves gradually from f_n to $3f$, that is from 0.28 to 0.3 Hz.

When the turbine is in power production (see Figure 2.15 lower panels) the transient behaviour quickly dissipates and the system responds at the loading frequencies. We only observe an augmentation of the energy at the frequencies $2f$ and $3f$ as the steepness increases.

The main results summarized above show that in regular waves, for very small wave steepnesses, the structure responds with the same amplitudes at the main wave frequency regardless of the wave theory used. As the steepness increases, the structure responds at the higher order loading frequencies as well. In this case, differences between linear, second-order and FNL models become relevant. From $ka = 0.01$ on, the linear and second-order theories become inaccurate. Under a constant wind speed, the structure reacts at the wave loading frequencies of f, $2f$ and $3f$ so that the PSDs between the parked and operating condition assume the same shape at high wave steepness. A more detailed discussion about the comparison of linear, second-order and FNL wave models for OWTs can be found in [78,79].

2.7 Conclusions

In this contribution, we have presented an overview of the current methods used in modelling wave loads on monopile supported OWTs. We have also summarized the main results obtained by the authors over the last 6 years. All results confirm that FNL wave kinematics play a crucial role in the assessment of loads as well as dynamic response of large wind turbines (5 MW). At the same time, it has been shown that linear and second-order wave theories are mostly unable to reproduce resonant vibrations and associated loads. This deficiency was dramatically evident in the case of turbines in parked state, while it was much less evident in the case of wind turbines in power production state. Such different behaviour is ascribed to the aeroelastic damping which, in case of power production, significantly concurs to increase the overall damping capabilities of the whole system. However, this evidence should not lead to the conclusion that nonlinear wave modelling is unnecessary when simulating the turbine in an operational state. Indeed, it has been observed that when wind and waves are not co-directional, even in power production state, the aeroelastic damping turns out to have a less beneficial effect.

We remark that in all the numerical studies discussed, the hydrodynamic forcing model was based on the standard Morison equation. The currently open discussion on the improvement of Morison equation or alternative approaches has been reported. Apparently, there is no simple model available capable of accounting for both FNL wave kinematics and nonlinear loads phenomena, such as the secondary load cycle and ringing loads. Additionally, at the preset time, it is not fully understood to what extent such nonlinear load phenomena may affect the global dynamic response of the system and the associated internal loads.

Acknowledgements

AM was supported by the European Commission's Framework Program "Horizon 2020", through the Marie Skłodowska-Curie Innovative Training Network (ITN) "AEOLUS4FUTURE – Efficient harvesting of the wind energy" (H2020-MSCA-ITN-2014: Grant agreement no. 643167). CL activity has been partially supported by the Research Council of Norway through the Centres of Excellence funding scheme NTNU/AMOS, project number 223254.

References

[1] GWEC. Wind in Numbers. Website: www.gwec.net/global-figures/wind-in-numbers/, 2017.

[2] GWEC. Global Wind Report 2016. Technical Report, 2017.

[3] IEC 61400-1 Wind Turbines – Part 1: Design Requirements, 2005.

[4] IEC 61400-3 Wind Turbines – Part 3: Design Requirements for Offshore Wind Turbines, 2009.

[5] D. P. Molenaar. *Cost-effective design and operation of variable speed wind turbines*. PhD thesis, Technische Universiteit Delft, 2003.

[6] J. M. Jonkman and L. B. Jr. Marshall. FAST User's Guide. Technical Report, NREL, 2005.

[7] R. C. T. Rainey. A new equation for calculating wave loads on offshore structures. *Journal of Fluid Mechanics*, 204:295–324, 1989.

[8] O. M. Faltinsen, J. N. Newman, and T. Vinje. Nonlinear wave loads on a slender vertical cylinder. *Journal of Fluid Mechanics*, 289:179–198, 1995.

[9] J. R. Chaplin, R. C. T. Rainey, and R. W. Yemm. Ringing of a vertical cylinder in waves. *Journal of Fluid Mechanics*, 350:119–147, 1997.

[10] T. Kristiansen and O. M. Faltinsen. Higher harmonic wave loads on a vertical cylinder in finite water depth. *Journal of Fluid Mechanics*, 833:773–805, 2017.

[11] M. E. Gurtin. *An Introduction to Continuum Mechanics*. Academic Press, Cambridge, MA, USA, 1981.

[12] M. S. Longuet-Higgins and E. D. Cokelet. The deformation of steep surface waves on water. I. A numerical method of computation. *Proceedings of the Royal Society of London. Series A, Mathematical and Physical Sciences*, 350(1660):1–26, 1976.

[13] J. W. Dold and D. H. Peregrine. An efficient boundary integral method for steep unsteady water waves. In K. W. Morton and M. J. Baines, editors, *Numerical Methods for Fluid Dynamics II*. Oxford University Press, Oxford, UK, pages 671–679, 1986.

[14] S. T. Grilli, J. Skourup, and I. A. Svedsen. An efficient boundary element method for nonlinear water waves. *Engineering Analysis with Boundary Elements*, 6(2):97–107, 1989.

[15] S. T. Grilli and I. A. Svendsen. Corner problems and global accuracy in the boundary element solution of nonlinear wave flows. *Engineering Analysis with Boundary Elements*, 7(4):178–195, 1990.

[16] T. Nakayama. Boundary element analysis of nonlinear water wave problems. *International Journal for Numerical Methods in Engineering*, 19(7):953–970, 1983.

[17] T. Nakayama. A computational method for simulating transient motions of an incompressible inviscid fluid with a free surface. *International Journal for Numerical Methods in Fluids*, 10(6):683–695, 1990.

[18] S. T. Grilli and J. Horrillo. Numerical generation and absorption of fully nonlinear periodic waves. *Journal of Engineering Mechanics*, 123(10):1060–1069, 1997.

[19] W. Tsai and D. K. P. Yue. Computation of nonlinear free-surface flows. *Annual Review of Fluid Mechanics*, 28:249–278, 1996.

[20] C. H. Kim. *Nonlinear Waves and Offshore Structures*. World Scientific, Singapore, 2008.

[21] G. X. Wu and R. Eatock Taylor. Finite element analysis of two-dimensional non-linear transient water waves. *Applied Ocean Research*, 16(6):363–372, 1994.

[22] C. A. Brebbia and C. Dominguez. *Boundary Elements, An Introductory Course*. WIT Press, Southampton, UK, 1998.

[23] Y. L. Shao and O. M. Faltinsen. A harmonic polynomial cell (HPC) method for 3d Laplace equation with application in marine hydrodynamics. *Journal of Computational Physics*, 274(Supplement C):312–332, 2014.

[24] F. C. W. Hanssen, A. Bardazzi, C. Lugni, and M. Greco. Free-surface tracking in 2d with the harmonic polynomial cell method: Two alternative strategies. *International Journal for Numerical Methods in Engineering*, 113:311–351, 2018.

[25] E. Marino. *An Integrated Nonlinear Wind-Waves Model for Offshore Wind Turbines*. Firenze University Press, Florence, Italy, 2011, ISBN: 978-88-6655-051-8.

[26] S. T. Grilli and R. Subramanya. Quasi-singular integrals in the modeling of nonlinear water waves in shallow water. *Engineering Analysis with Boundary Elements*, 13(2):181–191, 1994.

[27] R. Cointe. Hydrodynamic impact analysis of a cylinder. *Journal of Offshore Mechanics and Arctic Engineering*, 109:237–243, 1987.

[28] R. Cointe. Numerical simulation of a wave channel. *Engineering Analysis with Boundary Elements*, 7(4):167–177, 1990.

[29] T. Sarpkaya and M. Isaacson. *Mechanics of Wave Forces on Offshore Structures*. Van Nostrand Reinhold Company, New York, NY, USA, 1981.

[30] S. K. Chakrabarti. *Hydrodynamics of Offshore Structures*. WIT Press (UK), 1998.

[31] R. G. Dean and R. A. Dalrymple. *Water Wave Mechanics for Engineers and Scientists*. World Scientific, Singapore, 1984.

[32] M. M. Fenton and J. D. Rienecker. A Fourier approximation method for steady water waves. *Journal of Fluid Mechanics*, 104:119–137, 1981.

[33] C. A. Brebbia and S. Walker. *Dynamic Analysis of Offshore Structures*. Newnes-Butterworths, Boston, MA, USA, 1979.

[34] M. K. Ochi. *Ocean Waves: The Stochastic Approach*. Cambridge University Press, Cambridge, UK, 2005.

[35] Y. Goda. *Random Seas and Design of Maritime Structures*. World Scientific Advanced Series on Ocean Engineering: Volume 33, 3rd edition, World Scientific Publishing Co. Pte. Ltd., Singapore, 2010.

[36] Design of Offshore Wind Turbine Structures. Technical report, Offshore Standard Det Norske Veritas DNV-OS-J101, 2014.

[37] N. Sharma and R. G. Dean. Second-order directional seas and associated wave forces. *Society of Petroleum Engineers Journal*, 21(01):129–140, 1981.

[38] P. Agarwal and L. Manuel. Incorporating irregular nonlinear waves in coupled simulation and reliability studies of offshore wind turbines. *Applied Ocean Research*, 33(3):215–227, 2011.

[39] D. G. Dommermuth and D. K. P. Yue. A high-order spectral method for the study of nonlinear gravity waves. *Journal of Fluid Mechanics*, 184:267–288, 1987.

[40] B. J. West, K. A. Brueckner, R. S. Janda, M. Milder, and R. L. Milton. A new numerical method for surface hydrodynamics. *Journal of Geophysical Research*, 92:11803–11824, 1987.

[41] F. Fedele, C. Lugni, and A. Chawla. The sinking of the El Faro: predicting real world rogue waves during Hurricane Joaquin. *Scientific Reports*, 7:1–15, 2017.

[42] E. Marino, C. Lugni, and C. Borri. A novel numerical strategy for the simulation of irregular nonlinear waves and their effects on the dynamic response of offshore wind turbines. *Computer Methods in Applied Mechanics and Engineering*, 255:275–288, 2013.

[43] E. Marino, C. Lugni, and C. Borri. The role of the nonlinear wave kinematics on the global responses of an OWT in parked and operating conditions. *Journal of Wind Engineering and Industrial Aerodynamics*, 123:363–376, 2013.

[44] J. R. Morison, J. W. Johnson, and S. A. Schaaf. The force exerted by surface waves on piles. *Journal of Petroleum Technology*, 2(05):149–154, 1950.

[45] S. Malenica and B. Molin. Third-harmonic wave diffraction by a vertical cylinder. *Journal of Fluid Mechanics*, 302:203–229, 1995.

[46] Y. Ma, G. Dong, M. Perlin, S. Liu, J. Zang, and Y. Sun. Higher-harmonic focused-wave forces on a vertical cylinder. *Ocean Engineering*, 36:595–604, 2009.

[47] B. T. Paulsen, H. Bredmose, H. B. Bingham, and N. G. Jacobsen. Forcing of a bottom-mounted circular cylinder by steep regular water waves at finite depth. *Journal of Fluid Mechanics*, 755:1–34, 2014.

[48] R. C. T. Rainey. Slender-body expressions for the wave load on offshore structures. *Proceedings of the Royal Society A: Mathematical, Physical and Engineering Sciences*, 450(1939):391–416, 1995.

[49] J. N. Newman. Nonlinear Scattering of Long Waves by a Vertical Cylinder. In John Grue, Bjørn Gjevik, and Jan Erik Weber, editors, *Waves Nonlinear Process. Hydrodyn.* Springer, Netherlands, Dordrecht, pages 91–102, 1996.

[50] O. M. Faltinsen. Ringing loads on a slender vertical cylinder of general cross-section. *Journal of Engineering Mathematics*, 35:199–217, 1999.

[51] M. Huseby and J. Grue. An experimental investigation of higher-harmonic wave forces on a vertical cylinder. *Journal of Fluid Mechanics*, 414:75–103, 2000.

[52] J. Grue and M. Huseby. Higher harmonic wave forces and ringing of vertical cylinders. *Applied Ocean Research*, 24(4):203–214, 2002.

[53] Norwegian Petroleum Directorate. NPD Annual Report 1992. Technical Report, 1992.

[54] J. Grue, G. Bjørshol, and Ø. Strand. Higher harmonic wave exciting forces on a vertical cylinder. *Preprint series. Mechanics and Applied Mathematics.* Publisher: Matematisk Institutt, Universitetet i Oslo, Oslo, Norway, pages 1–28, 1993.

[55] C. T. Stansberg, H. Huse, J. R. Krokstad, and E. Lehen. Experimental study of non-linear loads on vertical cylinders in steep random waves. In *Fifth Int. Offshore Polar Eng. Conf.*, pages 75–82, 1995.

[56] C. T. Stansberg. Comparing ringing loads from experiments with cylinder of different diameters – an experimental study. In *BOSS '97 Behav. Offshore Struct.*, volume 2, 1997.

[57] K. R. Gurley and A. Kareem. Simulation of ringing in offshore systems under viscous loads. *Journal of Engineering Mechanics*, 124(5):582–586, 1998.

[58] K. R. Gurley, A. Kareem, F. Waisman, and M. Grigoriu. A stochastic model of ringing. In *Int. Conf. Offshore Mech. Arct. Eng.*, 1998.

[59] F. Waisman, K. Gurley, M. Grigoriu, and A. Kareem. Non-Gaussian model for ringing phenomena in offshore structures. *Journal of Engineering Mechanics*, 7:730–741, 2002.

[60] J. R. Krokstad and F. Solaas. Study of nonlinear local flow. In *Tenth Int. Offshore Polar Eng. Conf.*, volume 4, pages 449–454, 2000.

[61] J. Grue. On four highly nonlinear phenomena in wave theory and marine hydrodynamics. *Applied Ocean Research*, 24(5):261–274, 2002.

[62] R. C. T. Rainey. Weak or strong nonlinearity: the vital issue. *Journal of Engineering Mathematics*, Newman Hon:1–27, 2007.

[63] A. Mockutė, E. Marino, C. Lugni, and C. Borri. Comparison of hydrodynamic loading models for vertical cylinders in nonlinear waves. *Procedia Engineering*, 199:3224–3229, 2017.

[64] A. Mockutė, E. Marino, C. Lugni, and C. Borri. Comparison of hydrodynamic loading models for bottom-mounted slender cylinders in highly nonlinear waves. In *Proc. of the 9th European Seminar OWEMES*, 2017.

[65] E. Marino, H. Nguyen, C. Lugni, L. Manuel, and C. Borri. Irregular nonlinear wave simulation and associated loads on offshore wind turbines. *Journal of Offshore Mechanics and Arctic Engineering*, 137(2):21901, 2015.

[66] E. Marino, A. Giusti, and L. Manuel. Offshore wind turbine fatigue loads: the influence of alternative wave modeling for different turbulent and mean winds. *Renewable Energy*, 102:157–169, 2017.

[67] C. Swan, T. Bashir, and O. T. Gudmestad. Nonlinear inertial loading. Part I: accelerations in steep 2-D water waves. *Journal of Fluids and Structures*, 16(3):391–416, 2002.

[68] J. Jonkman, S. Butterfield, W. Musial, and G. Scott. Definition of a 5-MW Reference Wind Turbine for Offshore System Development. Technical Report, NREL, 2009.

[69] E. Marino, C. Borri, and U. Peil. A fully nonlinear wave model to account for breaking wave impact loads on offshore wind turbines. *Journal of Wind Engineering and Industrial Aerodynamics*, 99(4):483–490, 2011.

[70] J. Wienke and H. Oumeraci. Breaking wave impact force on a vertical and inclined slender pile – theoretical and large-scale model investigations. *Coastal Engineering*, 52:435–462, 2005.

[71] E. Marino, C. Borri, and C. Lugni. Influence of wind-waves energy transfer on the impulsive hydrodynamic loads acting on offshore wind turbines. *Journal of Wind Engineering and Industrial Aerodynamics*, 99(6–7):767–775, 2011.

[72] C. Borri, P. Biagini, and E. Marino. Large Wind Turbines in Earthquake Areas: Structural Analyses, Design/Construction & In-situ Testing. In C.C. Baniotopoulos, C. Borri, and T. Stathopoulos, editors, *Environmental Wind Engineering and Design of Wind Energy Structures*. Springer, Vienna, Austria, 2011.

[73] E. Marino, C. Lugni, and C. Borri. Aero-hydroelastic instabilities on an offshore fixed-bottom wind turbine in severe sea states. In *Proceedings of the 11th International Conference on Hydrodynamics (ICHD 2014)*, 2014.

[74] S. Schløer, H. Bredmose, and H. B. Bingham. The influence of fully nonlinear wave forces on aero-hydro-elastic calculations of monopile wind turbines. *Marine Structures*, 50:162–188, 2016.

[75] M. B. Van Der Meulen, T. Ashuri, G. J. W. Van Bussel, and D. P. Molenaar. Influence of nonlinear irregular waves on the fatigue loads of an offshore wind turbine. In *The Science of Making Torque from Wind*, 2012.

[76] A. Mockutė, E. Marino, and C. Borri. Effects of wind-wave misalignment on the dynamic response of an offshore wind turbine in highly nonlinear waves. In *Proc. of the 7th European-African Conference on Wind Engineering*, 2017.

[77] H. Bredmose, J. Mariegaard, B. T. Paulsen, *et al.* The Wave Loads Project. Technical Report, DTU Wind Energy, 2013.

[78] E. Marino, G. Stabile, C. Borri, and C. Lugni. A comparative study about the effects of linear, weakly and fully nonlinear wave models on the dynamic response of offshore wind turbines. In *Proceedings of the 5th International Conference on Structural Engineering, Mechanics and Computation, SEMC*, pages 983–988, 2013.

[79] E. Marino, C. Lugni, G. Stabile, and C. Borri. Coupled dynamic simulations of offshore wind turbines using linear, weakly and fully nonlinear wave models: The limitations of the second-order wave theory. In *Proceedings of the 9th International Conference on Structural Dynamics, EURODYN 2014*, pages 3603–3610, 2014.

[80] B. M. Lake and H. C. Yuen. A new model for nonlinear wind waves. Part 1. Physical model and experimental evidence. *Journal of Fluid Mechanics*, 88:33–62, 1978. (cited in: J. R. Chaplin, R. C. T. Rainey, and R.W.Yemm. Ringing of a vertical cylinder in waves. *Journal of Fluid Mechanics*, 350: 119–147, 1997.)

[81] L. W. Schwartz. Computer extension and analytic continuation of Stokes' expansion for gravity waves. *Journal of Fluid Mechanics*, 62:553–578, 1974. (cited in: J. R. Chaplin, R. C. T. Rainey, and R.W.Yemm. Ringing of a vertical cylinder in waves. *Journal of Fluid Mechanics*, 350:119–147, 1997.)

[82] N. Atkins, R. A. Lyons, and R. C. T. Rainey. Summary of findings of wave load measurements on the Tern platform. HSE Rep. OTH 95 467. HSE Books, London, 1997. (cited in: J. R. Chaplin, R. C. T. Rainey, and R.W.Yemm. Ringing of a vertical cylinder in waves. *Journal of Fluid Mechanics*, 350: 119–147, 1997.)

Chapter 3

Numerical and experimental tools for small wind turbine load analysis

Karczewski Maciej[1], *Sobczak Krzysztof*[1], *Lipian Michal*[1], *and Jozwik Krzysztof*[1]

3.1 Introduction

The designing process of wind turbines is a multidisciplinary problem of mechanical and electrical engineering. Each kind of wind turbines requires methods that generally could be recognized as similar. In the field of mechanics, aerodynamic loads have to be carefully recognized. A small diffuser augmented wind turbine (DAWT), a kind of turbine being of great interest on the prosumer market, was designed within the STOW project carried out by the Military University of Technology, Warsaw, the Lodz University of Technology and the University of Agder, specifically for modest wind speeds prevailing in Poland. The main object of the activities conducted at the Institute of Turbomachinery, Lodz University of Technology (IMP TUL) within this project was an aerodynamic design of the turbine. Numerical and experimental tools for the load analysis are indispensable in this process.

Figure 3.1 presents a modified approach to a small wind turbine (SWT) design process adapted from the International Electrotechnical Commission (IEC) 61400-2 standard. The process is divided into several steps. The criteria (listed in white boxes) that help to define and later to evaluate and judge the performance of the machine and the durability of proposed construction elements are the central axis of the decision path. A decision on the design quality must be informed; hence the feedback is provided from a number of tests: measurement campaigns, functional tests and inspections (green colour). Furthermore, the decisions are supported with supplementary fact sheets and models that, through calculations, provide a complementary set of data for analysis. The models are indispensable in the design process as they help to cut costs. In the small turbine design process, the models can be of various origins from country-wide and region-wide wind data, the IEC-based load model, manufacturer-provided fact sheets about materials, measurement equipment, to end with own very

[1] Faculty of Mechanical Engineering, Institute of Turbomachinery, Lodz University of Technology, Poland

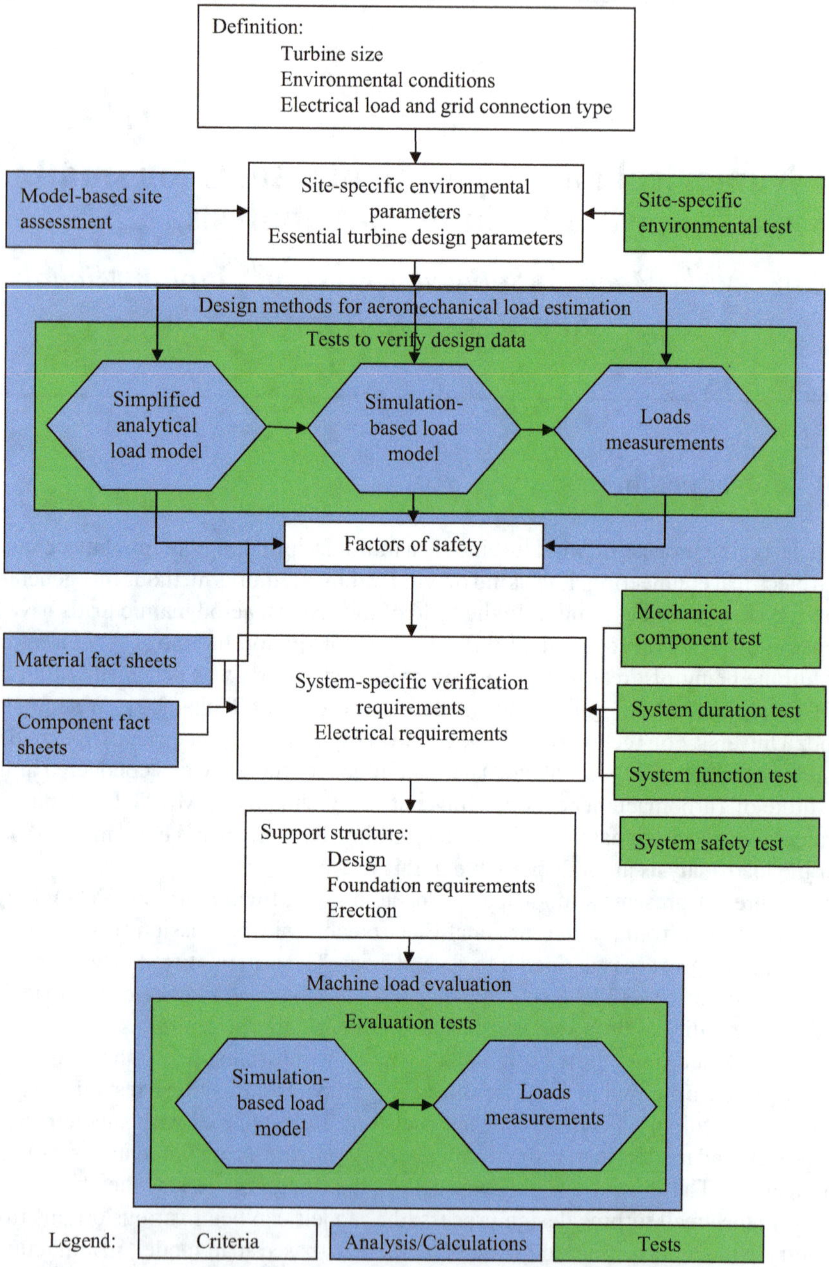

Figure 3.1 Proposed decision path leading to a successful design of the small wind turbine adapted from the IEC 61400-2 standard [7]

sophisticated computational fluid dynamics (CFD) models, experiments in scale and lengthy open field measurement campaigns of a prototype.

In the designing process, one must begin with defining local environmental conditions. In absence of site-specific data, measurements from a near-by site can be used. For the presented examples, the wind speed records gathered by the airport authorities at the Lodz city airport [1] were initially used. Later site-specific wind conditions were measured via cup anemometers installed on the top of the building. With the mean wind velocity established, an outline design of a ducted wind turbine was possible. The duct (diffuser with a brim) was to increase the mass flow rate through the turbine working section, providing thus a higher power production rate. To design the turbine blade, an algorithm proposed by [2] for the DAWT was implemented in a two-dimensional CFD (2D CFD) study and supplemented by own design optimization strategy [3]. Then, the design investigations were split into two paths: experimental tests and 3D simulations. The first measurement campaign in scale 1:6 was conducted in a low-speed wind tunnel and included separate tests on a diffuser size. The first campaign helped to size the objects allowable in an open test section. The second experimental campaign focused on the turbine open rotor. The third visit to the tunnel ended with a set of data for the frozen geometry configuration – a ducted wind turbine. Each campaign was supplemented with a complementary 2D/3D CFD study [4–6]. A simplified load model (SLM) was prepared to assist in the assessment of specific loads under normal and extreme operating conditions. At the same time, 3D CFD simulations of the full scale turbine were performed. They provided data for design verifications and modifications.

With the models and designs validated in scale, the process moved a level down to the construction of a prototype in 1:2 scale. A detailed design of mechanical and electrical components began. A number of functional tests was prepared in cooperation with electrical engineers, leading to the first design and then to the execution of the complicated measurement, regulation and safety systems. The construction was completed with a trussed tower prepared to withstand the extreme operating conditions defined by the standard. The prototype was set atop the building, and the load estimation under actual operating conditions was possible. The chapter introduces this path, beginning with a look on the CFD methodology, normative requirements, to finish with the experimental assessment of loads where some comparisons are provided.

3.2 SWT design and development

3.2.1 Analytical methods

Any efficient designing process that includes aerodynamic optimization of an object cannot be accomplished without an analytical approach. The so-called momentum theory approach was one of the first methods offered for wind turbine rotor studies. This method, proposed originally at the end of the nineteenth century by Froude and Rankine, incorporates a 1D flow model. The rotor is represented as a frictionless thin disk of constant load that induces a uniform pressure drop (Figure 3.2).

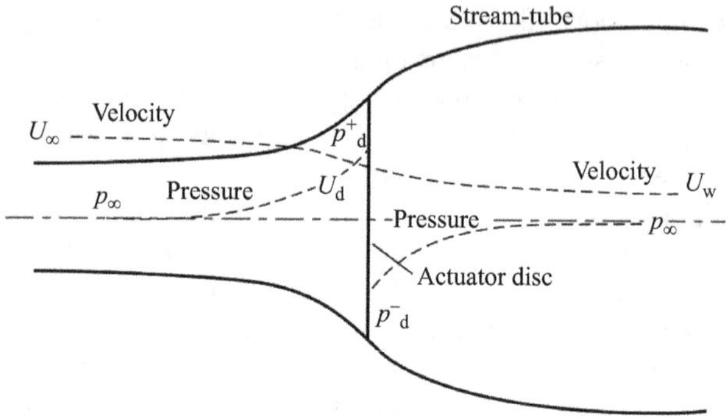

Figure 3.2 Illustration of the 1D rotor theory, [8]

Thus, the model ignores the wake rotation effect. The pressure drop Δp is related to a velocity field variation and can be determined by employing the Bernoulli theorem:

$$\Delta p = \frac{\rho(U_\infty^2 - U_w^2)}{2} \tag{3.1}$$

The pressure variation creates the thrust force T on the rotor of a diameter D, which can be translated onto the rotor power as

$$T = \Delta p \cdot \pi \cdot \left(\frac{D}{2}\right)^2 \tag{3.2}$$

$$P_{\text{rotor}} = T \cdot U_d \tag{3.3}$$

A streamwise variation of the axial velocity component is described by the so-called axial induction factor (a), dependent on the rotor geometry and operating conditions. It relates velocity magnitudes in three characteristic locations: far upstream (free-stream velocity U_∞), in the rotor plane (U_d) and far downstream (U_w):

$$\begin{aligned} U_d &= (1 - a) \cdot U_\infty \\ U_w &= (1 - 2a) \cdot U_\infty \end{aligned} \tag{3.4}$$

Two coefficients playing the main role in the wind turbine analysis are related to power and axial force (thrust). The wind turbine power coefficient Cp (3.5) can be interpreted as the aerodynamic efficiency of the rotor. The thrust coefficient Ct (3.6) is a measure of the axial effort of the rotor exerted on the flow, and thus, on the resulting pressure drop.

$$\text{Cp} = \frac{P_{\text{rotor}}}{P_{\text{wind}}} = \frac{P_{\text{rotor}}}{\rho A U_\infty^3/2} \tag{3.5}$$

$$\text{Ct} = \frac{T_{\text{rotor}}}{T_{\text{wind}}} = \frac{T_{\text{rotor}}}{\rho A U_\infty^2/2} \tag{3.6}$$

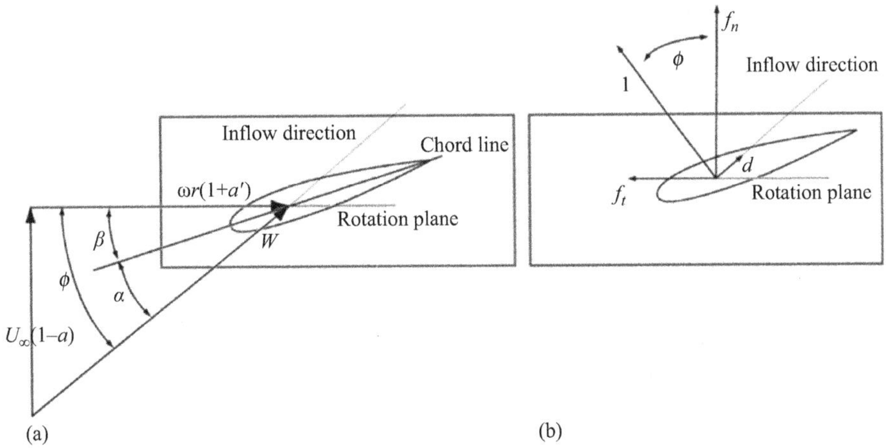

Figure 3.3 (a) velocity components, angles and (b) forces for wind turbine blade cross-section, own materials after [11]

P_{wind} and T_{wind} correspond, respectively, to the power comprised in the wind stream of a circular section and the diameter D. When (3.1)–(3.4) are substituted into (3.5) and (3.6), we obtain after some transformation:

$$\text{Cp} = 4a(1-a)^2 \tag{3.7}$$

$$\text{Ct} = 4a(1-a) \tag{3.8}$$

Relationship (3.7) is fundamental from the viewpoint of the blade geometry oimization, as it relates directly the rotor efficiency to the axial induction factor. Equation (3.7) is a third-order polynomial of a and it attains its local maximum at $a = 1/3$. The corresponding value $\text{Cp}(a = 1/3) = 16/27 \approx 0.593$ is known as the Betz or Lanchester–Betz limit. It follows, therefore, that no wind turbine rotor can actually achieve a higher aerodynamic efficiency than about 60%. It has been observed that relationships (3.7) and (3.8) hold for a value lower than about 0.4 only. With a further increase in the induction factor (and, thus, the efforts exerted by the rotor on the flow), the wake model assumed in the momentum theory is no longer valid. Notably, following formula (3.4), the wind velocity in the wake would become negative for $a > 0.5$. This problem was studied and addressed by Glauert in his works concerning originally helicopter rotors (see, e.g. [9]).

To compute aerodynamic properties of a rotor with the given geometry, Drzewiecki [10] proposed the so-called Blade Element Theory (BET). He considered the rotor as an ensemble of aerofoils studied independently from one another. Its performance was unsatisfactory, as the model did not initially take into account the induced velocity component. The vortex studies by Prandtl at the beginning of the twentieth century enabled Joukowski and Betz to address this issue. A reconstructed velocity triangle over the rotating aerofoil is shown in Figure 3.3(a). Similarly, as a refers to a modification of the axial velocity component, a' (tangential induction factor) applies to the tangential velocity component. Knowing the relative velocity magnitude W at a certain radial position r, one can determine the local aerodynamic forces, a lift (l) and a drag (d) per the unit blade span, as

$$\begin{aligned} l &= \frac{\rho W^2}{2} \cdot c \cdot C_l(\alpha, Re, Ma) \\ d &= \frac{\rho W^2}{2} \cdot c \cdot C_d(\alpha, Re, Ma) \end{aligned} \tag{3.9}$$

In equation system (3.9), c is an aerofoil chord, and C_l C_d are the lift and force coefficients, correspondingly. Their values are functions of the angle of attack α and local Reynolds and Mach numbers (the latter essentially for high-speed flows). Aerodynamic force coefficients are evaluated both experimentally and numerically and are a guideline for determination of aerodynamic profile applicability considering expected flow conditions. For example, for SWTs, the aerofoils have to exhibit good performance at low values of Reynolds numbers, as the Re magnitude is usually around 100,000 for such devices.

Lift and drag forces are perpendicular and parallel to the relative velocity vector, respectively [see Figure 3.3(b)]. In the global coordinate system, they can be expressed as

$$\begin{aligned} f_n &= l \cdot \cos(\phi) + d \cdot \sin(\phi) \\ f_t &= l \cdot \sin(\phi) - d \cdot \cos(\phi) \end{aligned} \tag{3.10}$$

where f_n and f_t denote the normal and tangential aerodynamic force components, correspondingly, in the considered wind turbine section. Integrating f_n along the blade span makes it possible to compute the total normal force exerted on the rotor by the flow. Similarly, integrating f_t multiplied by the local radius r gives the total rotational torque, Q (for complete formulae, refer, e.g. to [11]). Similarly, as for power and thrust, the torque coefficient Cq can be introduced:

$$Cq = \frac{Q_{\text{rotor}}}{Q_{\text{wind}}} = \frac{Q_{\text{rotor}}}{(\rho A U_\infty^2/2) \cdot R} \tag{3.11}$$

In both the presented methods, the axial induction factor a is the remaining unknown. Thus, Betz proposed to consolidate the actuator disk approach and the BET into the Blade Element Momentum (BEM) theory. The iterative procedure that permits to calculate the induced velocity magnitude, and thus the rotor performance, was employed and further developed by Glauert in 1935. The theory was also further modified to take into account the finite number of blades (Prandtl tip-loss correction) and different wake behaviour for high values of a (Glauert's correction). The BEM approach is currently the basic tool in industrial applications of the wind turbine blade design and optimization. The model is appraised for its favourable relationship between the quality of results and the demand of resources. Ready-made computational scripts est and can be consulted, for instance in [12].

An assumption of independent operation of neighbouring blade sections is an important disadvantage of the BET. Addressing this issue, Lanchester and Prandtl developed the so-called lifting-line theory. The lift force over a unitary wing section is related to the circulation Γ, as follows from the Kutta–Joukowski theorem [11]:

$$l = \rho \cdot U_\infty \cdot \Gamma \tag{3.12}$$

Thus, a spanwise lift distribution can be determined if the spanwise circulation distribution is known. The latter is directly related to the vortex sheet forming downstream of the trailing edge and can be found, providing that the wing or blade geometry is known. Therefore, the lifting line approach is capable of predicting the rotor properties without the need of additional corrections. This method is, however, less popular than the BEM.

3.2.2 BET–CFD coupling

In some applications, the momentum theory cannot be applied directly due to a local velocity/pressure field modification. This is the case for high values of a, when a backflow occurs downstream of the rotor and is compensated by the aforementioned Glauert's correction. As another example, ducted rotors are encased in a special structure meant to increase the flow velocity through the wind turbine rotor. In such a case, the local flow fields must be determined otherwise, for example by applying CFD. A development in these methods makes it possible to perform high-quality, yet rapid determination of velocity and pressure fields. In this manner, the local velocity vector is readily available and can be directly used in equation system (3.9). The forces determined from system (3.10) are supplied to the Navier–Stokes equations in the form of source terms (with a negative sign, bearing in mind Newton's 3rd law of motion). This, in turn, modifies the local pressure and velocity fields, which are then again used to determine aerodynamic forces. The iterative procedure is continued until the desired level of solution convergence is reached.

The simplest BET–CFD approach (referred to as actuator models further on) is a 2D (or pseudo-2D for 3D-exclusive solvers like ANSYS CFX) model [Figure 3.4(a)]. The rotor, similarly as in the momentum theory, is represented as a section of the axisymmetric thin disk (hence its name: Actuator Disk Model, ADM). Axial forces are added to the Navier–Stokes equations in the cells adjacent to the disk surface. Such an approach can be favoured when a rapid and simple model is required. That could include applications such as initial optimization of the wind farm layout, in which modelling of a wind turbine interaction in full 3D models could be computationally unjustifiable. An optimization of the aerodynamic design of a blade or other components of the wind turbine (hub, shrouding in the case of the ducted rotor) might be another application. Since the model readily provides a pressure distribution along the blade span, it is also possible to estimate the resulting stresses and ensure proper blade mechanical resistance.

Among the limitations of the 2D ADM model, its incapability to depict effects of wake rotation, since only the axial forces are supplied to the flow equations, is possibly the most important one. To take into account the tangential loads, it is necessary to consider the case as a full 3D problem [Figure 3.4(b)]. This is the classic 3D ADM approach, widely used and acclaimed for studies of the flow around wind turbines.

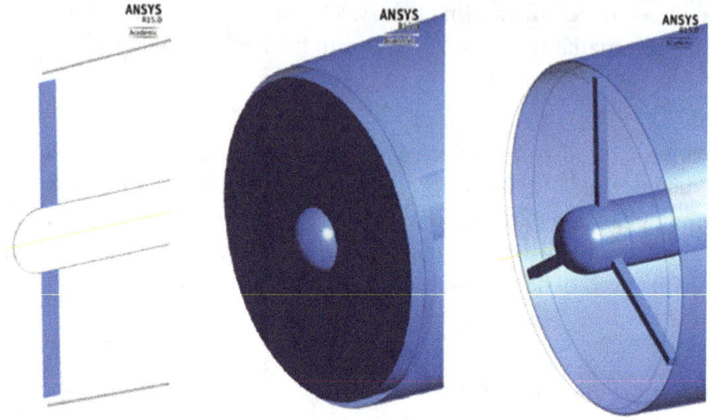

(a) (b) (c)

Figure 3.4 2D AMD (a), 3D ADM (b) and 3D ALM (c) BET–CFD models for the shrouded wind turbine study [13]

Disk loads are considered as axisymmetric – all cells situated at the same radial position r attain the same value of the applied force. Hence, the wake behind the actuator disk remains essentially axisymmetric.

In reality, the rotor is a structure composed of a finite number of blades. It seems natural to consider an actuator model in which forces are applied in discrete regions, corresponding to the actual blade positions. Such an approach is known as the actuator line model [ALM, Figure 3.4(c)]. Usually considered as unsteady, the model permits to reproduce the helical character of the wake and examine its meandering. In the basic approach, loads are distributed along the lines that represent blades. An application of concentrated loads in singular cells imposes high pressure gradients that disturb locally pressure fields. This phenomenon can lead to solution instability, hence the need to distribute loads onto a wider range of cells. For this purpose, a Gaussian distribution can be used (e.g. [14]), usually in the tangential direction. The load magnitude in a cell is determined as a sum of distributed loads coming from each blade. It depends on the cell angular position and the blade position at a given timestep. Helicoidal vortex formation downstream of the rotor in the unsteady ALM is presented in Figure 3.5. Since the solver chooses appropriate cells on its own, it is possible to use relatively simple meshes. Own experience proves, however, that a structured regular mesh provides a better control over the model parameters and operation. An alternative approach (see [13]) involves grouping together cells representing blades at the level of the meshing process. The zones thus created can have a simple form of cylinder or cuboid or may resemble the actual blade geometry. Hence, this approach is sometimes referred to as the Actuator Shape Model in some investigations. To depict unsteady phenomena and rotor movement, typically it is necessary to apply a sliding mesh approach, which can introduce additional errors due to flow fields averaging at interfaces. In return, the model offers a better control over the location of cells affected by the loads.

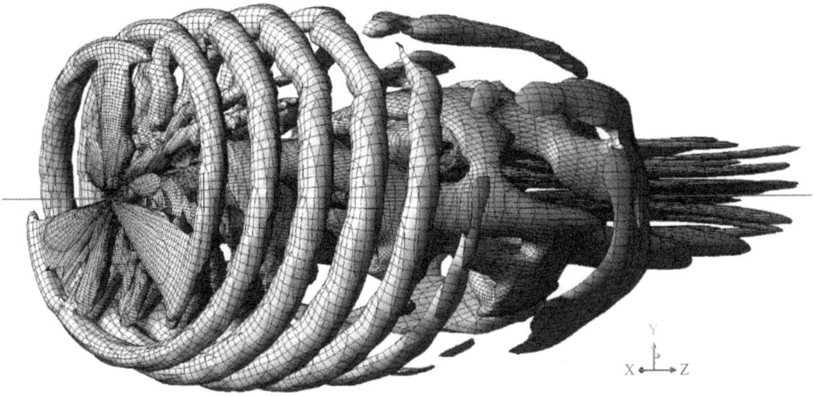

Figure 3.5 Swirling strength in the helicoidal wake downstream of the wind turbine, unsteady ALM with a Gaussian load distribution

There is an open question concerning an application of the Prandtl tip-loss correction in the BET–CFD coupling models. Although it was developed for modifying the momentum theory limitations, its physical interpretation is to address a tip leakage. In the tip region of the wing or blade, the pressure gradient between its intra- and extrados provokes a transversal movement of air. This violates the basic assumption of the BET, as the flow cannot be considered as 2D in that region. The tip leakage provokes a dramatic decrease in the blade end region contribution to an overall lift production. The Prandtl tip-loss correction enables taking this into consideration by reducing gradually the applied loads towards the blade end.

3.2.3 Numerical methods for the WT development

In diffuser augmented wind turbines (DAWT), a special structure is used to increase the flow velocity through the wind turbine rotor. It is achieved by an application of a shroud in the form of a diffuser with the rotor placed at its inlet (throat). What is more, a brim (flange) can be placed at the diffuser outlet as proposed in [15] and further investigated experimentally in [2] and numerically in [16]. Flow separations downstream of the brim decrease the pressure in this region and additionally increase the flow rate through the turbine rotor (Figure 3.6). As mentioned above, a strong modification of the local velocity/pressure field due to shroud presence makes a direct application of the classical momentum theory impossible. Therefore, the BET–CFD approaches described above can be used.

A design stage in the DAWT development requires numerous (hundreds or thousands) CFD simulations to optimize an aerodynamic design of rotor blades and a diffuser as well. Therefore, the rapid determination of velocity and pressure fields is necessary, which can be achieved by an application of 2D ADM models. Despite the fact that this numerical approach is the least accurate of all the above mentioned BET–CFD approaches, it has one undeniable advantage – a solution of the particular

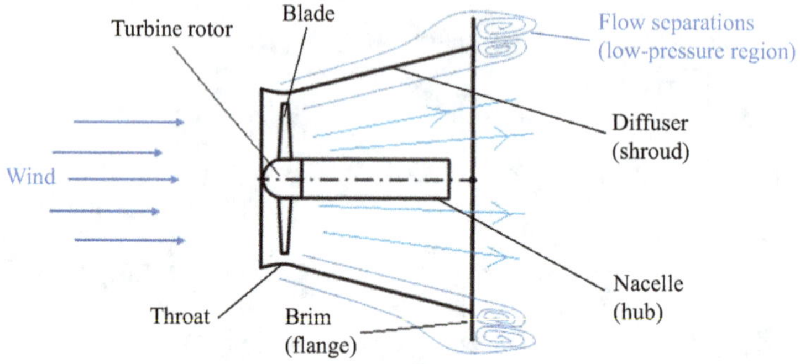

Figure 3.6 Scheme of the DAWT with a brim [3]

configuration is attained in less than half an hour on a PC. Other 3D methods (especially the ALM) can provide more precise prediction; however, the individual solution requires hours of computations on multiprocessor/multicore machines.

A 2D ADM model was applied to design a small DAWT of the rotor diameter equal to 2 m. The optimal rotor blade geometry (distribution of the blade pitch angle and the aerofoil chord along the blade height) was determined for each diffuser arrangement under consideration for selected values of the tip speed ratio (TSR), defined as

$$\lambda = \text{TSR} = \frac{\omega D}{2U_\infty} \tag{3.13}$$

An influence of the most important geometric parameters of the shroud (diffuser angle and length, brim height) and hub (length, diameter) on the turbine performance was investigated. Taking into account the aerodynamic criterion, an optimal design of the diffuser was determined [angle (2θ) – 18°, length – 2.5 m, inlet lip length – 0.2 m, brim height – 0.3 m].

In the next stage of analysis, full 3D numerical models (FRM – fully resolved rotor model) of the optimal turbine configurations were developed and their performance was verified with 3D simulations. In contrast to the previously described methods (3D ADM and ALM), the rotor geometry was fully represented; thus, its influence on the flow structure in the diffuser was taken into account in a straightforward manner without any extra modelling [beside intrinsic turbulence modelling in Reynolds Averaged Navier–Stokes (RANS) methods]. The models were prepared, and simulations conducted for the two blade configurations corresponding to TSR = 6 (a massive blade) and 8 (the slender one).

A comparison of 2D ADM and full 3D predictions of the velocity distribution in the diffuser of the DAWT for these two blade configurations was presented in Figure 3.7. The circumferentially averaged velocity distributions in the diffuser from 3D simulations were compared to 2D predictions. For slander blades (TSR = 8), the velocity fields from both solutions were very similar; however, some decrease in

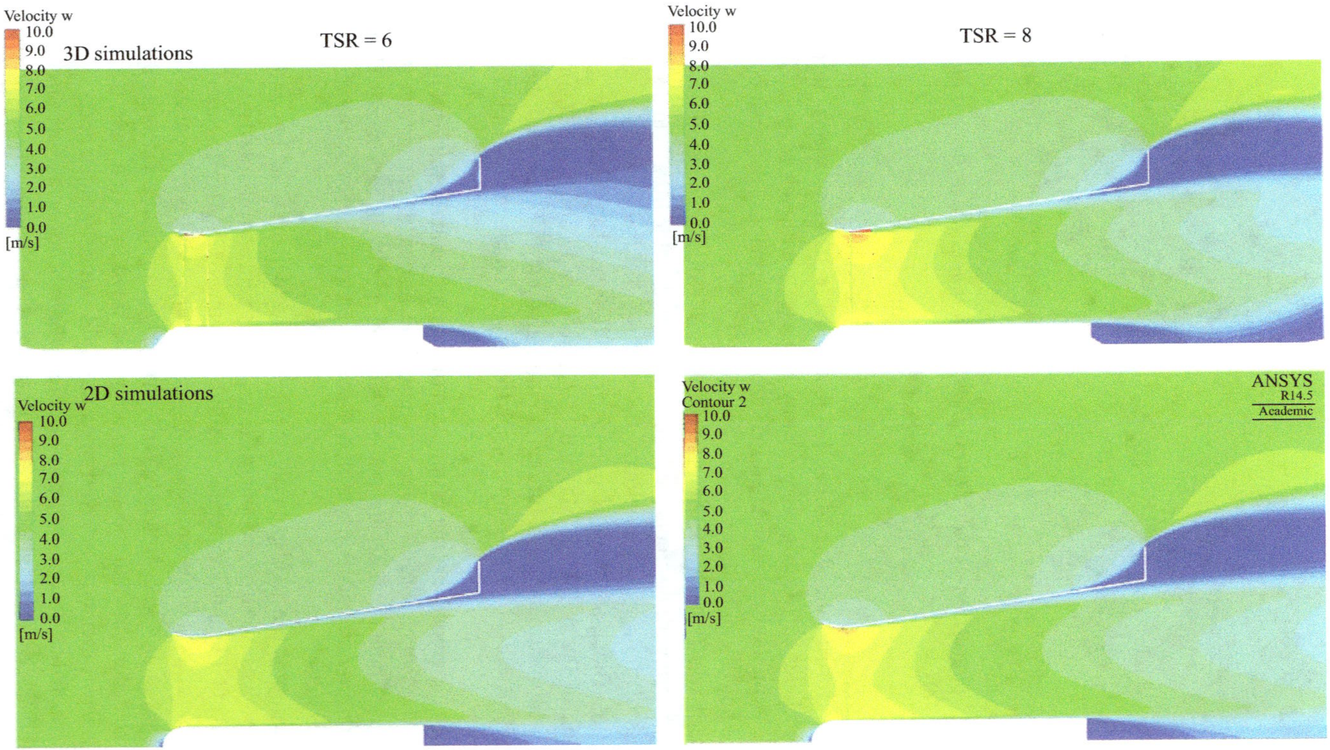

Figure 3.7 Comparison of 2D ADM and full 3D predictions of the velocity distribution in the diffuser of the DAWT [3]

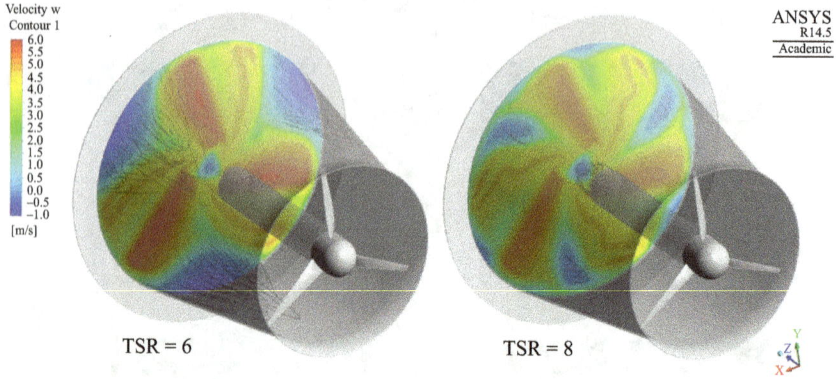

Figure 3.8 Comparison of the 3D flow field prediction at the diffuser outlet for TSR = 6 and 8 [3]

the velocity close to the shroud can be observed for 3D simulations. In the case of massive blades (TSR = 6), this velocity decrease was much higher. Generally, the 2D ADM method overestimated the turbine performance. For both considered blades, the power coefficient was the same – 1.12. Unfortunately, for the full 3D prediction, it was 0.95 for TSR = 8 and 0.85 for TSR = 6, respectively. In both cases, the performance discrepancies can be explained by the flow field disturbances downstream of the rotor. The velocity fields at the diffuser outlet were presented in Figure 3.8. Black streamlines were additionally used to present the regions of recirculation in the diffusers. Massive blades in the case of TSR = 6 intensified the flow field disturbances. A further strength analysis [17] revealed that slander blades (for TSR = 8) are not strong enough to sustain combined centrifugal and wind loads.

It can be concluded that the assumption of circumferentially uniform flow fields in the 2D ADM model leads to far too optimistic results. It seems that the ALM would be more beneficial as far as the prediction quality is concerned. However, the computing requirements made it still unaffordable as a tool at the design and optimization stage.

A wind turbine is a good example of objects for which the fluid flow and strength aspects have to be taken into account to design an efficient and reliable device. Numerical methods of CFD and computational solid mechanics (CSM) can be used to assess the fluid flow and the structure strength, respectively. However, separate CFD and CSM analyses cannot be sufficient to verify properly designs with a strong interaction of deformable elements and the surrounding fluid flow. In such a case, a combination of these two methods, referred to as the fluid–structure interaction (FSI), has to be applied. It has become popular in practical applications as software producers (e.g. ANSYS, Abaqus) introduced automatic procedures enabling the appropriate data transfer between solvers. If one of the solution fields strongly affects other fields but is not affected by any other field, the coupling between fields is essentially unidirectional [18]. In such a case, the one-way FSI can be sufficient to assess the design. If the aerodynamic loads obtained from CFD simulations of the flow field are used in the CSM prediction of the structure in contact with the fluid, then it is a typical application of this method. However, in this approach, the deformation of the structure is not taken into account in the CFD analysis. In the case of wind turbine blades which can deform considerably, this approach has a limited scope of application. It can be used to assess initially the blade durability, but precise simulations require full coupling of the solvers. Therefore, if there is a strong interaction between fluid and structure fields, the two-way FSI has to be applied [18]. In this method, the loads obtained from CFD are used in CSM solutions, and the deformations of the structure obtained from the structural solver are transferred to the flow solver modifying the shape of the flow channel. Usually, the procedure is time consuming because a significant number of data exchange loops are necessary to obtain the solution, i.e. the finally deformed structure and the corresponding fluid flow field. However, it can provide reliable data on the efficiency and durability of a device.

As mentioned above, in the project of the small DAWT, the CSM simulations were performed for two designs of the blades: the massive (optimized for TSR = 6) and slender (optimized for TSR = 8) one [17]. The one-way FSI revealed that for the rated velocity of 12 m/s, the slender blade deformed too much, and there was a problem of its fatigue under the variable velocity (and load) operation. Therefore, despite its poorer performance, the massive blade was selected.

Later, two-way FSI simulations were conducted to verify the strength of the blade and the turbine performance for deformed blades for nominal (5 m/s) and rated (12 m/s) wind speeds. The FSI simulations were performed in the ANSYS Workbench, which integrates all the products of the ANSYS company and facilitates an appropriate data transfer between CFD and CSM solvers. The fluid flow part of the FSI is based on the previous CFD analysis with ANSYS CFX, the results of which were presented in Figure 3.7 and Figure 3.8. A structure analysis was performed in the ANSYS solver. At that stage of the project, the blade made of aluminium alloy was considered. The fix support was applied at the blade root at its connection to the rotor hub. Beside the aerodynamic loads obtained from the CFD solution, the centrifugal load due to blade rotation was considered. The maximal blade deformation (at its tip) for the nominal wind speed was very low (0.5 mm) and had no impact on the turbine performance. Prior to performing FSI simulations for the rated wind speed, there was a doubt if a significant blade deformation deteriorates its aerodynamic performance. Fortunately, even though the maximal deformation was quite high – 3.5 mm, the blade was deflected downstream but not twisted (Figure 3.9). Therefore, the angle of attack was not changed, and the turbine performance remained at the level predicted in the pure CFD simulation. What is more, the maximal von Mises stress was far beneath the yield strength of the aluminium alloy (Figure 3.10). Therefore, in further stages of the turbine development, some other blade materials (E-glass, polyamide) were considered. It is interesting to notice the highest stress concentration in the middle of the blade height, which indicates that the place of the blade aerofoil was changed from thicker into thinner.

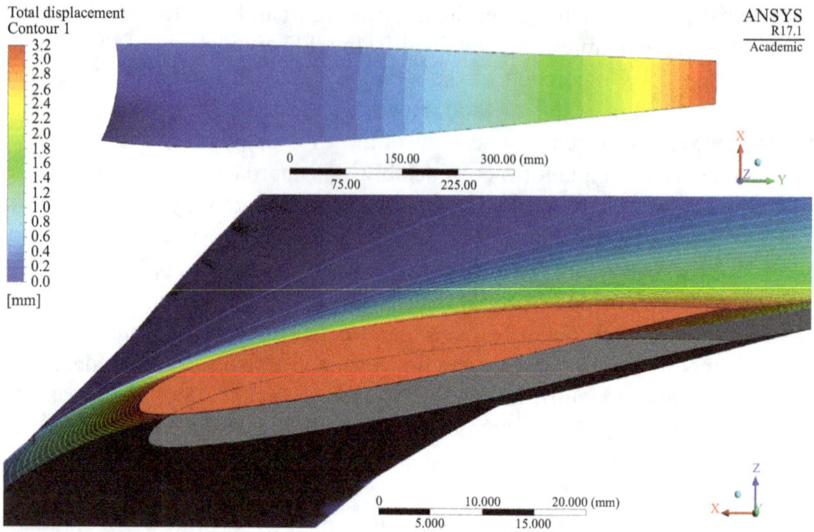

Figure 3.9 Total blade deformation: top – upstream view of the blade, bottom – top view of the blade – non-deformed blade is shown in grey

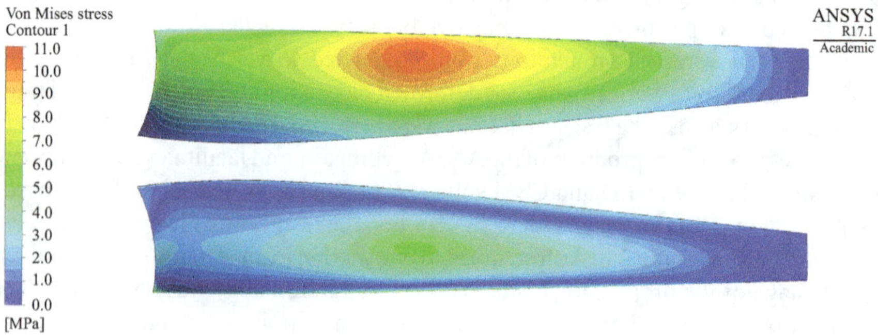

Figure 3.10 Von Mises stress of the blade: top – upstream, bottom – downstream views of the blade

3.2.4 CFD simulations of the WT

CFD simulations of the wind turbine were carried out for the task configuration which took into account the most important elements of the final design. The geometry of the numerical model is presented in Figure 3.11. Due to the mesh generation, it was necessary to simplify the geometry of the diffuser support. Additionally, all small elements used for the diffuser assembly were neglected. The simulations were carried out for idealized conditions of the free, steady, axial wind flow. It was assumed that the turbine was placed far away from the ground and buildings. Therefore, the computational space was in the form of a cylinder (45 m long and 15 m in radius)

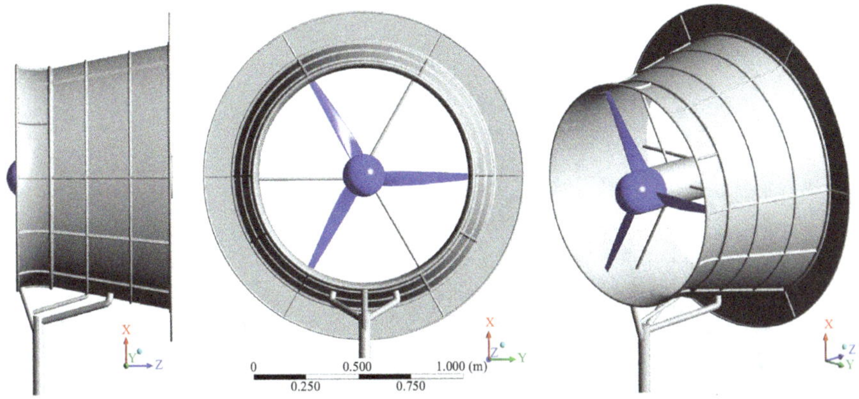

Figure 3.11 Side, front and general views of the wind turbine numerical model

coaxial with the turbine rotor axis with no extra elements inside, beside the turbine. The incoming wind direction was along the turbine axis, with a low intensity of turbulence.

The computational mesh used in the simulations was composed of several parts. On the basis of the experience gained during simplified design simulations, it was decided to use a block-structured hexahedral mesh in the region inside and downstream of the diffuser. However, the geometry of the diffuser support did not allow one to use the preferable hexahedral mesh outside the diffuser. Therefore, in this region, a tetrahedral mesh with prismatic elements in the boundary layer regions was generated. The same mesh was also used for a far field, i.e. the part of the domain away from the turbine. The computational mesh in the diffuser symmetry plane is shown in Figure 3.12. Due to the fact that the most important elements of the turbine were taken into account, the computational mesh was relatively large. It was composed of 56 million of nodes (124 million of elements). The mesh was refined enough to solve fully the boundary layer for the nominal wind speed (5 m/s) case. In the case of the rated (12 m/s) and maximal (40 m/s) velocities, the automatic wall treatment (wall function) was applied in the regions where the dimensionless wall distance of the first element $y+$ was higher than 2.

Steady-state flow simulations were carried out in ANSYS CFX 17.1 for the turbine operation under the average site wind speed (5 m/s), rated (12 m/s) and extreme (40 m/s) wind speeds. The Shear Stress Transport (SST) turbulence model was used to close the system of Reynolds RANS equations. Due to the fact that the flow velocity was relatively low ($Ma < 0.3$), the air was treated as incompressible ($\rho = 1.185\ kg/m^3$). The computational space was divided into two domains: the rotational domain for the region inside the diffuser around the rotor and the stationary one for the rest of the space. In the case for the nominal and rated speeds, the angular velocities of the rotor (60 and 144 rad/s) resulted from its optimal TSR ($\lambda = 6$). Frozen rotor [18] interfaces were used to transfer data between the stationary and rotational domains. In the case of the extreme wind velocity (40 m/s), it was assumed

Figure 3.12 Computational mesh in the turbine vicinity in the symmetry plane of the diffuser

that the turbine was stopped and the domain around the rotor was specified as the stationary one. The General Grid Interface was used to transfer data between different meshes. This approach permits connections between regions with non-matching node locations [18]. The nodal values of parameters on one side are interpolated on the nodes of the second side of the interface. In the case of significant differences in the mesh refinement on two sides, some numerical problems can appear. However, in the presented cases, interpolation mistakes were not observed. At the inlet to the cylindrical computational domain, the axial wind velocities (5, 12 and 40 m/s) were imposed as the boundary conditions with low (1%) turbulence intensity. At the outlet, the atmospheric pressure (100,000 Pa) was applied. At the lateral surface of the computational domain, the free slip wall condition directed the flow axially without friction effects. All the turbine walls were assumed to be smooth.

The second order discretization schemes were used to approximate the governing equations. The flow in and, especially, downstream of the diffuser was not perfectly stationary. Therefore, the root mean square residuals of the momentum equations were just below the 1e–4 and 1e–5 level for the mass equation, which indicates an acceptable level of convergence (solution precision) [18]. Nevertheless, some unsteady flow structure appeared, and additional transient simulations could provide more interesting information, e.g. data on time-dependent loads for the wind turbine fatigue analysis. Such simulations were not conducted yet as they are much time consuming.

The numerical simulations for the wind turbine provided data of the flow field and allowed one to analyse the flow pattern. The velocity distributions on the diffuser

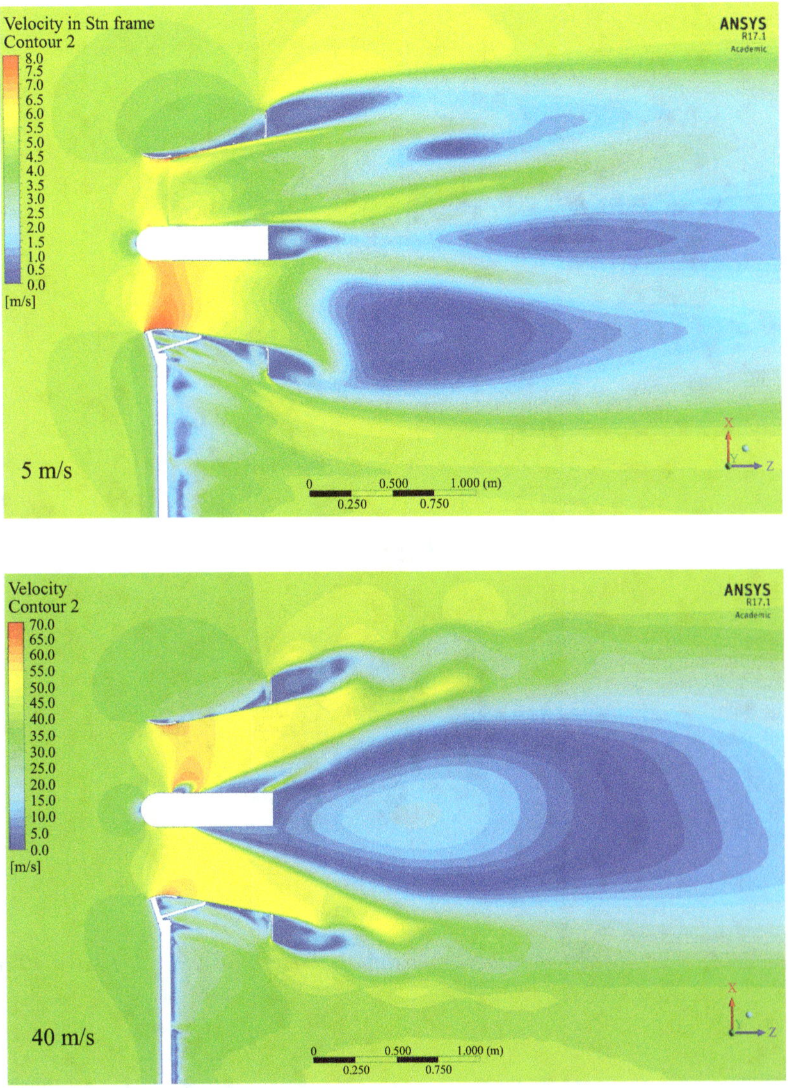

Figure 3.13 Velocity distributions in the diffuser symmetry plane for 5 and 40 m/s wind speeds

symmetry plane presented in Figure 3.13 show a separation zone downstream of the diffuser generated by the brim and nacelle. One can see that when the rotor is stopped (case for 40 m/s), the flow pattern changes drastically and the flow destabilizes. This can be also observed in Figures 3.14 and 3.15, where velocity distributions at the outlet and downstream of the diffuser are shown. In these figures, the disturbances of the flow generated by the mast and the ribs of the diffuser are observed. They deteriorate the suction effect of the diffuser and reduce the turbine performance.

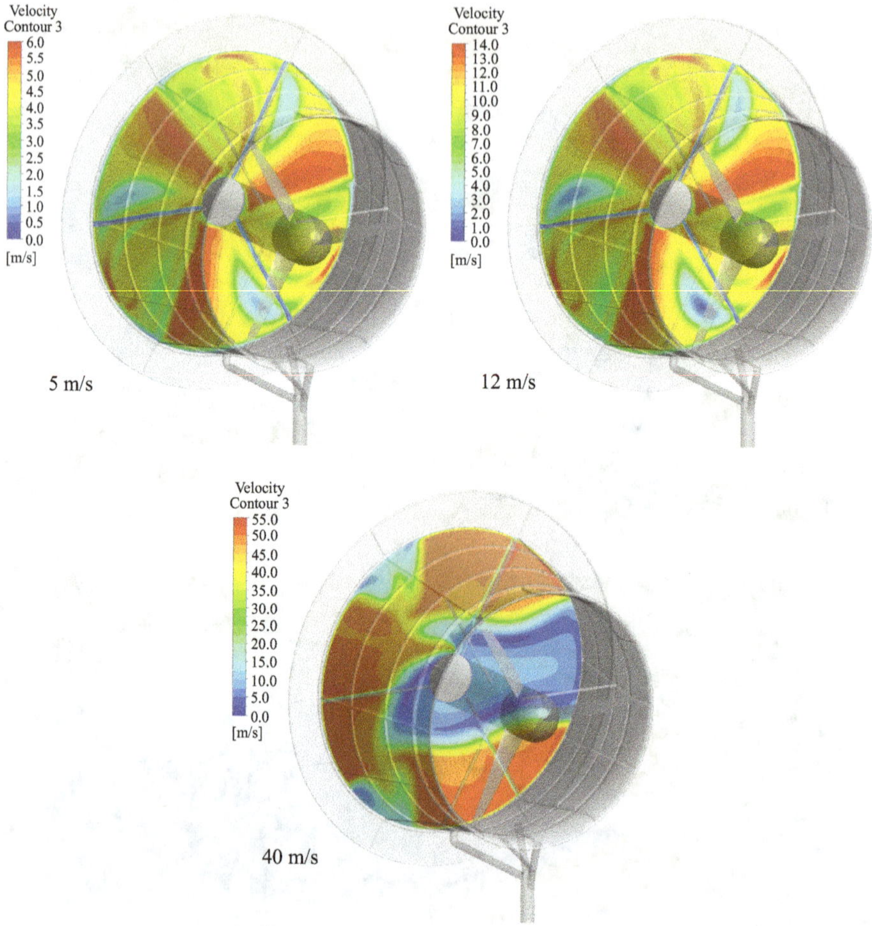

Figure 3.14 Velocity distributions at the outlet of the diffuser for 5, 12 and 40 m/s wind speeds

The loads exerted by the flowing air on the elements of the turbine (rotor blades and a diffuser) were one of the important aspects. The pressure distribution on the turbine elements (Figure 3.16) can be used as a load in the one-way FSI to determine the deformation and stresses in the blades or the diffuser. The software can also provide global data, whose example is presented in Table 3.1.

3.2.5 BET–CFD simulations of the turbine model in scale

The presented numerical data have been obtained from a hybrid BET–CFD model, within the approach described previously. The commercial ANSYS Fluent solver is coupled with an in-house BET code to form a 3D ADM flow simulation. An overview

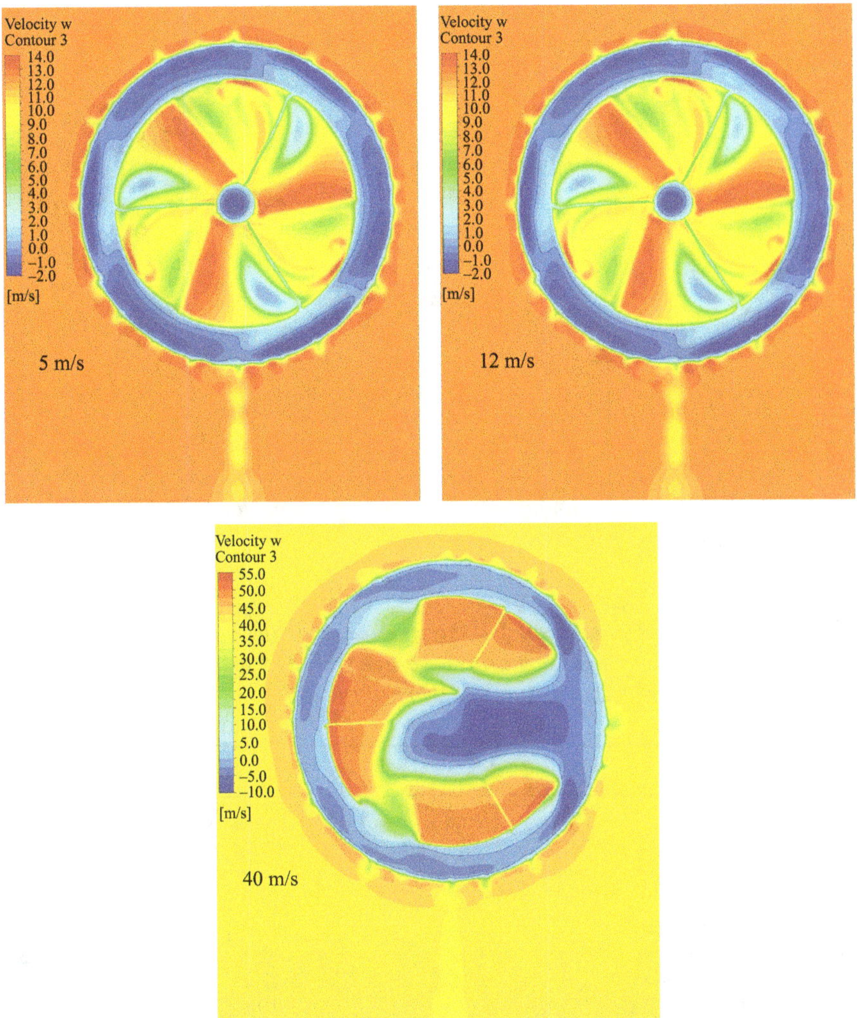

Figure 3.15 Velocity distributions 50 mm downstream of the diffuser for 5, 12 and 40 m/s wind speeds

of the discussed simulation model is presented in Figure 3.17. The domain depicts the geometry of an open-test section of the subsonic wind tunnel located at the Institute of Turbomachinery, Lodz University of Technology. The wind turbine rotor is represented by a red circular surface. The velocity fields provided by the flow solver are used by the BET model to compute (local) aerodynamic loads, introduced back to the solver in a form of source terms. In this way, the forces acting on the wind turbine blades can be estimated without complete rotor modelling.

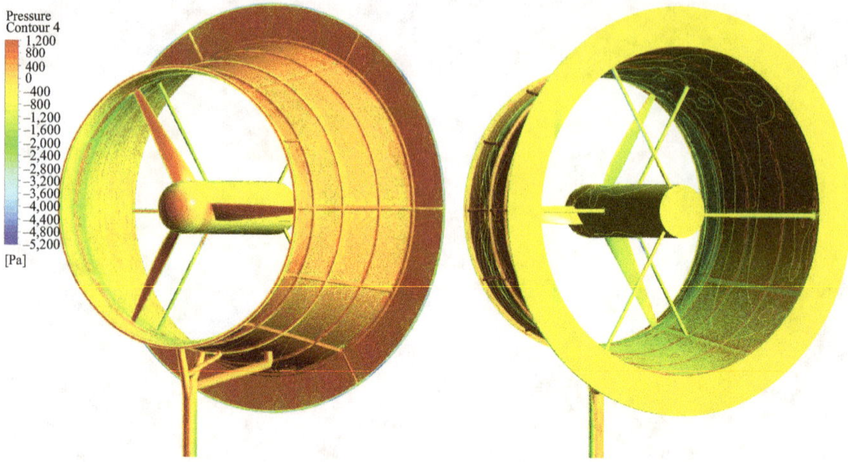

Figure 3.16 Pressure distribution on the turbine elements for 40 m/s wind speed, front and rear views

Table 3.1 Axial force component and three moments acting on the rotor blades at three distinct load scenarios: average site wind speed (5 m/s), rated wind speed (12 m/s), extreme wind (40 m/s)

Wind speed (m/s)	**5**	**12**	**40**
Angular velocity	60 rad/s	144 rad/s	0 m/s
Tip speed ratio	6	6	0
	Data for the first blade		
Axial force (N)	2.76	16.4	35.1
Flapwise moment (N m)	0.944	5.61	9.59
Torsional moment (N m)	0.0074	0.052	0.525
Rotational torque (N m)	−0.240	−1.432	−2.639
Power (three blades)	43.2 W	619 W	0
Cp	0.743	0.770	0

A correct depiction of the wind turbine rotor in the ADM simulation is influenced by numerous factors. The profile characteristics, dependant in turn on the flow character, is among the most important ones. The local Reynolds number and the turbulence intensity change the aerodynamic properties of an aerofoil. The SWTs have a tendency to operate in the transitional flow regime, with Re based on the chord length (see also [19]). In the model under consideration, the local Re is of the order of 10^4–10^5. The wind turbine model operates at a low turbulence level (the order of magnitude is 1%), with velocities of 8 and 12 m/s as noted in Figures 3.20–3.22. The

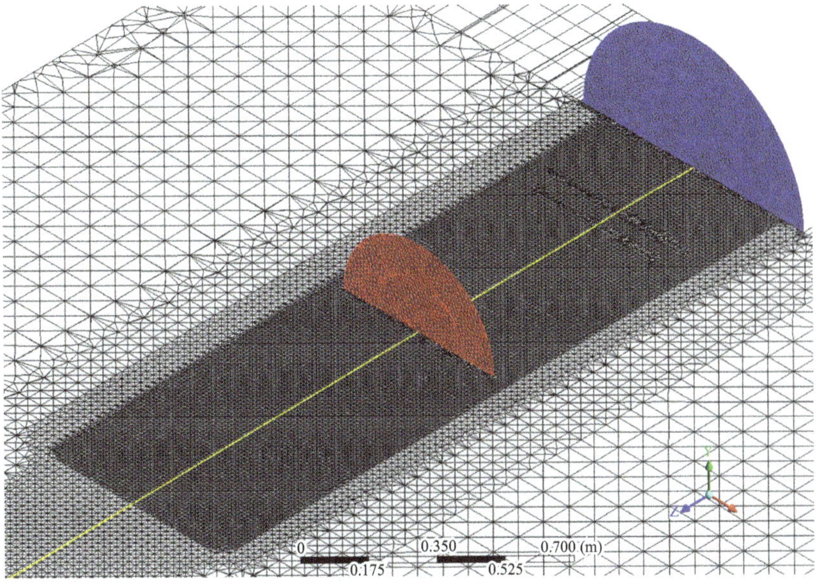

Figure 3.17 *Sectional view of the ADM simulation domain; rotor marked in red, domain inlet in blue*

SST turbulence model was applied in this particular case. A tetra-dominant unstructured mesh was composed of 1.4 million of nodes (8.3 million of elements), and it was refined downstream and upstream of the rotor to ensure a proper solution in the high-gradient region and modelling of turbulent phenomena.

3.3 Experimental tools for the small wind turbine load analysis

The majority of standards used in energy converters of renewable sources specify two important operational conditions for wind turbines. These are normal and extreme operating conditions. The normal conditions are associated with long-term loads that can have an impact on performance, durability and device control system functions. They occur very frequently during operation and thus are ideal for machine-health monitoring. The extreme operating conditions list a number of load cases critical to structural safety and integrity. Hence, these load cases will be rare and are used to design and scale an energy converter via safety factors, which are defined typically with a period of recurrence of 1 year up to even 50 years.

The IEC standard [7] specifies a number of load scenarios to be considered in the design of SWTs. Under the normal operating conditions, the turbine should be tested for yawing errors and the maximum thrust occurrence. Mechanical faults leading to the maximum rotational speed, short circuit, emergency shutdown are also important.

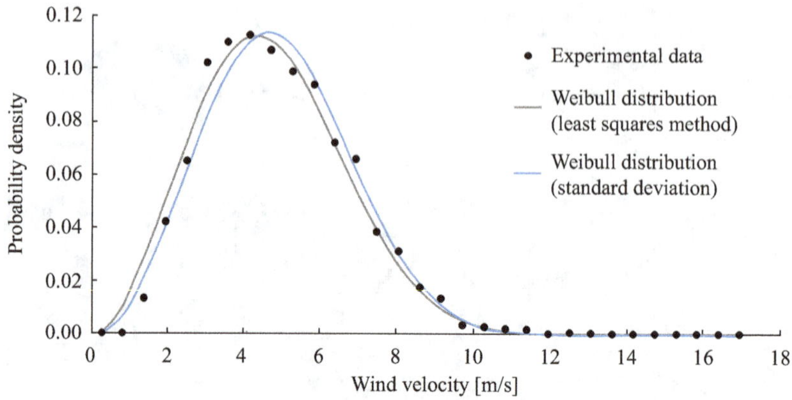

Figure 3.18 Weibull distribution for wind conditions at the campus B of the Lodz University of Technology [20]

The load cases in the extreme conditions demand an analysis of wind loading on the parked turbine. Even the loads that can impact on the structure during transportation and assembly are to be considered. However, from the power generation point of view, the most frequent load scenario is a fatigue type load resulting from the site-specific wind profile and turbulence changes. Hence, thrust and rotational torques acting on turbine blades are going to be mainly considered. The extreme operating and coherent gusts, wind direction changes are difficult to test in wind tunnel conditions and remain the object of simulations, discussed in the previous section. Finally, the electrical loads associated with surge, circuits, earthing and conductors are usually used as a part of functional and system safety assessment.

According to the methodology presented in Figure 3.1, a wind turbine design starts with site-specific wind conditions. Ideally, the site where a turbine is to be located needs a thorough evaluation prior to the erection of the machine structure. The measurements are averaged in 10-min bins and are advised over a number of years to take out the seasonality effects. In the absence of detailed site data, the conditions can be assessed from other nearest sites or via models such as the wind resource forecasting.

The site evaluation typically ends with a wind speed distribution. Figure 3.18 presents a probability density function for wind speeds measured by a pair of cup anemometers placed on the roof of the IMP TUL, representative of urban wind conditions in Poland. The data is derived from a-year-long measurements. It is accompanied by two curves, which can be further used to model the site annual energy production more precisely. The key thing that these measures enable is establishment of the annual average wind speed. According to the distribution provided, the average wind speed is equal to 5 m/s at this site. Although the plot does not show it, the same campaign helped to establish another important load parameter – a site-specific maximum wind velocity of 31 m/s [20].

Table 3.2 Characteristic normal and extreme operating conditions defined for the presented site-specific study

Short name	**SWT class** **Parameter definition**	**III**	**IV**	**S DAWT**
Vref	Reference wind velocity [m/s]	37.15	30	31.1
Vave	Annual average wind speed [m/s]	7.5	6	5
TI_{15}	Turbulence intensity at 15 m/s	0.18	0.18	0.18
a	Slope for standard deviation equation	2	2	2
Vhub	Hub-height wind velocity [m/s]	8.24	6.59	6.50
α	Wind shear power law exponent [–]	0.2	0.2	0.2
$\sigma 1$	Hub-height longitudinal wind velocity standard deviation [m/s]	1.89	1.69	1.60
Ve50	Expected extreme wind speed with 50-year recurrence [m/s]	49.85	39.88	41.3
Ve1	Expected extreme wind speed with 1-year recurrence [m/s]	37.39	29.91	31

On the basis of the available wind conditions, the turbine class can be chosen in order to arrive at a list of load scenarios necessary to consider. For the turbine example under discussion, the characteristic parameters were derived in agreement with the IEC standard and are presented in Table 3.2. The reader is advised that the hub-specific velocity values in the table incorporate an effect of the wind shear.

In the absence of similar standard conditions, one may decide to adopt the S turbine class and name other conditions specific to the site. Here, they are stated in the last column. For convenience, regular class III and class IV, as defined by the standard, are also provided. It must be noted that in the prepared S (special) class, the hub-height velocity is much higher than the average wind speed. This is a direct result of using a diffuser that increases the throughput velocity for the analysed turbine by a factor of 1.25. The survival wind speed is computed to be just over 41 m/s. Thus, as presented before, the loads exerted onto this turbine were evaluated via CFD modelling at 40 m/s in a parked standstill mode. Overall, the derived S class values situate the design to be more demanding than class IV.

With the information on the normal and extreme aerodynamic loads, the development path moves onto to design of a rotor–nacelle assembly. In the case of the ducted wind turbines, a diffuser is also designed at this stage. To aid the simulation tools in the process, a wind tunnel test campaign has to be planned. Figure 3.19 presents a view of the test section of the low speed wind tunnel at the IMP TUL. At the early stage, a model of the ducted wind turbine in scale 1:6 can be prepared. For this size, the blockage coefficient was 0.3. In previous works [4], we demonstrated how the blockage effect was controlled. Simultaneous measurements of ambient pressure, temperature and humidity, dynamic pressure of the airstream, rotational speed of the shaft, torque of the shaft were performed. The position of the measuring equipment is shown in Figure 3.19. The power produced was stored in supplied batteries.

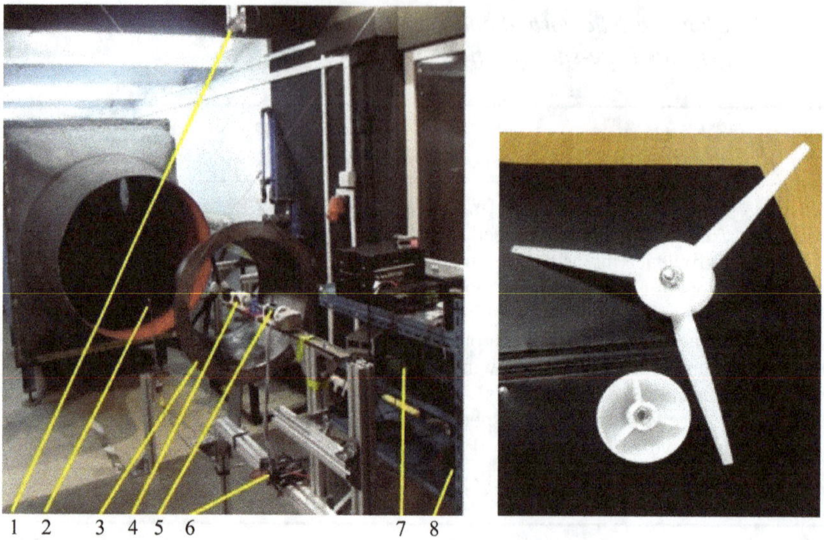

Figure 3.19 View of the test section with the scaled DAWT model – low speed wind tunnel at IMP TUL (1 – rotational speed sensor, 2 – Prandtl tube, 3 – diffuser, 4 – torque meter, 5 – generator, 6 – generator servo controller, 7 – power unit for measuring devices, 8 – batteries), a 3d printed rotor model in scale 1:6 is seen on the right (FDM, fused deposition modelling)

Three series of measurement sessions were performed. The first one focused on loads exerted on the diffuser in three different scales to evaluate the blockage effect, and this measurement allowed one to select the scale (1:6) for further tests. In the second measurement, a bare wind turbine rotor (open rotor wind turbine) was evaluated in the airstream velocity ranging from 5 to 18 m/s. At each freestream velocity, the rotor rotational speed was regulated via a servo controller. As a result, wind turbine characteristics at a number of reference velocities were obtained. Two of the curves, measured at 8 and 12 m/s, can be seen in Figure 3.20.

The measurements helped to validate the numerical model presented in the previous section of this chapter. A very good agreement was achieved for this scale. The rotor was designed to operate in the duct, hence its nominal TSR (TSR = 6) decreases to a lower value, when it operates as a bare wind turbine rotor. At this scale, the turbine blade load is under influence of friction forces and experiences the so-called Reynolds number effect. One can notice that an increase in velocity from 8 to 12 m/s, i.e. an increase in the Reynolds number by 50% (e.g. the Re based on the diameter of the rotor from $1.8 \cdot 10^5$ to $2.7 \cdot 10^5$) led to a significant increase in the power coefficient. Nevertheless, Reynolds numbers in this scale are low and, hence, the turbine is not able to achieve a power coefficient higher than 0.2.

In the third campaign, measurements of the ducted wind turbine version were conducted at 8 and 12 m/s (Figure 3.21). The plots show that the diffuser is able to

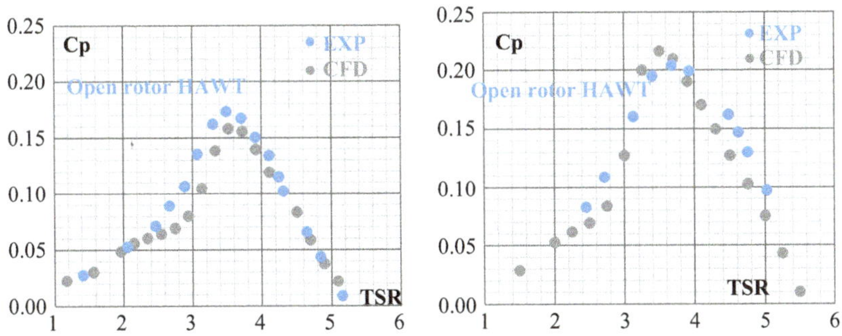

Figure 3.20 *Power coefficient as a function of the tip speed ratio for the open rotor wind turbine in scale 1:6 at 8 m/s (left) and at 12 m/s (right); plots compare the results of the derived numerical model to the experimental measurement*

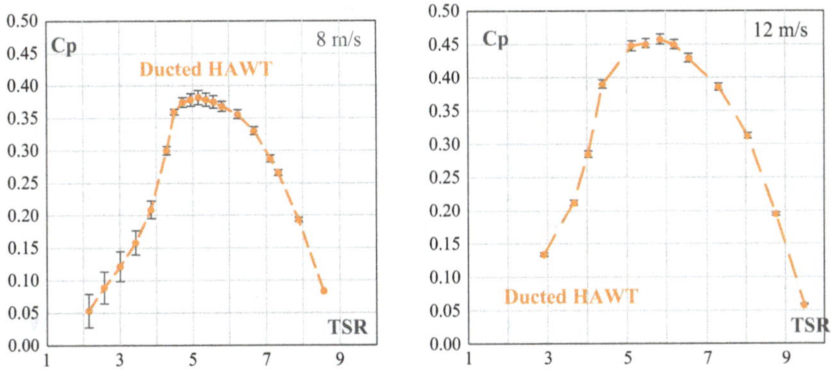

Figure 3.21 *Ducted wind turbine characteristics in scale 1:6, measurements in the IMP TUL wind tunnel at 8 and 12 m/s*

increase the mass flow through the turbine section and, thus, increase the power production capacity at the same rotor size. The theoretical turbine power equation based on the rotor swept area is used to derive the power coefficient for a ducted converter. With the duct even at lower velocities, the turbine power production increases two times or more. This is well illustrated in Figure 3.22, in which the bare rotor turbine performance is compared to the performance of a ducted wind turbine.

The validated numerical model was then used to design the turbine in scale 1:1, where its diameter reaches 2 m and the nominal power is equal to 2,000 W. In the next step, it was decided to build a prototype in scale 1:2. This scale can diminish the Reynolds number effect observed for 1:6 and provide more reliable data on full-scale design loads. The data can be also used in further verification of the numerical model.

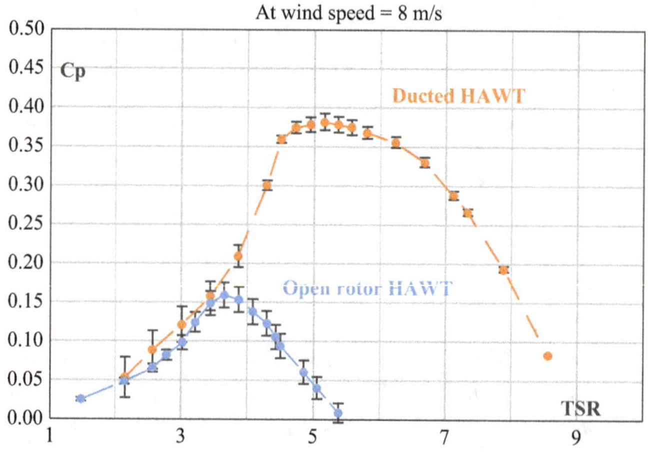

Figure 3.22 Power coefficient as a function of the tip speed ratio measured in the wind tunnel facility, turbine without and with the diffuser in scale 1:6

3.4 Analytical methods for estimation of aeromechanical rotor loads

The first round of the design tests on aerodynamic loads supports the concept and gives a chance to validate the simulation models against one another as well as to compare them with the experiment. The simulation models can now be used on a larger scale prototype and to detail the design in a higher degree.

The SLM is one of convenient ways to verify the structural integrity of the SWT. It can be used for certain types of wind turbines: horizontal axis rotors with no teetering, possessing at least two cantilever-like blades with no pitch control. The SLM is based on key aeromechanical formulae helping to estimate the axial, radial and tangential loads such as forces and moments acting on machine components. The load scenarios to consider were enumerated in the previous sections. The turbine blades and the shaft are the major turbine components to be considered from the aeromechanical point of view. In this part of the chapter, a few basic equations will be stated for both above-mentioned elements, however, only the blade structural integrity will be analysed.

Within a number of load case scenarios, at least three deserve special attention. They are normal operation in the power production mode, the maximum thrust developed on the blade and the extreme wind loading. The first load case endangers the blade and the shaft with fatigue loads that decrease the material strength over time. This puts especially the rotor structural integrity at risk since the safety factor decreases over time with the blade ability to carry loads, whereas the design load remains relatively unchanged within the assumed peak-to-peak amplitudes. The second scenario creates unfavourable operating conditions on the shaft, where the highest possible axial load from rotor blades is now added to the shaft torsional load. These loads, together with an eccentricity of the shaft, can create a substantial reaction force on the bearing closest to the plane of rotation. The third scenario is important from the structure design point of view. The extreme loading is a theoretical event that the designer must be prepared for. Hence, all machine components, and rotor blades in particular, have to be designed to survive under such conditions.

In the normal operating conditions, the fatigue loads are results of the centrifugal force and varying aerodynamic loads. The variation comes from stochastic changes in turbulence scales present in wind as well as from an interaction of blades with the tower and other supporting elements. This variation impacts the instantaneous design load, which cannot be considered as static. To counter these effects, it is suggested to consider a range of design loads manifested by varying rotational velocity and torque. This produces a certain offset to the considered forces and moments which can be estimated and added to the static load.

Typically, the stress concentration happens on the blade root section, hence the axial (due to the centrifugal force) and bending loads (due to a torque couple) are derived (shear stresses are negligible):

- centrifugal force:

$$\Delta F_{zB} = 2m_B R_{cog}\omega_{n,\text{design}}^2 \tag{3.14}$$

- edge-wise blade root bending moment:

$$\Delta M_{xB} = \frac{Q_{\text{design}}}{N} + 2m_B g R_{cog} \tag{3.15}$$

- flapwise blade root bending moment:

$$\Delta M_{yB} = \frac{\lambda_{\text{design}} Q_{\text{design}}}{N} \tag{3.16}$$

The fatigue loads resulting from the normal operation have to be translated now into design stresses in the principal axis:

$$\sigma_{zB} = \frac{\Delta F_{zB}}{A_B} \tag{3.17}$$

The equivalent stress formula for bending of the rectangular-shaped blade root is in the form:

$$\sigma_{MB} = \frac{\Delta M_{xB}}{W_{xB}} + \frac{\Delta M_{yB}}{W_{yB}} \tag{3.18}$$

where the section moduli W is computed for the blade root aerofoil. The equivalent combined (axial + bending) stress can be approximated by

$$\sigma_{eqB} = \sigma_{zB} + \sigma_{MB} \tag{3.19}$$

To assess the blade fatigue strength at the machine lifespan, an *S–N* curve or a plot of the magnitude of the alternating stress versus the number of cycles to failure for a given material is needed. The *S–N* curve functions as a 'lookup table' between the alternating stress level and the number of cycles to failure. This knowledge can be used to pick adequately safety factors and avoid an under-design of the turbine and the supporting tower. On the other hand, a lack of exact data concerning fatigue performance for the chosen blade material implies a use of simplified relations. The fatigue limit may be therefore estimated in the most conservative way as an order of the magnitude lower than the material ultimate strength. With the fatigue limit stress known, Miner's rule is applied for fatigue safety check. According to this rule, the limit state is reached when the accumulated damage value within the SWT lifetime exceeds 1 [7].

The second scenario, i.e. the maximum thrust loading, is typically associated wh operation of a turbine at the rated point. Variable rotational speed wind turbines, as the one presented in this chapter, are regulated to diminish the thrust and maintain their rotor power. The rated wind speed associated with an occurrence of the highest axial force is typically two to three higher than the average wind speed. Thus, under these conditions, the shaft loading is associated with the freestream dynamic pressure and the rotor thrust coefficient:

$$F_{Bshaft} = \frac{1}{2} C_T \rho (2.5 V_{ave})^2 \pi R^2 \tag{3.20}$$

whereas for the extreme wind loading, the third scenario under consideration, the designed SWT remains stationary and is exposed to a 50-year extreme wind speed in the form of a 3-s-long gust. The out-of-plane blade root bending moment in this case is dominated by drag forces and, thus, is defined as

$$M_{yB} = C_D \frac{1}{4} \rho V_{e50}^2 A_{proj,B} R \tag{3.21}$$

The shaft thrust force for a parked rotor (along the rotation axis) and the associated equivalent stress is expressed by

$$F_{B-shaft} = B C_D \frac{1}{2} \rho V_{e50}^2 A_{proj,B} \tag{3.22}$$

$$\sigma_{eq} = \frac{M_{yB}}{W_{yB}} \tag{3.23}$$

It is worth stating that the calculated equivalent stresses provide only a simplified and superficial insight into the dynamic response of the designed SWT blades under the operational conditions under consideration. A summary of all results for the blade structural integrity is presented in Table 3.3 and the design proves safe in all load cases. Whenever possible, less conservative safety factors were used, especially for cases in which a high fidelity 3D CFD simulation was carried out (D and H).

The calculations allowed one to qualify safely the design and begin the assembly. With a turbine prototype laid out, the rotor balance was checked (Figure 3.23).

The rotor blades were entirely 3D printed from a polyamide in the SLS (selected layer sintering) technology. The rotor balancing was conducted in two stages. In the first stage, the turbine top section, above the connection to the tower, was investigated alone. A CSI 2130 vibration monitor from Emerson was used to assess the vibration amplitudes and load induced by the spinning rotor on the diffuser. To rotate the turbine blades, a Kinco electrical engine was used. The tests were conducted up to the maximum rotational speed of 1,669 rpm, associated with the load case E as determined by the SLM.

Figure 3.25 *DAWT prototype in scale 1:2 on the top of the building – IMP TUL, a cup anemometer is visible on the diffuser*

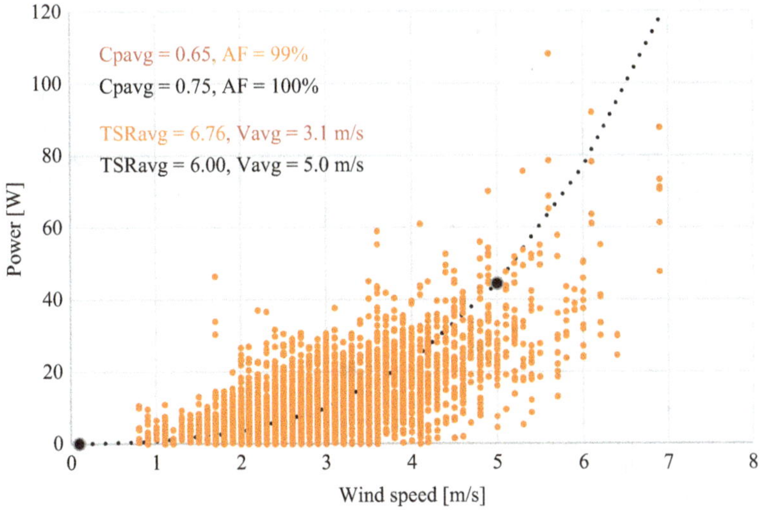

Figure 3.26 *Power measured on the shaft as a function of the wind speed – measurements in 30 min span (February 2017) – compared to the theoretical power curve derived from the accurate 3D CFD model*

idealized geometrical model which did not take into account some details of the real construction and the flaws in the prototype manufacturing and assembly. What is more, the simulations were conducted for ideal wind conditions (constant velocity and flow along the turbine axis), whereas significant fluctuations in the velocity magnitude were observed during the tests.

The average value of the power coefficient (defined on the basis of the rotor diameter) obtained from the experiment is much higher than for typical wind turbines of the same dimensions. It was also proven that the designed turbine, thanks to the duct, was able to begin generating power at very low wind speeds. Despite a more complex construction and investment costs, it can be considered as an interesting alternative to traditional designs of SWTs, specifically for modest wind speeds prevailing in Poland.

3.5 Summary

The practical application of the proposed decision path specified at the beginning of the chapter allowed one to design successfully a small diffuser augmented wind turbine (DAWT). Multiple methods were applied to design and verify the construction, ranging from an application of simple analytical methods through complex numerical simulations, to finish with the experimental tests of models in different scales.

An influence of the turbine duct (diffuser) made it impossible to use standard methods of turbine design and forced one to combine analytical methods with CFD predictions of the flow. A set of different numerical models was used, ranging from the 2D CFD ADM up to 3D FRM, which took the most important elements of the final construction into account. Those simulations provided data on aerodynamic loads, further used to verify the design with analytical methods. Some trials of the load analysis employing the FSI were carried out. They proved to be useful for the design load analysis. However, significant requirements of computer resources for simulations of the complex turbine model as well as high deformations of polyamide blades limited the practical application of that method within the project so far. Nevertheless, continuous progress in the computer performance and numerical methods, combining the fluid flow and structural solvers, should provide tools for detailed design analysis in the near future. At the same time, the Fourier transform analysis can be performed on the rotor power signal to provide the fault tolerance and to increase the turbine load factor. Combined with the high frequency measurement of wind velocity, a link between dangerous loading due to environmental conditions and the resulting rotor power fluctuations can be established early enough. In this case, a potential fault could be avoided, thus keeping the turbine uptime to its maximum.

Nomenclature

ADM actuator disk model
AF availability factor
ALM actuator line model
FDM fused deposition modelling

FRM		fully resolved rotor model
BEM		Blade Element Momentum (theory)
BET		Blade Element Theory
CFD		computational fluid dynamics
DAWT		diffuser augmented wind turbine
SLM		simplified load model
rpm		revolutions per minute
SLS		selective laser sintering
SWT		small wind turbine
TSR		tip speed ratio
IMP TUL		Institute of Turbomachinery, Lodz University of Technology
a	–	axial induction factor/slope of a standard deviation equation
a'	–	tangential induction factor
c	m	chord
d	N/m	drag force per unit blade length
f_n	N/m	axial component of aerodynamic force per unit blade length
f_t	N/m	tangential component of aerodynamic force per unit blade length
g	kg m/s^2	gravitational acceleration
l	N/m	lift force per unit blade length
m	kg	mass
p	Pa	pressure
r	m	radial position
A	m^2	area (e.g. of the rotor)
B	–	number of blades
$C_{d,l}$	–	drag, lift coefficient
Cp	–	power coefficient
Cq	–	torque coefficient
Ct	–	thrust coefficient
D	m	rotor diameter
F	N	force
M	N m	moment
Ma	–	Mach number
N	–	blade number
P	W	power
R	m	radius
Q	N m	torque
Re	–	Reynolds number
T	N	thrust force
TI_{15}	–	turbulence intensity at 15 m/s wind
U	m/s	axial velocity
V	m/s	reference wind velocity
Vave	m/s	average wind speed
Ve1	m/s	extreme wind speed with the recurrence of 1 year
Ve50	m/s	extreme wind speed with the recurrence of 50 years

Vhub	m/s	wind velocity at the hub height
W	m/s, m^3	relative velocity, section modulus for blade root cross-section
α	rad/–	angle of attack/power law exponent
β	rad	pitch angle (blade section twist)
λ	–	tip speed ratio
μ	Pa s	dynamic viscosity ($\mu = \nu \cdot \rho$)
ν	m^2/s	kinematic viscosity
ρ	kg/m^3	density
σ	N/m^2	stress
$\sigma 1$	m/s	standard deviation of longitudinal wind velocity
ϕ	rad	inflow angle (between the relative wind and the rotation plane)
ω	rad/s	rotational speed
Γ	m^2/s	circulation
(r, φ, z)		coordinates in a cylindrical reference system
(x, y, z)		coordinates in a Cartesian reference system

Subscripts

1, 2, 3	pertaining to principal axes for stress calculations
all	allowable
B	pertaining to blade
cog	centre of gravity
design	design
eq	equivalent
n	nominal
proj	projected
shaft	pertaining to shaft
x, y, z	in the direction of x, y, z

References

[1] M. Karczewski, P. Baszczynski, P. Wiklak, K. Sobczak, K. Jozwik, "Economic analysis of small wind turbines," in *Techniki Komputerowe w Inżynierii*, Teresin, Poland, 2016.

[2] K. Abe, M. Nishida, A. Sakurai, *et al.*, "Experimental and numerical investigations of flow fields behind a small wind turbine with a flanged diffuser," *Journal of Wind Energy and Industrial Aerodynamics*, vol. 93, pp. 951–970, 2005.

[3] K. Sobczak, K. Kacprzak, R. Wieczorek, M. Lipian, M. Karczewski, K. Jozwik, "Design and performance numerical verification of a small Diffuser Augmented Wind Turbine," in *Techniki Komputerowe w Inżynierii*, Teresin, Poland, 2016.

[4] M. Kulak, M. Karczewski, K. Olasek, K. Jozwik, "CFD Analysis of Diffuser Augmented Wind Turbine Model for Wind Tunnel Investigation," in *Proceedings of IECON 2016 – 42nd Annual Conference of the IEEE Industrial Electronics Society*, 2016.

[5] M. Lipian, M. Karczewski, J. Molinski, K. Jozwik, "Numerical simulation methodologies for design and development of diffuser-augmented wind turbines – analysis and comparison," *Open Engineering*, vol. 6, no. 1, pp. 235–240, 2016.

[6] K. Olasek, M. Karczewski, M. Lipian, P. Wiklak, K. Jozwik, "Wind tunnel experimental investigations of a diffuser augmented wind turbine model," *International Journal of Numerical Methods for Heat & Fluid Flow*, vol. 26, no. 7, pp. 2033–2047, 2016.

[7] IEC 61400-2:2013, Wind Turbines – Part 2: Small Wind Turbines. http://www.iec.ch/standardsdev/resources/draftingpublications/directives/principles/referencing.htm

[8] T. Burton, N. Jenkins, D. Sharpe, E. Bossanyi, Wind Energy Handbook, 2nd ed., West Sussex: John Wiley & Sons, 2011.

[9] W. F. Durand, Aerodynamic Theory, vol. 4, Berlin: Springer Berlin Heidelberg, 1935.

[10] S. Drzewiecki, Théorie générale de l'hélice; hélice aériennes et hélices marines. Paris: Gauthier-Villars et Cie, 1920.

[11] M. O. L. Hansen, Aerodynamics of Wind Turbines, 2nd ed., London: Earthscan, 2008.

[12] D. Wood, Green Energy and Technology: Small Wind Turbines, London: Springer, 2011, p. 270.

[13] M. Lipian, M. Karczewski, K. Jozwik, "Analysis and comparison of numerical methods for design and development of small Diffuser-Augmented Wind Turbine (DAWT)," in *Proceedings of IECON 2016 – 42nd Annual Conference of the IEEE Industrial Electronics Society*, Florence, 2016.

[14] N. Troldborg, J. N. Sorensen, R. F. Mikkelsen, "Actuator line simulation of wake of wind turbine operating in turbulent inflow," *Journal of Physics: Conference Series*, vol. 75, p. 012063, 2007.

[15] Y. Ohya, T. Karasudani, T. Sakurai, K. Abe, M. Inoue, "Development of a shrouded wind turbine with a flanged diffuser," *Journal of Wind Energy and Industrial Aerodynamics*, vol. 96, pp. 524–539, 2008.

[16] S. Takahashi, Y. Hata, Y. Ohya, T. Karasudani, T. Uchida, "Behavior of the blade tip vortices of a wind turbine equipped with a brimmed-diffuser shroud," *Energies*, vol. 5, pp. 5229–5242, 2012.

[17] J. Bukala, K. Damaziak, M. Krzeszowiec, J. Malachowski, K. Sobczak, "Analysis approach for diffusor augmented small wind turbine rotor," in *Advances in Mechanics: Theoretical, Computational and Interdisciplinary Issues*, London, Taylor & Francis Group, 2016, pp. 109–112.

[18] ANSYS Help, Release 17.1, ©ANSYS, Inc., Canonsburg, PA, USA, 2016.

[19] J. Bartl, L. Sætran, "Blind test comparison of the performance and wake flow between two in-line wind turbines exposed to different atmospheric inflow conditions," *Wind Energy Science Discussions*, vol. 2, pp. 55–76, 2017.

[20] K. Kacprzak, *"Numerical and Experimental Analysis of the Flow in Savonius Wind Turbines" (in Polish) – PhD Thesis*, Lodz University of Technology, 2016.

[21] ISO 1940-1, "Mechanical vibration — Balance quality requirements for rotors in a constant (rigid) state – Part 1: Specification and verification of balance tolerances," 2003.

Chapter 4

Structural control concept for load reduction of offshore wind turbines

Yulin Si[1], *Dahai Zhang*[1], *and Hamid Reza Karimi*[2]

4.1 Offshore wind energy development

Most current large wind turbines around the world are installed on land with sparse population and vast land. However, in many countries, inhabitants are concentrated in places along coastlines where land is scarce while power is in high demand. Therefore, utilizing offshore wind resources is more beneficial, which will both reduce electricity transmission loss and reserve more land space for people, animals, and plants. More importantly, offshore wind quality has been evaluated to be much better than that onshore. According to [1], a wind farm located offshore could experience wind speeds that are, on average, 90% greater than that over land. Therefore, global wind energy exploitation has been gradually moving to offshore areas [2]. Many countries start developing large offshore wind farms over the last 10 years, among which Europe leads the role, with the United Kingdom currently having the largest installed offshore capacity >5,000 MW. China has also made large investment on offshore wind with more than 1,500 MW OWTs already installed, and a large expansion is expected in the near future due to its large south-eastern coastal electricity load and well-developed grid there. The United States has also started to consider the opportunities on its eastern coastlines (Figure 4.1).

Near offshore wind farms in shallow water have been extensively built in recent years, but they often get affected by fishery and transportation conflicts, and their foundations may also leave relatively large seabed footprints [4]. In contrast, with less space constraints and more consistent wind, deep sea wind energy is more promising for those coastal cities without enough ideal shallow water areas. According to the experience borrowed from offshore oil and gas industry, floating foundations can be regarded as a possible way of wind turbine deployment when water depth is over 60 m, see Figure 4.2. Deep offshore wind energy from floating windmills has many advantages compared to its onshore and near offshore counterparts. First, the quality of wind in deep sea areas is better with higher wind speeds and less turbulence. Second,

[1]Department of Ocean Engineering, Ocean College, Zhejiang University, China
[2]Department of Mechanical Engineering, Politecnico di Milano, Italy

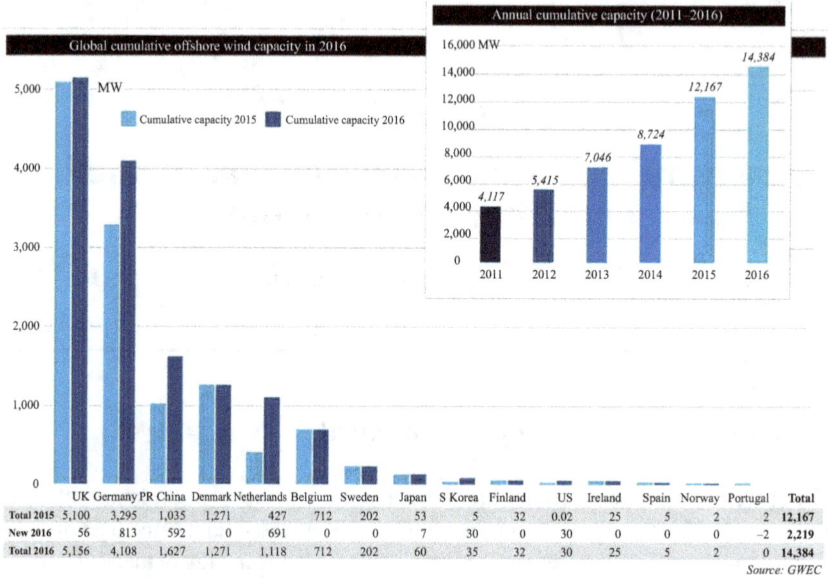

Figure 4.1 Global cumulative offshore wind capacity in 2016 and annual cumulative capacity 2011–16 [3]

Figure 4.2 Offshore wind turbine foundations [5]

the size of floating wind turbines is not limited by road or rail logistical constraints if they can be manufactured near the coastline, and seabed quality also does not restrict the turbine weight, so they can possibly grow bigger with theoretically higher efficiency. Third, the visual and noise annoyances of wind turbines can be avoided if

Figure 4.3 Full-scale floating offshore wind turbine demonstrators (left: Hywind, middle: WindFloat, right: FukushimaFORWARD)

they are installed in an enough distance from shore. Fourth, vast uninterrupted open sea areas are available, and the installations will not occupy land or interfere with other land uses. Last but not the least, since floating platforms can be towed by boats, the wind turbines could be brought to shore for maintenance instead of expensive operation in field.

Several full-scale floating offshore wind turbines (FOWTs) have been installed as pilot projects to demonstrate the technical feasibility for deep sea wind energy. In 2009, one Norwegian company Statoil developed the world first full-scale experimental FOWT Hywind with a Siemens 2.3-MW wind turbine mounted on a spar floater. Making use of the operational experience, Statoil has started to build the world's first offshore floating wind farm in Scotland in 2017. WindFloat is another full-scale FOWT prototype constructed in 2011. Different from Hywind, WindFloat uses a semisubmersible platform to support a Vestas 2-MW wind turbine. In 2013, Japan initiated the FukushimaFORWARD FOWT demonstration project, which serves as a symbol of Fukushima's recovery from the nuclear disaster in 2011. In one demonstration of this project, a Mitsubishi 7-MW turbine was installed on semi-subtype floater, resulting in the world's largest FOWT until now (Figure 4.3).

4.2 Offshore wind turbine design challenges

The main difference between onshore and offshore wind turbines are their foundations. OWTs use extra supporting structures ranging from fixed bottom types (monopile, tripod, jacket, etc.) to floating ones (spar, semisubmersible, TLP), which will introduce extra cost in the installation and O&M process. Economic analyses of offshore wind sites to date have shown that the wind turbine installation cost is approximately 100% more than onshore, whilst the O&M cost is 18%–23% more than onshore, depending upon the offshore wind location. Moreover, integrated design of OWTs is also different from onshore design due to the complex dynamics of these supporting platforms, where the wind turbine dynamic stability and structural integrity should be governed in various wind and wave environment. Particularly, motions of FOWTs even have six extra degrees of freedom (DoF), i.e., surge, sway, heave, pitch, yaw, and roll, brought by their floating platforms, such that more ultimate and fatigue loadings are imposed on wind turbine critical structures. Studies have shown that sea–land ratio of ultimate loads on tower base bending moments has reached 4.4, and the value could even be as much as over 7 for fatigue loadings (Figure 4.4).

The IEC 61400-3 standard [6] requires to run a number of design load case (DLC) simulations to verify the structural integrity of an offshore wind turbine design. The results from each DLC are analyzed to determine the ultimate and fatigue loads expected over the lifetime of the turbine. For instance, several DLCs in Table 4.1 are selected in the simulation process of this chapter. Here, NTM, EWM, NSS, and

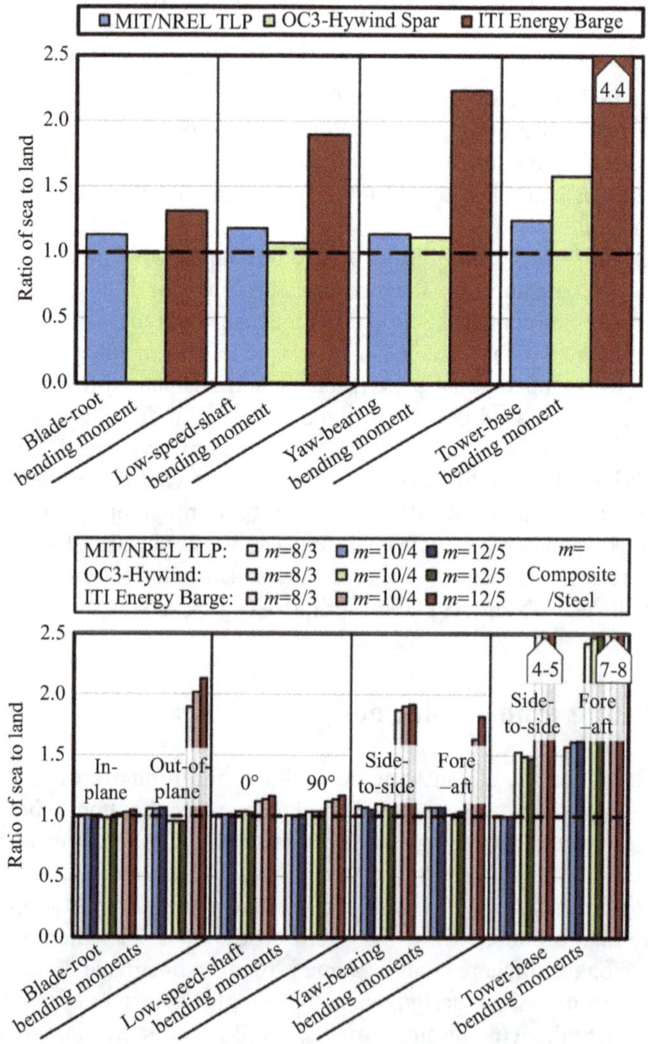

Figure 4.4 Sea-to-land ratios of ultimate and fatigue loads on various wind turbine components [7]

Table 4.1 Selected design load cases

DLC	Wind conditions		Wave conditions		Events
	Model	Wind speed (m/s)	Model	Wave height (m)	
1.1a	NTM	$V_{hub}^{mean}=10$	NSS	$H_s=2.3$	Normal operation
1.1b	NTM	$V_{hub}^{mean}=18$	NSS	$H_s=3.7$	Normal operation
6.1a	EWM	$V_{hub}^{mean}=34$	ESS	$H_s=13.8$	Parked

ESS denote normal turbulence model, turbulent extreme wind model, normal sea state, and extreme sea state, respectively. For wind condition, the mean value of the turbulent wind at hub height is defined as 10 and 18 m/s, representing below-rated and above-rated conditions, respectively. Kaimal spectra and the power law exponent of 0.14 are used for the wind field. The normal turbulence intensity is set as level B, i.e., 18% (10 m/s case) and 15% (18 m/s case). For wave condition, JONSWAP spectrum is utilized to generate the stochastic wave inputs. The significant wave height is set as 2.3 (10 m/s case) and 3.7 m (18 m/s case), and the peak spectral period is both defined as 14 s. Besides, the parked case with 50-year extreme wind and wave is also considered. Mean value of the extreme turbulent wind is defined as 37 m/s with 11% intensity and 0.11 power law exponent, and the significant wave height and wave period are defined as 13.8 m and 19 s. In this case, the generator torque and blade pitch controller are turned off, and all the blades are feathered to 90 degree to minimize the aerodynamic loading. For each case, at least two sets of random seeds are used to generate wind and wave data.

4.2.1 Ultimate loads

Ultimate loads stand for absolute maximum loads that specific wind turbine structure can bear without failing, and they should be first checked when designing a wind turbine in case of structural failure of critical structures. Normally, 95th percentile of the loads are seen as the ultimate loads on wind turbine critical components, and the ultimate loads to be governed for FOWTs include fore–aft and side–side tower base bending moments, flapwise bending moment at the blade root, and tension of anchors.

The DLCs selected for ultimate load design must cover realistic combinations of a wide range of external wind conditions and machine states. For example, this work examines the ultimate loads with normal wind conditions in combination with normal machine states and extreme wind and wave conditions in combination with normal machine states. Besides, the occurrence of fault should also be examined in the ultimate loads evaluation process while it is not analyzed here.

4.2.2 Fatigue loads

In fact, the design of most wind turbine components is not governed by ultimate loads but fatigue loads [8]. According to [9], the rotor of a 2-MW onshore wind turbine will rotate some 10^8 times during a 20-year life, with each revolution causing a complete gravity stress reversal in the tower base, in the low speed shaft, and in each blade, together with a cycle of blade out-of-plane loading due to the combined effects of wind shear, yaw error, shaft tilt, tower shadow, and turbulence. This number will be much higher with extra hydrodynamic loads for an OWT. Therefore, special attention should be drawn on the fatigue load evaluation of OWT structures.

The damage equivalent loads (DELs) are a measure of equivalent fatigue damage taking into account the material properties, namely, the *S–N* named curves (Wöhler exponents). Since usually only a limited number of DLCs are used in the evaluation simulations in the initial design process, the short-term DEL, instead of lifetime DEL, for each time-series can be used as the indicator of fatigue load. For the calculation of DEL, rainflow counting [10] should be first performed, where the fluctuating loads are broken down into individual hysteresis cycles by matching local peaks with local valleys. Normal rainflow counting procedure is listed below.

1. Reduce the time-series to a sequence of peaks and valleys.
2. Imagine that the time-series is a template for a rigid sheet, and rotate it clockwise 90 degree.
3. Each peak is imagined as a source of water that drips down.
4. Count the number of half-cycles by looking for terminations in the flow occurring when either:
   a. it reaches the end of the time history,
   b. it merges with a flow that started at an earlier peak or
   c. it flows when an opposite peak has greater magnitude.
5. Repeat step (4) for valleys.
6. Assign a load range to each half-cycle equal to the magnitude difference between its start and termination.
7. Pair up half-cycles of identical magnitude (but opposite sense) to count the number of complete cycles.

Figure 4.5 illustrates an example of how to perform the rainflow counting for a time-series, and the results are shown in Table 4.2.

Based on the rainflow counting numbers, a DEL can be calculated, representing a constant-amplitude fatigue load that occurs at a fixed load-mean and frequency and produces the equivalent damage as the variable spectrum loads such that:

$$
\begin{aligned}
D_j^{ST} &= \sum_i \frac{n_{ji}}{N_{ji}} = \frac{n_j^{STeq}}{N_j^{eq}}, \\
n_j^{STeq} &= f^{eq} T_j, \\
N_j^{eq} &= \left(\frac{L^{ult} - |L^{MF}|}{0.5DEL_f^{STF}} \right)^m, \\
N_{ji} &= \left(\frac{L^{ult} - |L^{MF}|}{0.5L_{ji}^{RF}} \right)^m,
\end{aligned}
$$

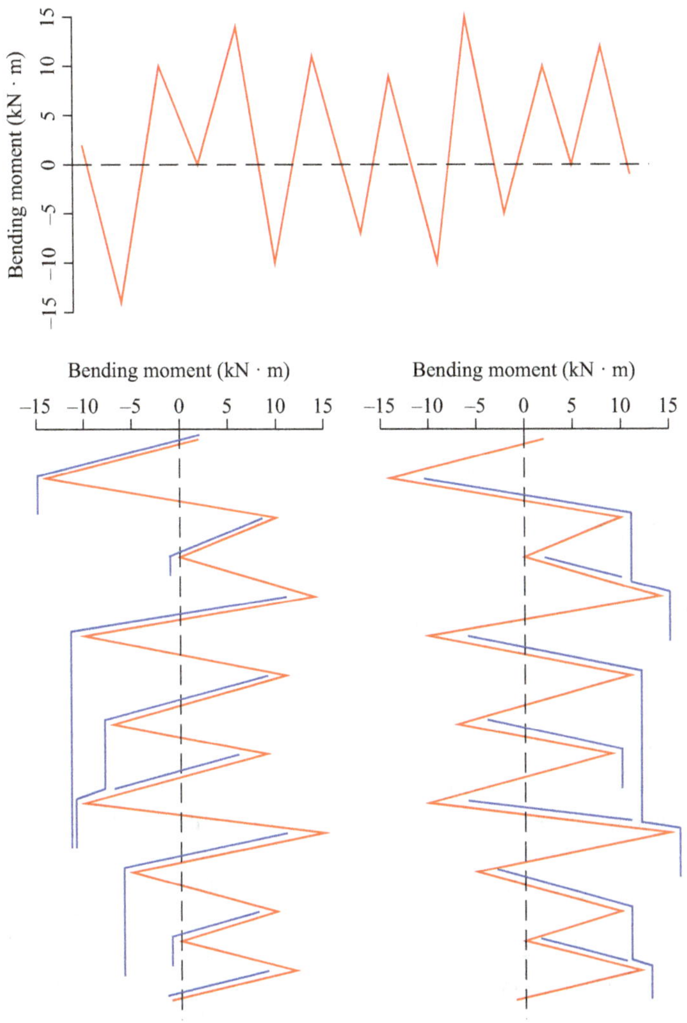

Figure 4.5 Example for rainflow counting

where D_j^{ST} is short-term damage rate for time-series j. N_{ji} and n_{ji} denote the cycle number to failure for load range L_{ji}^{RF} and its rainflow counting number, respectively. f^{eq} is the DEL frequency, T_j is the elapsed time of time-series j. L^{ult} is the ultimate design load of the component, L^{MF} is the fixed load-mean. n_j^{STeq} is the total equivalent fatigue counts for time-series j, DEL_f^{STF} is the short-term DEL for time-series j about a fixed mean, and N_j^{eq} is the equivalent number of cycles until failure for time-series j. m denotes the Whöler exponent, which is specific to the component. In this work,

Table 4.2 Rainflow counting results

Load range (kN m)	Whole cycles	Half cycles
10	2	0
13	0	1
16	0	2
17	0	2
19	1	0
20	0	1
22	0	1
24	0	1
27	0	1

m is set as 3 for steel structures, such as tower, and 10 for fiberglass components, like blades. Solving these equations will yield

$$DEL_f^{STF} = \left(\frac{\sum_i (n_{ji}(L_{ji}^{RF})^m)}{n_j^{STeq}} \right)^{1/m}.$$

4.2.3 Load reduction solutions

One idea for wind turbine load mitigation is to improve the blade pitch control strategy. There have been many promising numerical results, see published [11–14], while these methods will possibly increase the control complexity or decrease the power quality and control stability. At the same time, a more direct approach to reduce loading is to utilize structural control devices, which have been applied successfully in large civil structures in the past few years, such as long bridges and tall buildings. For example, as shown in Figure 4.6, Taipei World Financial Center installed a huge tuned mass damper (TMD) in its 87th–92rd floors to dissipate the vibration from seismic earthquake and strong wind. Therefore, this chapter is going to introduce the structural control concept for load reduction of OWTs.

4.3 Structural control methods

Structural control is generally a concept in civil engineering discipline, which refers to the techniques that are used to reduce acceleration and loading in buildings and bridges due to wind, wave and earthquake. In the past decades, numerous structural control devices have been implemented for large civil structures, and there are many different types of devices or designs, ranging from massive pendulums to motor controlled mass dampers. Following is a brief introduction of several structural control devices (Figure 4.7).

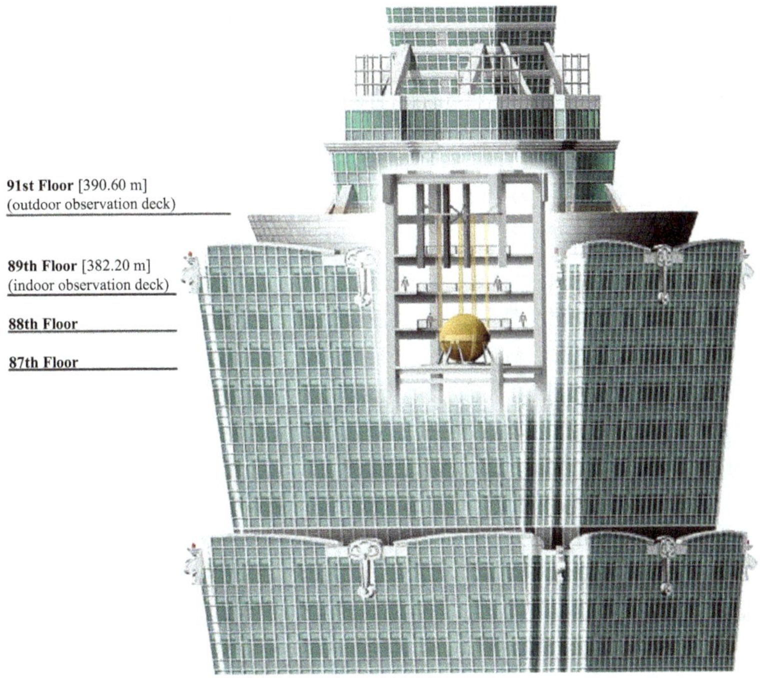

Figure 4.6 Location of Taipei 101's tuned mass damper[1]

4.3.1 Tuned mass dampers

The most common structural control device is called the TMD. This device utilizes a mass on an ideally frictionless track. The TMD mass and the main structure are connected via a spring and damper, providing stiffness and damping. In an ideal TMD, both of these components are linear and have a constant spring and damping coefficient, which are also the assumptions in this work. Following is the sketch for an ideal TMD on top of the target system for horizontal acceleration mitigation, and the mathematical model can be established as:

$$\begin{aligned} m_s\ddot{v}_s + D_s\dot{v}_s + K_s v_s &= f_w + Kx + D\dot{x}, \\ m\ddot{x} + Kx + D\dot{x} &= 0, \end{aligned}$$

where m_s, D_s, K_s denote structural mass, damping coefficient, and stiffness of the target system, respectively, while m, D, K represent corresponding properties of the TMD. $\ddot{v}_s$ and $\ddot{x}$ are the horizontal acceleration of the target and the TMD. f_w is the external force imposing on the target (Figure 4.8).

[1] https://en.wikipedia.org/wiki/Tuned_mass_damper/

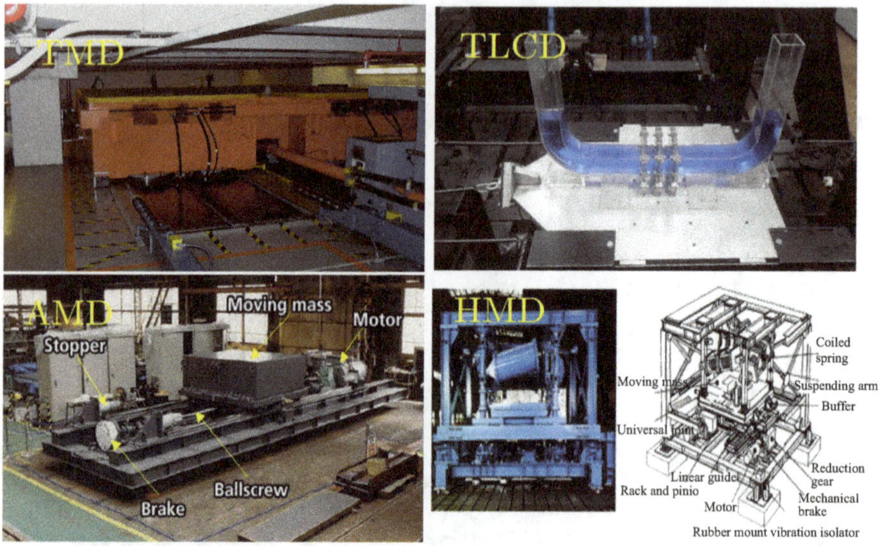

Figure 4.7 Different types of structural control devices[2,3,4]

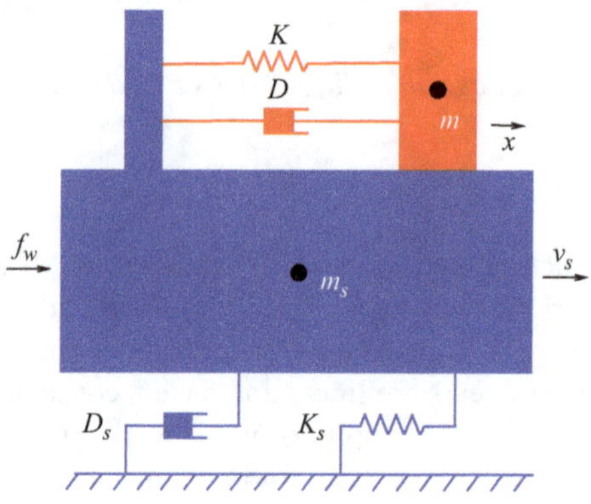

Figure 4.8 Sketch of a tuned mass damper on top of the target system

The mass and spring are usually tuned to be consistent with the main system frequency regarding load reduction, and this will lead to the TMD mass vibrating at the same frequency. At the same time, the damper dissipates energy from the whole system in the form of heat. However, in a real TMD design, optimizing the spring and damping constants can be difficult due to space constraints and parameter

[2]http://www.lemessurier.com/john_hancock_tower
[3]http://www.mdpi.com/1996-1073/8/1/111
[4]http://www.ihi.co.jp/iis/english/products/damper_mass.html

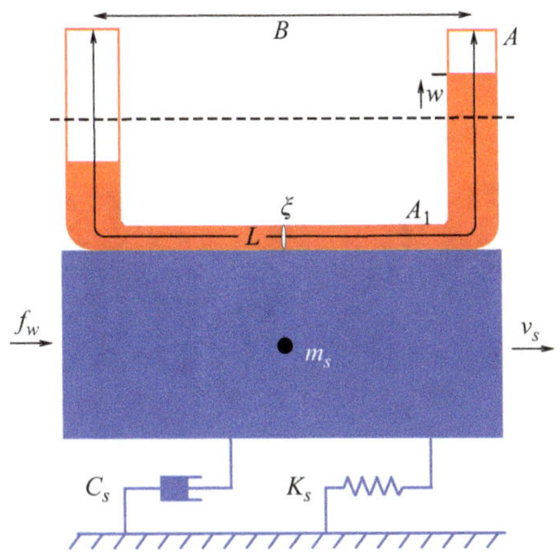

Figure 4.9 Sketch of a tuned liquid column damper on top of the target system

nonlinearities. For structures with more DoF and nonlinearities like an OWT, this is much more difficult.

4.3.2 Tuned liquid column dampers

Tuned liquid column damper (TLCD) uses two attached vertical columns of liquid with an orifice between them. The difference between the heights of the two liquid columns provides an equivalent spring force, and the fluid passing through the orifice provides a damping force. Figure 4.9 gives a sketch for a TLCD on top of the target, and the equations of motion can be described as:

$$\begin{aligned} m_s\ddot{v}_s + C_s\dot{v}_s + K_s v_s &= f_w - \rho AB\ddot{w} - \rho AL_{em}\ddot{v}_s, \\ \rho AL_{ee}\ddot{w} + 0.5\rho A\xi|\dot{w}|\dot{w} + 2\rho Agw &= -\rho AB\ddot{v}_s m, \\ |w| &< \frac{L-B}{2}, \end{aligned}$$

where w represents the liquid relative displacement. A and A_1 denote cross-sectional area of liquid column vertical and horizontal sections, respectively. B is the horizontal length, and L is the total length of the liquid. $\alpha = A/A_1$ is the area ratio, and $L_{ee} = L - B + \alpha B$. ρ is the liquid density, and ξ is the coefficient of head loss. $L_{em} = B/\alpha + (L - B)$ is the length of an equivalent uniform cross-sectional area liquid column with area A which has the same mass as the TLCD.

Compared with TMD, TLCD will significantly reduce the cost as it only uses liquid instead of huge concrete or steel. This can be seen as an advantage over other structural control methods, since cost of energy is the main topic in wind industry. Similar with TMD, how to define the dimension and position of TLCD needs to be investigated. Also, TLCD provides a semiactive control channel because of the tunable orifice, which is promising for further performance improvement.

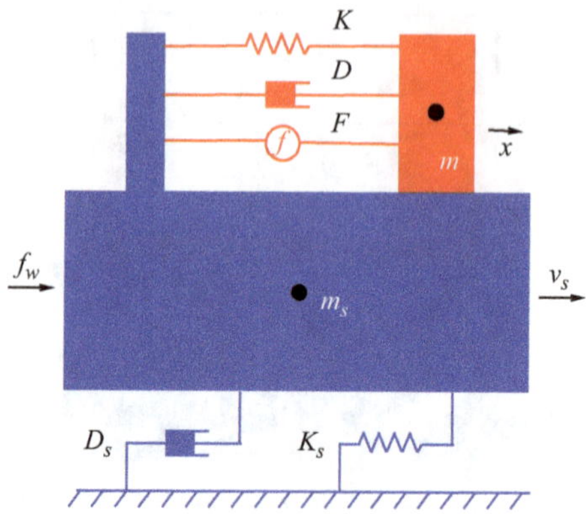

Figure 4.10 Sketch of a hybrid mass damper on top of the target system

4.3.3 Hybrid mass dampers

An active mass damper (AMD) consists of a mass and an actuator, which can be actively controlled to apply a force to the mass and an equal and opposite force on the structure. Since there is no physical spring and damper in this system, the actuator must provide all of the forces to the mass. There is also the potential to destabilize the system if the control scheme is not well designed.

Based on this, the hybrid mass damper (HMD) combines the TMD and AMD. It usually consists of a tuned mass, spring, and damper system as well as an actuator, such as servomotor or hydraulic actuator. Figure 4.10 sketches an HMD on top of the target, and its dynamics are similar to that of TMD, except a force F from the actuator is added.

$$\begin{aligned} m_s\ddot{v}_s + D_s\dot{v}_s + K_s v_s &= f_w + Kx + D\dot{x}, \\ m\ddot{x} + Kx + D\dot{x} &= F. \end{aligned}$$

With the added actuator, the HMD gains the potential for improved performance over a passive system. However, HMD can add energy to the system, thus instability will be possibly introduced, and how to design a proper controller for the actuator is one of the topics in this research. Besides, the HMD includes a passive system, so it can still provide load reduction with no actuation power.

4.4 Structural control of offshore wind turbines

Structural vibration control have been successfully applied in civil engineering structures, such as skyscrapers and bridges [15]; thus, it is also expected to be a promising solution for extending the fatigue life of OWTs. The above-mentioned structural control devices can be possibly installed at the locations of wind turbine nacelle for load

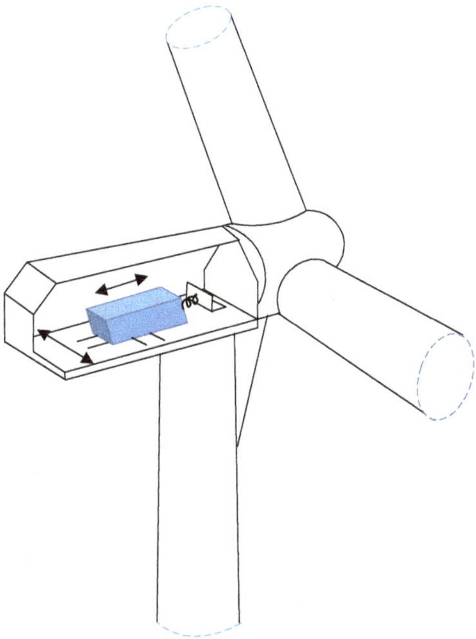

Figure 4.11 Wind turbine structural control concept

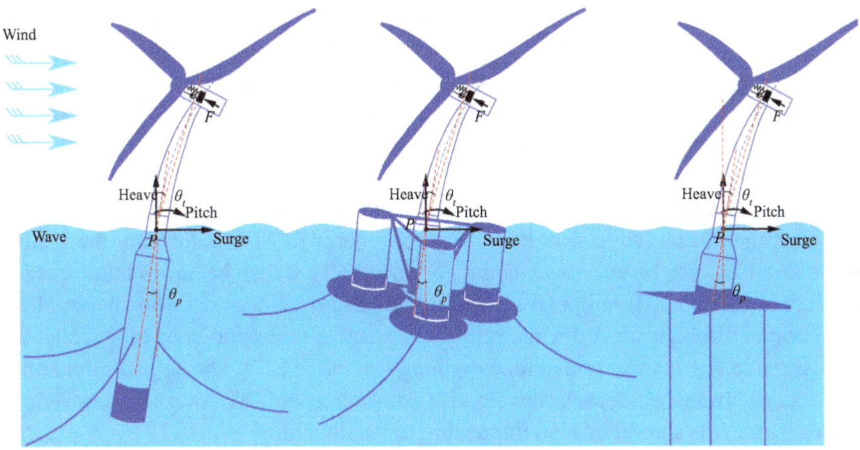

Figure 4.12 Structural control concept for FOWTs

reduction. As shown in Figure 4.11, the TMD can either move in the *x* or *y* axis to mitigate the wind turbine loads in the for–aft and side–side directions. The structural control systems can even be installed in the supporting platforms for FOWTs shown in Figure 4.12.

There have been many published results discussing the structural control idea for load mitigation of wind turbines recent years. In [16], the authors investigated the use of a TMD placed at the tower top of a simplified wind turbine model for vibration mitigation. Following the same installation idea, Colwell *et al.* explored the structural responses of a fixed-bottom offshore wind turbine with a TLCD [17]. Later, the reliability of this idea was assessed in [18]. Moreover, Li *et al.* performed an experimental study on an offshore wind turbine with a ball vibration absorber fixed on top of the nacelle [19]. However, these discussions are about vibration mitigation of fixed-bottom wind turbines, while their motion dynamics are quite different from that of FOWTs. Besides, these works are not based on the cutting-edge high-fidelity codes for wind turbine models, which may not capture the comprehensive coupled nonlinear dynamics of wind turbines.

Based on the aero–hydro–servo–elastic wind turbine numerical simulator FAST (fatigue, aerodynamics, structures, and turbulence) [20], Lackner *et al.* implemented a new simulation tool, called FAST-SC, for passive, semiactive, and active structural control design of wind turbines [21]. It has incorporated TMDs into the nacelle or platform of wind turbines for load mitigation. Utilizing this code, Lackner *et al.* presented more realistic simulation results by installing a TMD in the nacelle of both a barge-type and a monopile supported wind turbines, and a simple parametric study was also performed to determine the TMD parameters [21]. In order to perform a more comprehensive parametric study, the authors in [22,23] established a 3-DOF dynamic model for different types of floating wind turbines based on first principles, and TMD parameters are designed under different optimization methods. This limited-DOF model has greatly facilitated parametric analysis, but the coupling between platform surge and pitch motion was not captured. This effect can be ignored for the barge model but might be a strong mode for other platforms [24,25]. In addition, TMD was also proposed to be installed in the platform of TLP or spar-type floating wind turbines [22], so that extra attention can be drawn on this idea since bigger mass becomes possible. Besides, the author in [26] investigated the effectiveness of TLCD for offshore wind turbine load reduction, but the results were not based on the cutting-edge simulator with fully coupled TLCD–turbine interaction. Since FAST-SC can be possibly customized for TLCD, it then becomes interesting to conduct further code development and simulation study. It was also shown more load reduction could be achieved when introducing active structural control, such as the multivariable H_∞ control with a loop-shaping technique [27]. The actuator dynamics and control-structure interaction were also considered in [28]. Alternative advanced control strategies should be also interesting to be studied.

Two different structural control scenarios are considered in this chapter, which both uses the OC3-Hywind model as a study object. One suggests to install a TMD into the spar platform as a passive structural control design, and the other proposes to install an HMD onto the wind turbine nacelle as an active structural control case. Instead of a specific device, an ideal TMD or HMD is used in this study, which is assumed to move on a linear nonfriction track, and its stiffness and damping coefficients are kept constant during operation (Figure 4.13).

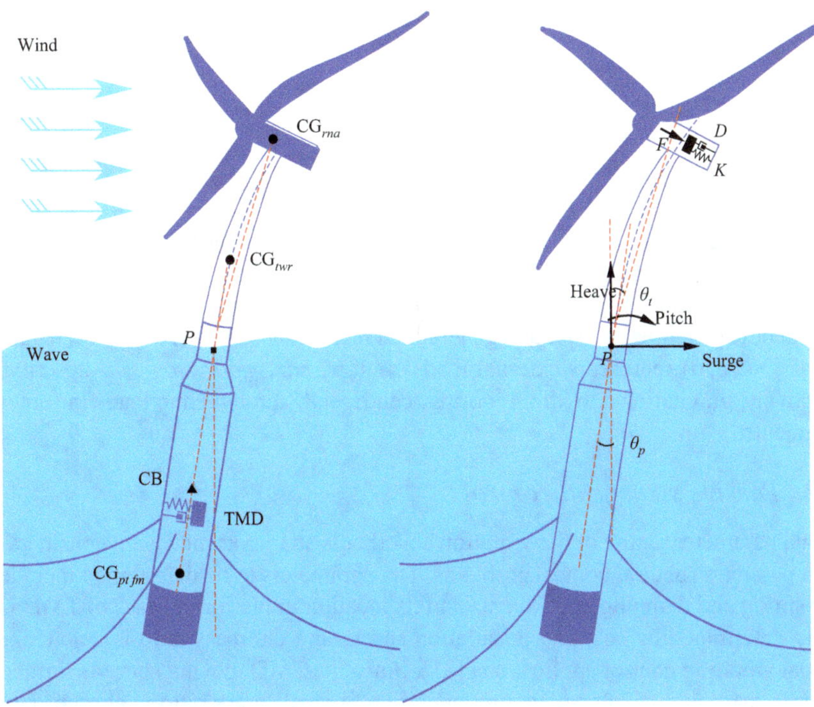

Figure 4.13 Passive and active structural control scenarios for OC3-Hywind FOWT [29]

4.4.1 Dynamic modeling

In order to investigate the parameters of structural control systems, optimize system performance, or further design an active controller, establishing one dynamic mathematical model is very helpful. Based on the D'Alembert's principle of inertial forces, the following longitudinal model for OC3-Hywind FOWT with an HMD installed in nacelle can be established as follows, see [30]:

$$\mathbf{M}(\mathbf{q})\ddot{\mathbf{q}} + \mathbf{L}(\mathbf{q}, \dot{\mathbf{q}}) = \mathbf{F}, \tag{4.1}$$

where

$$\mathbf{q} = \begin{bmatrix} x_{sg} \\ x_{hv} \\ \theta_p \\ x_{hmd} \\ \theta_t \end{bmatrix}, \quad \mathbf{M}(\mathbf{q}) = \begin{bmatrix} M_{sg}^{sg} & 0 & I_{sg}^{p} & M_{sg}^{hmd} & 0 \\ 0 & M_{hv}^{hv} & I_{hv}^{p} & M_{hv}^{hmd} & 0 \\ M_{p}^{sg} & M_{p}^{hv} & I_{p}^{p} & M_{p}^{hmd} & 0 \\ M_{hmd}^{sg} & M_{hmd}^{hv} & I_{hmd}^{p} & M_{hmd}^{hmd} & 0 \\ M_{t}^{sg} & M_{t}^{hv} & 0 & M_{t}^{hmd} & I_{t}^{t} \end{bmatrix},$$

$$\mathbf{L}(\mathbf{q},\dot{\mathbf{q}}) = \begin{bmatrix} F_{sg}^{hdr} + F_{sg}^{moor} + F_{sg}^{ctr} + F_{sg}^{wnd} \\ F_{hv}^{gr} + F_{hv}^{hdr} + F_{hv}^{moor} + F_{hv}^{ctr} + F_{hv}^{wnd} \\ \tau_{p}^{gr} + \tau_{p}^{hdr} + \tau_{p}^{moor} + \tau_{p}^{ctr} + \tau_{p}^{wnd} \\ F_{hmd}^{gr} + F_{hmd}^{hmd} \\ \tau_{t}^{gr} + \tau_{t}^{p} + \tau_{t}^{ctr} + \tau_{t}^{wnd} \end{bmatrix}, \quad \mathbf{F} = \begin{bmatrix} 0 \\ 0 \\ 0 \\ F \\ 0 \end{bmatrix}.$$

In this model, $\mathbf{q}$ is the state vector, and sg, hv, p, hmd, t represent, respectively, the enabled 5 DOFs, i.e., platform surge, heave, pitch motion about P, HMD translation, and tower deflection. $\mathbf{M}(\mathbf{q})$ is the system inertial matrix, which is positive definite. M_i^j and I_i^j denote generalized mass and generalized inertial tensor for DOF i with regard to DOF j. $\mathbf{L}(\mathbf{q},\dot{\mathbf{q}})$ represents external loads, and gr, hdr, $moor$, ctr, wnd describe, respectively, gravitational, hydro, mooring, centripetal, aerodynamic loads in forces or moments.

4.4.2 Passive structural control

Optimal parameter tuning of the vibration absorber is an important design consideration in passive structural control problems. The parameters to be determined include TMD spring and damping coefficients and its location in platform. Mass of TMD is usually determined by cost and installation space, and the mass ratio is below 5% for most vibration control applications [31]. Since the TMD moving space is limited and large mass is difficult for marine operation, a small mass ratio is chosen, i.e., 100,000 kg, about 1.25% of the whole platform-tower-RNA weight.

In fact, the most convincing optimization solution here is to try all possible values of these parameters in FAST-SC. However, this global searching process will take tens of thousands of calls from FAST-SC, and it even takes minutes to run it for only one time. Therefore, exhaustive search (ES) based on FAST-SC is almost impossible with ordinary computers, and appropriate optimization methods are needed.

4.4.2.1 Frequency and damping analysis

In engineering applications, the natural frequency of TMD is usually tuned to be near to that of the target system, so that it will effectively dissipate the undesirable system vibration energy. In order to systematically describe this phenomenon, Den Hartog [32] analyzed the response of undamped main system with TMD subjected to harmonic external forces and derived an explicit expression to determine the optimal TMD natural frequency and damping ratio for vibration inhibition, which has been widely adopted in many practical applications. The optimal solution is given by

$$\begin{aligned} f_{tmd} &= \frac{f}{1+\mu}, \\ \xi_{tmd} &= \sqrt{\frac{3\mu}{8(1+\mu)}}, \end{aligned} \tag{4.2}$$

where μ denotes the mass ratio m_{tmd}/m. f and ξ are the natural frequency and damping ratio of target system. f_{tmd} and ξ_{tmd} represent, respectively, the optimal natural

Table 4.3 Natural frequencies and damping ratios of modes for OC3-Hywind surge-heave-pitch motion

DOF	f (Hz)	ξ
Platform surge	0.0080	0.1364
Platform heave	0.0324	0.0384
Platform pitch	0.0342	0.1418
1st tower fore–aft bending	0.4732	0.0087
2nd tower fore–aft bending	3.7505	0.0102

frequency and damping ratio of TMD. However, all systems contain damping, which should not be neglected in the parameter tuning process. It is also found that the optimum tuning frequency is strongly influenced by the system damping level [33]. In face of this, Tsai *et al.* presented a numerical searching procedure to determine the optimum parameters with target system damping considered, and a curve fitting scheme was then used to produce the equations for the optimal tuning parameters [33], given as

$$f_{tmd} = \left(\left(\frac{\sqrt{1+0.5\mu}}{1+\mu} + \frac{1}{\sqrt{1-2\xi^2}} - 1 \right) - (0.288 - 0.661\sqrt{\mu} + 1.12\mu)\sqrt{\mu}\xi \right.$$

$$\left. - (2.298 - 6.739\sqrt{\mu} + 8.316\mu)\sqrt{\mu}\xi^2 \right) f, \tag{4.3}$$

$$\xi_{tmd} = \sqrt{\frac{3\mu}{8(1+\mu)}} + 0.151\xi - 0.187\xi^2 + 0.238\xi\mu.$$

Their results also showed that the higher the system's damping is, the more the optimum parameters deviate from those in undamped systems. In the OC3-Hywind dynamics, there exist high structural damping (above 0.1) for the main translational and rotational modes; thus, the solution (4.3) is more appropriate than (4.2) for this optimization problem.

In order to adopt Tsai's method, eigenanalysis based on model linearization result should be performed first to obtain system natural frequencies and damping ratios for the modes of interest. In the FAST linearization process, all the platform translation and rotation DOFs, all tower bending DOFs, drivetrain DOF, and blade flexibility DOFs are enabled. Rotor speed and wind speed are both set as 0. Incident wave and hydro radiation damping effects are disabled. Part of the eigenanalysis result is presented in Table 4.3, which agrees with the data in [8].

Among these DOFs for OC3-Hywind surge–heave–pitch motion, the most critical modes that affect system performance are the 1st tower fore–aft bending mode and the low frequency platform pitch mode, which is also the dominant cause of tower bending. Therefore, the natural frequency and damping ratio of these two modes are used here for TMD parameter optimization. Based on formula (4.3), the TMD spring and damping coefficients are determined and listed in Table 4.5. It should be noted

Table 4.4 Performance indices

Index	Description
$J_1 = \sqrt{(1/T)\int_0^T (x_{tt} - \tilde{x}_{tt})^2 dt}$	Standard deviation of tower top displacement under its equilibrium point
$J_2 = \sqrt{(1/T)\int_0^T (\theta_p - \tilde{\theta}_p)^2 dt}$	Standard deviation of platform pitch angle under its equilibrium point
$J_3 = \max(x_{ttd}) - \min(x_{ttd})$	Maximum range of tower top displacement
$J_4 = \max(\theta_p) - \min(\theta_p)$	Maximum range of platform pitch angle

that the TMD location in platform is not analyzed, and it is intuitively specified as 100 m below the reference point P.

However, the nonlinearity ofTMD stops due to space constraints is not considered in this design process, which has been shown to have strong influence on TMD load reduction effectiveness according to the nonlinear FAST-SC simulation results. Besides, parametric study of TMD location in platform cannot be performed based on this method. Therefore, a more thorough method should be proposed to find the best combination of TMD parameters.

4.4.2.2 Exhaustive search

In the last section, we have obtained a mathematical model describing OC3-Hywind surge–heave–pitch motion, which manages to capture most of system structural dynamics, hydro and mooring effects. More importantly, the time for solving this dynamic equation is less than 1 s; thus, ES, a global parameter tuning method, becomes a possible solution to determine the optimal TMD parameters.

Before presenting the result, the performance indices are introduced which are used in the performance evaluation process. In fact, the fore–aft tower top deflection is the best indicator of tower bottom bending moments, and the author in [22] used standard deviation (STD) of the tower top displacement as the performance index, which is also adopted in this work as the first index J_1. Similarly, the STD of platform pitch angle is another criterion J_2, since platform pitch motion is the source of extra tower fatigue load, and it will also introduce extra loading on the important devices in nacelle such as drivetrain and generator. Second, we also care about load reduction effectiveness of the proposed method in extreme events; thus, the range of tower top displacement and platform pitch angle in the free decay test are treated as another two evaluation indices J_3 and J_4.

ES is performed based on these indices. The parameter searching range and interval are chosen when both time consumption and accuracy are considered. The slice plot for J_1 is illustrated in Figure 4.14, and the optimization results are listed in Table 4.5.

Although ES could be regarded as a global optimization method, it is still limited by predefined parameter searching range and interval. Besides, it is very

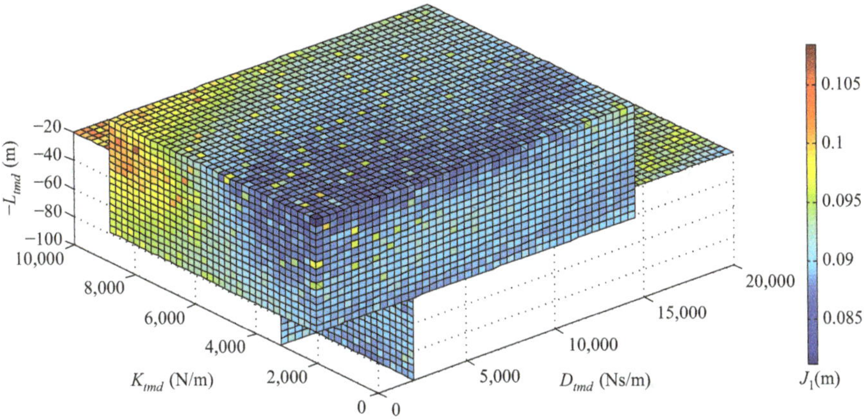

Figure 4.14 Slice plot subjected to performance index J_1 with TMD installed in platform

computationally expensive, which will take hours or days long to finish one optimization process. Moreover, there might exist better solution if the parameter interval is not small enough. Therefore, more intelligent and efficient optimization algorithms are demanded.

4.4.2.3 Simplex coding genetic algorithm

In the past few years, genetic algorithm has been widely applied in a broad spectrum of real-world systems [34]. This approach starts with randomly generated population, and individuals with better fitness will be selected as the basis of next generation. The improved population will keep evolving after inheritance, mutation, selection, and crossover procedures until it meets the final requirement. As a global optimization method, genetic algorithm is based on stochastic variables and does not require the derivatives of object function, which brings the advantages of global evaluation and objective tolerance when compared with other gradient-based local optimization methods. It usually helps to obtain a better result in optimization problems with nonsmooth objective functions, thus is suitable for the optimization problem in this work.

However, the genetic algorithm may still suffer from the slow convergence that brings about high computational cost. Considering this problem, several researchers tried to furnish genetic algorithm with the ability to simulate the fast convergence of local search methods. The simplex coding genetic algorithm (SCGA) proposed by Hedar *et al.* is adopted in this work for efficient parameter optimization [35]. SCGA combines the genetic algorithm and simplex-based local optimization algorithm called Nelder–Mead method [36], which is one of the most efficient derivative-free nonlinear optimization approaches. In SCGA, each chromosome in the population is a simplex, and the gene is a vertex of this simplex. Nelder–Mead method is applied to improve the population in the initial stage and every intermediate step when new children are

Table 4.5 Parameter optimization result (m_{tmd} = 100,000 kg)

Method	Performance index	K_{tmd} (N/m)	D_{tmd} (N s/m)	L_{tmd} (m)
Tsai	Tower bending (Tsai1)	868,010	40,134	N/A
Tsai	Platform pitch (Tsai2)	4,660	3,663	N/A
ES	$J_1 = 0.0812$ m	2,000	3,200	20
ES	$J_2 = 1.2327$ degree	0	6,400	20
ES	$J_3 = 0.8880$ m	3,000	0	20
ES	$J_4 = 8.8926$ degree	0	14,800	20
SCGA	$J_1 = 0.0826$ m	719	3,389	20
SCGA	$J_2 = 1.2312$ degree	0	11,733	20
SCGA	$J_3 = 0.7600$ m	95,9526	5,733	20
SCGA	$J_4 = 8.8924$ degree	0	14,595	20

generated. Detailed steps of SCGA are not presented here for brevity, while only the values of important parameters are given in the following. In the simplex searching parameters, number of simplices per coordinate direction is set as 5, and maximum number of local iterations is 2. For the GA initialization, probability of the roulette wheel uniform crossover is chosen as 0.6, and the mutation probability 0.01 is used.

Based on the same evaluation indices in Table 4.4, the optimization results are shown in Table 4.5. First, it can be easily noticed that SCGA gives a better result with respect to J_3, but the exhausted searching range is limited by its searching scope. Second, compared with GA, the SCGA searching range is not strictly restricted by the initial lower and upper limits, since the reflection and expansion steps in the simplex searching process will possibly cross the predefined bounds. This feature also brings the advantage of wider searching range, where the optimal values may exceed the predetermined limits in the J_3 case. Third, SCGA is more efficient than GA, which only takes 10 min to finish one optimization process.

4.4.2.4 Simulation and analysis

Based on the optimization result, fully nonlinear simulations are performed in FAST-SC with all wind turbine DOFs enabled. Each test runs 630 s, and the output data in first 30 s are not recorded, waiting for generator torque and blade pitch motion arriving normal operation state. The modified generator torque and blade pitch controller from National Renewable Energy Laboratory (NREL) is used in the form of a dynamic link library for all tests [25].

As mentioned above, three different simulation scenarios are considered. The wind and wave conditions in [27] are adopted as two cases in this experiment. For wind condition, the mean value of the turbulent wind is defined as 10 and 18 m/s, respectively. The wind field is generated by TurbSim [37], where Kaimal spectra and the power law exponent of 0.14 are used according to the IEC61400-3 offshore wind turbine design standard. The normal turbulence intensity is set as level B, i.e., 18% (10 m/s case) and 15% (18 m/s case). For wave condition, JONSWAP spectrum is utilized to generate the stochastic wave inputs. The significant wave height is set as 2.3 (10 m/s case) and 3.7 m (18 m/s case), and the peak spectral period is both defined as

Table 4.6 Percentage of load reduction with different TMD parameters

Case (m/s)	Evaluation index	Tsai1	Tsai1′	Tsai2	Tsai2′	J_1	J_2	J_3	J_4
10	DEL TwrBsMyt	5.40	−0.28	2.21	−0.35	1.71	1.32	3.30	1.33
	DEL TwrBsMxt	17.59	−0.90	−1.36	−2.16	−2.05	−2.64	18.73	−2.54
	DEL RootMyc1	0.31	−1.32	0.77	−0.01	0.51	0.36	0.56	0.32
	DEL Anch1Ten	5.25	2.75	5.25	2.75	0.51	0.36	0.56	0.32
	95th TwrBsMyt	1.12	2.57	0.40	1.58	0.27	0.05	0.68	0.05
	95th TwrBsMxt	4.44	1.60	1.33	1.70	0.81	0.33	3.11	0.34
	95th RootMyc1	0.42	0.37	0.47	0.42	0.47	0.51	0.42	0.47
	95th Anch1Ten	4.09	4.07	4.06	4.05	4.06	4.07	4.09	4.07
	95th PtfmPitch	1.92	7.91	0.49	5.46	0.25	0.02	1.96	0.04
	95th PtfmRoll	0.90	2.11	2.74	3.53	2.49	2.11	0.81	2.11
18	DEL TwrBsMyt	0.80	0.06	14.26	6.77	13.89	13.24	−0.38	13.1
	DEL TwrBsMxt	18.70	5.46	−4.76	4.52	−6.88	1.25	9.17	2.93
	DEL RootMyc1	−0.89	4.47	8.05	7.14	7.80	7.28	−0.98	7.29
	DEL Anch1Ten	4.81	1.41	−3.52	2.40	−2.01	−0.97	4.82	−0.97
	95th TwrBsMyt	−0.78	2.00	12.84	6.54	13.17	12.16	−1.03	11.95
	95th TwrBsMxt	3.82	2.40	1.90	3.61	0.64	3.03	1.54	3.65
	95th RootMyc1	−0.90	−0.61	7.11	2.41	7.56	7.07	−0.75	6.97
	95th Anch1Ten	4.55	4.60	2.96	4.38	3.15	3.39	4.55	3.41
	95th PtfmPitch	−0.38	7.12	18.80	13.78	18.53	17.50	−0.40	17.27
	95th PtfmRoll	−3.64	8.42	13.92	12.60	12.97	12.10	−3.66	12.07
	RMS GenPwr	−8.32	12.63	46.90	37.29	46.47	42.45	−8.29	41.86
37	DEL TwrBsMyt	3.73	−2.50	−7.14	−1.98	−7.75	−7.39	2.69	−6.12
	DEL TwrBsMxt	−0.56	0.53	−0.61	0.73	−0.42	−0.49	−0.67	−0.27
	DEL RootMyc1	0.93	−2.51	−8.66	−3.28	−8.41	−5.70	0.45	−4.40
	DEL Anch1Ten	3.29	3.87	3.13	3.38	3.25	3.66	3.14	3.25
	95th TwrBsMyt	2.27	−3.70	2.42	−2.76	1.21	0.39	3.09	1.26
	95th TwrBsMxt	0.36	1.33	−0.02	1.44	0.15	0.17	0.34	0.47
	95th RootMyc1	1.01	−1.12	−4.73	−1.00	−4.40	−0.62	1.14	−1.57
	95th Anch1Ten	3.67	3.82	3.52	3.67	3.54	3.53	3.66	3.55
	95th PtfmPitch	2.15	−8.16	4.30	−5.23	4.00	2.87	2.17	2.98
	95th PtfmRoll	1.32	6.22	1.22	6.37	1.06	0.59	1.32	0.37

14 s. Besides, the parked case with 50-year extreme wind and wave is also considered. Mean value of the extreme turbulent wind is defined as 37 m/s with 11% intensity and 0.11 power law exponent, and the significant wave height and wave period are defined as 13.8 m and 19 s. In this case, the generator torque and blade pitch controller are turned off, and all the blades are feathered to 90 degree to minimize the aerodynamic loading. For each case, at least two sets of random variables are used to generate wind and wave data.

Percentage of load reduction with different TMD parameter choice is shown in Table 4.6. In the cases Tsai1 and Tsai2, TMD is placed near to the top of the spar, i.e., 20 m from P, while L_{tmd} is 100 m for Tsai1′ and Tsai2′. Regarding J_1–J_4, the better design parameters are chosen when comparing the index values between ES and SCGA in the optimization results. In order to measure the fatigue and extreme loading, DEL and the 95th percentile of fore–aft and side–side tower base bending moments (TwrBsMyt and TwrBsMxt), flapwise bending moment at the first blade root (RootMyc1), and tension of the first anchor (Anch1Ten) are calculated. STD and the 95th percentile of platform pitch and roll rotation angle are also calculated. In the above-rated situation, the root mean square (RMS) of generated power is considered as another index. Three time series simulation results based on the J_1 design are shown in Figures 4.15–4.17, and our remarks on these data are presented below.

- First, when comparing the results of Tsai1–Tsai2 and Tsai1′–Tsai2′, it is interesting to find out that this TMD is more effective when placed at the upper side of the spar instead of the bottom. This agrees with the other optimization results, and it could also be indicated from the fact that most tall buildings have TMDs installed on top floors.
- Second, it can be seen that the design of TMD with small spring coefficient (J_1, J_2, and J_4 cases) achieves much load reduction in above rated condition, and the RMS of generated power is even improved by over 40%. Also, the load reduction effectiveness can be clearly noticed from Figure 4.16, where the TMD is taking effect when the platform is in resonant motion. Besides, from the other point of view, only a damper is needed for the above-rated case to achieve such performance based on the value of optimized parameters.
- Third, with these designs, however, it does not bring much load reduction when the turbine is working below rated in nonresonant motion. This ineffectiveness is due to the constant wind force component acting on the turbine rotor so that the TMD is leaning on one side most of time [22], which can be seen from Figure 4.15. Besides, in parked condition, the system performance is also worse since external load mainly comes from wave, and the platform pitch frequency is dominated by wave, so that the proposed TMD design loses effect.
- Fourth, the design with large spring and damping constants (J_3 case), in contrast, will produce moderate load reduction in all three working conditions, no matter the platform is in resonant motion or not. At this time, the TMD behaves like a high-frequency load absorber, working around its original position and never exceeding its motion constraints. Therefore, it will not contribute too much when the wind turbine platform is working above rated or in low frequency resonant motion.

4.4.3 Active structural control

Passive structural control systems have fixed properties and do not require an external source of energy, so it is not possible to change a passive control system dynamically to compensate for unexpected loadings. In contrast, an active structural control system has the ability to determine the present state of the target structure, synthesize the control strategies in order to change this state to a more desirable one, and carry out these control actions. Active structural control methods are promising to mitigate more loads but at the expense of more energy consumption and instability risk.

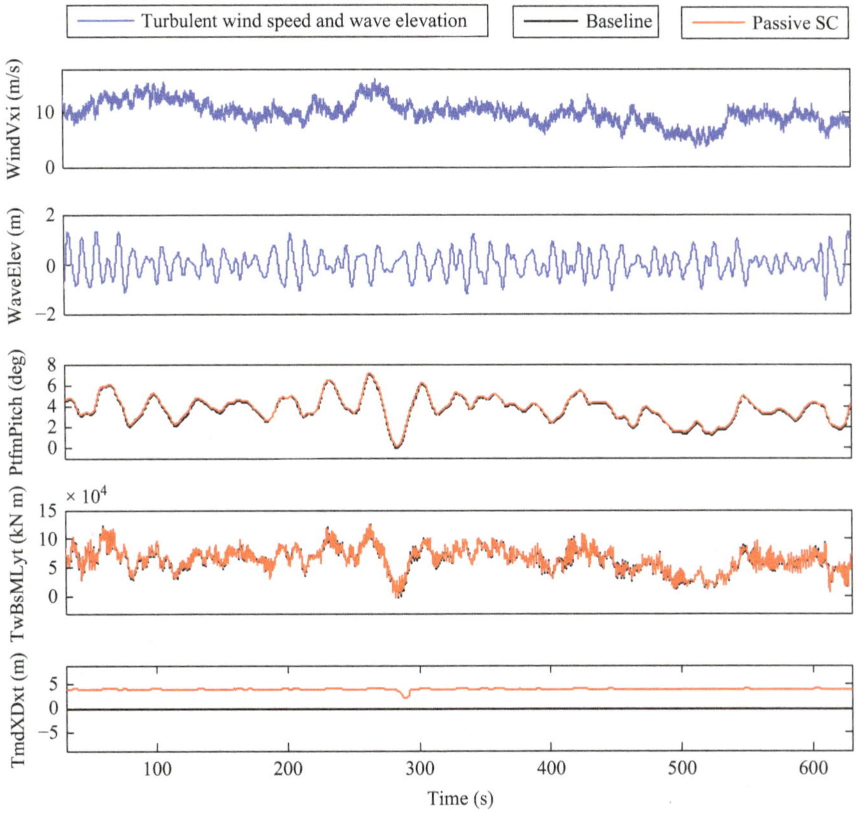

Figure 4.15 FAST-SC simulation results with 10 m/s turbulent wind and 2.3 m wave height

This section discusses the active structural control design for load mitigation of OC3-Hywind FOWT, where a HMD is installed in the nacelle. Based on the derived dynamic model of OC3-Hywind longitudinal motion, model linearization is first performed. Then, a gain scheduling state-feedback H_2/H_∞ structural controller is designed. Numerical simulation results based on the proposed control design is given in the end.

4.4.3.1 Model linearization

Based on small deviation approximation, the model around each equilibrium point can be linearized into the following state-space representation:

$$\begin{cases} \dot{x} = Ax + Bu \\ z = Cx \end{cases}, \tag{4.4}$$

where $x = \left[\dot{x}_{sg}, \dot{x}_{hv}, \dot{\theta}_p, \dot{x}_{hmd}, \dot{\theta}_t, x_{sg}, x_{hv}, \theta_p, x_{hmd}, \theta_t \right]^T$ is the state vector. It includes the velocity and displacement of platform surge, heave, pitch, HMD mass translation,

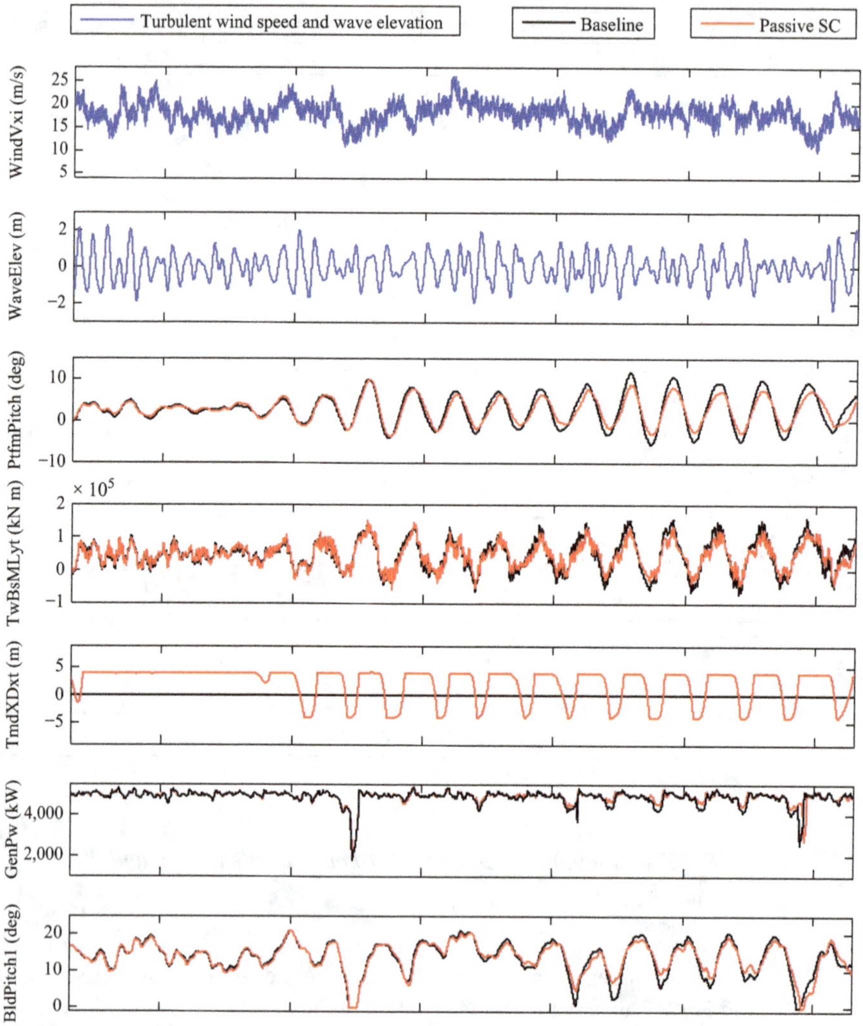

Figure 4.16 FAST-SC simulation results with 18 m/s turbulent wind and 3.7 m wave height

as well as tower pitch motion. $u = F$ is the control input, and $z = \theta_p - \theta_t$ is the controlled output.

4.4.3.2 H_2/H_∞ control design

Similar to (4.4), consider the linear system around a certain setpoint

$$\begin{cases} \dot{x} = Ax + Bu + B_w w \\ z = Cx \end{cases}, \tag{4.5}$$

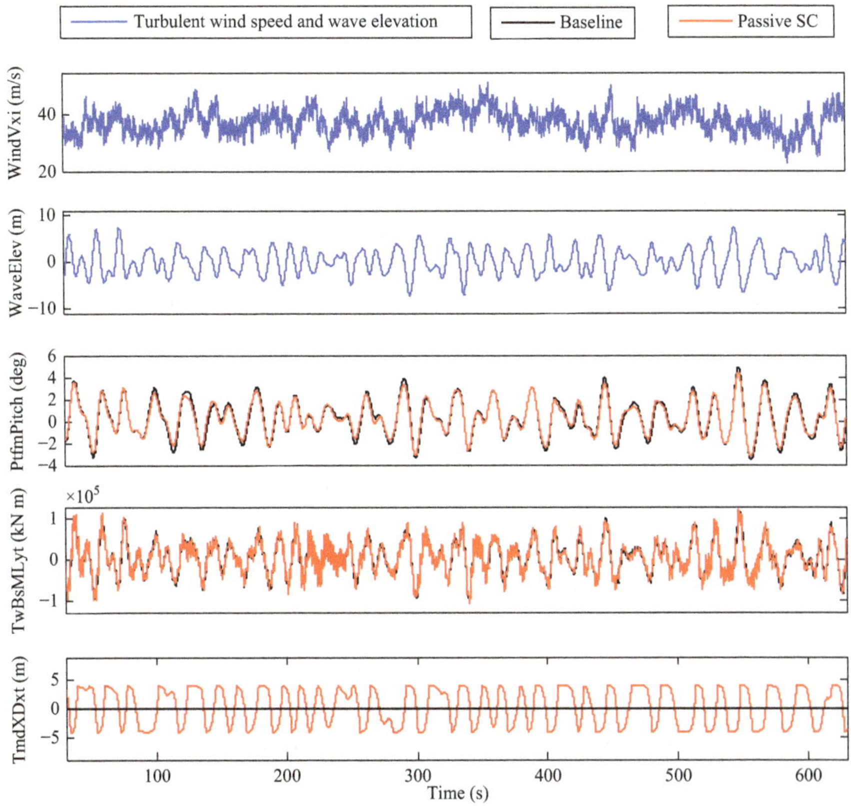

Figure 4.17 FAST-SC simulation results with 37 m/s 50-year extreme turbulent wind and 13.8 m wave height

where w is the aerodynamic disturbance acting on the rotor. We would like to design a state feedback controller $u = Kx$ that keeps the closed-loop system

$$\begin{cases} \dot{x} = (A + BK)x + B_w w \\ z = Cx \end{cases} \tag{4.6}$$

asymptotically stable and improves the dynamic performance of the closed-loop system simultaneously. More specifically, regarding performance improvement, the controller should keep the closed-loop system robust to disturbance w, i.e., the H_∞ norm of the transfer function T_{wz} in the closed-loop system does not exceed a given upper bound γ_1. More importantly, the H_2 norm of T_{wz} should be as small as possible (e.g., less than γ_2) so that the vibration energy of tower top deflection will be reduced.

Therefore, this problem is equivalent to a mixed H_2/H_∞ control design, see [38,39], and the design objective is to determine a desired state feedback gain K

such that the closed-loop system is asymptotically stable and γ_2 is minimal for the controllers such that

$$\|T_{wz}\|_\infty < \gamma_1, \|T_{wz}\|_2 < \gamma_2.$$

The following theorem helps to convert the H_2/H_∞ control design problem into an optimization process for several linear matrix inequalities (LMIs). It is then more convenient to solve by using well developed LMI toolbox.

Theorem 4.1. *For the closed-loop system (4.6), if there exists a given $\gamma_1 > 0$ and the following LMIs have an optimal solution,*

$$\begin{aligned} &\min \gamma_2 \\ s.t. &\begin{bmatrix} AX + BW + (AX + BW)^T & B_w & (CX)^T \\ B_w^T & -\gamma_1 I & 0 \\ CX & 0 & -\gamma_1 I \end{bmatrix} < 0 \\ &AX + BW + (AX + BW)^T + B_w B_w^T < 0, \\ &\begin{bmatrix} -Z & CX \\ (CX)^T & -X \end{bmatrix} < 0 \\ &Trace(Z) < \gamma_2 \end{aligned} \tag{4.7}$$

where $X = X^T > 0$, $Z = Z^T > 0$ and W are matrices of appropriate dimensions, then the state feedback H_2/H_∞ control design is feasible, and the control law is

$$u = WX^{-1}x. \tag{4.8}$$

4.4.3.3 Gain scheduling

It is possible to design a controller for each steady-state point, but controller switching will be frequent and when to switch becomes a problem. Therefore, these setpoints are categorized in seven intervals, which are determined by rotor speed and blade pitch angle, which are possible to obtain from a wind turbine.

4.4.3.4 Low pass filter

The control force has to pass a second-order low-pass filter,

$$G(s) = \frac{\omega^2}{s^2 + 2\zeta\omega s + \omega^2},$$

which represents the actuator dynamics. Here, $\omega = 10$ rad/s and $\zeta = 0.5$.

4.4.3.5 Simulation analysis

In this section, based on the control design, fully nonlinear simulations are performed in FAST-SC with all wind turbine DOFs enabled. Each test runs 630 s, and the

Table 4.7 Property of the hybrid mass damper in simulation

Mass *m*	Spring constant *K*	Damping constant *D*
20,000 kg	120,000 N/m	16,000 N/(m/s)

Table 4.8 Percentage of load reduction for passive and active structural control (%)

Case (m/s)	Term	Passive	Active
10	DEL tower fore–aft bending	9.7	12.7
	DEL tower side–side bending	35.1	40.3
	95th tower fore–aft bending	4.1	2.57
	95th tower side–side bending	11.7	11.4
18	DEL tower fore–aft bending	4.1	6.5
	DEL tower side–side bending	32.4	42.25
	95th tower fore–aft bending	0.3	0.2
	95th tower side–side bending	14.9	19.3

output data in first 30 s are not recorded, waiting for generator torque and blade pitch motion arriving normal operation state. The modified generator torque and blade pitch controller from NREL is used in the form of a dynamic link library for all tests [25].

Here, we consider two different simulation scenarios. The wind and wave conditions in [27] are adopted as two cases in this experiment. For wind condition, the mean value of the turbulent wind is defined as 10 and 18 m/s, respectively. The turbulent wind file is generated by TurbSim, where Kaimal spectra and the power law exponent of 0.14 are used according to the IEC61400-3 offshore wind turbine design standard. The normal turbulence intensity is set as level B, i.e., 18% (10 m/s case) and 15% (18 m/s case). For wave condition, JONSWAP spectrum is utilized to generate the stochastic wave inputs. The significant wave height is set as 2.3 (10 m/s case) and 3.7 m (18 m/s case), and the peak spectral period is defined as 14 s. For each case, at least two sets of random variables are used to generate wind and wave data.

According to the parameter study in [23,30], the property of the HMD on tower top is chosen as follows, which matches first tower fore–aft vibration mode (Table 4.7).

Nonlinear simulation results for tower bottom load reduction can be seen in Table 4.8. Compared with the passive case, more load reduction can be achieved with the designed controller. One simulation comparison is illustrated in Figure 4.18. TwrBsMxt and TwrBsMyt denote side–side and fore–aft tower base bending moment, respectively. TmdXDxt is the HMD displacement, and TmdXFext is the actuating force. However, this load reduction improvement is based on more energy consumption in HMD, and it could also be risky for instability.

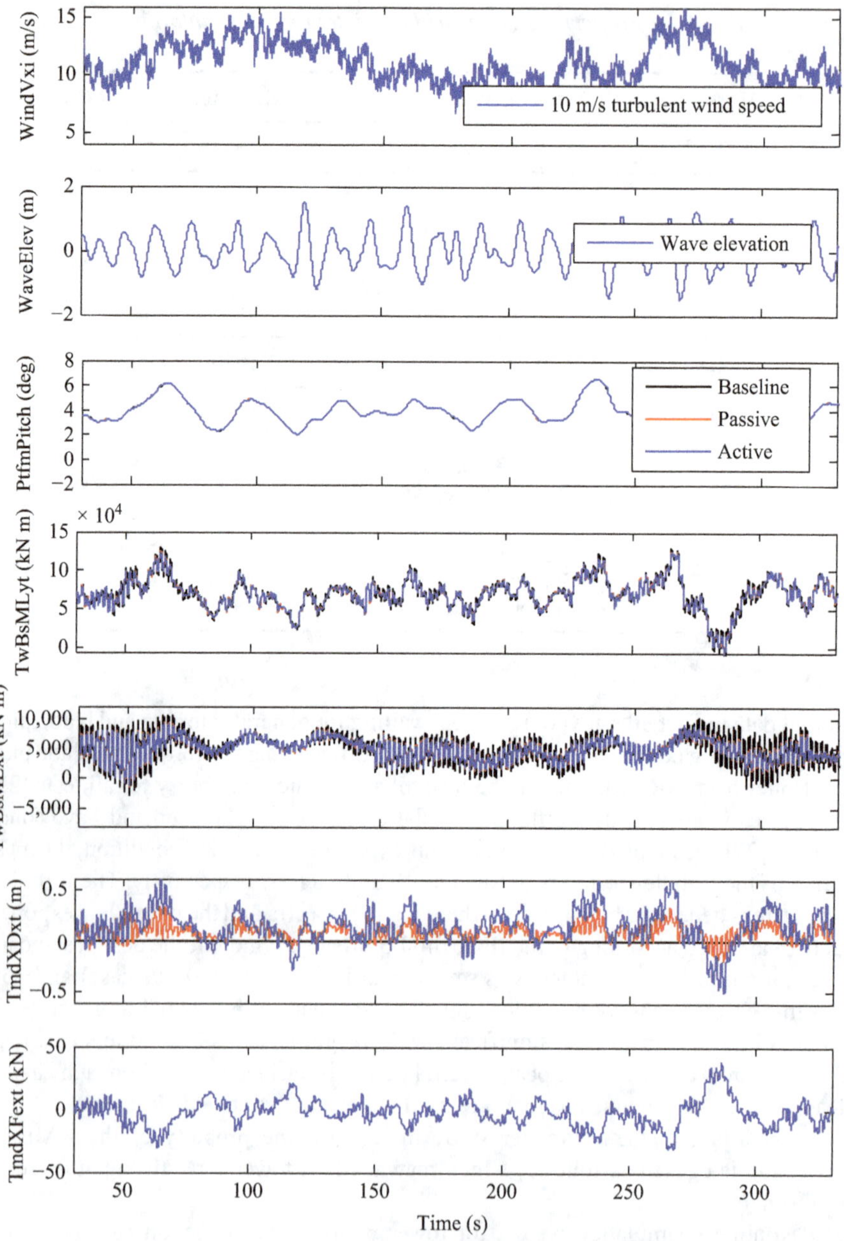

Figure 4.18 Nonlinear simulation comparison under 10 m/s turbulent wind and 2.3 m wave

4.5 Conclusions

This chapter introduces the structural control concept for load reduction of OWTs. Theories for both passive and active structural control are introduced. Particularly, a spar-type FOWT is used as the study case to demonstrate the load mitigation effectiveness of the proposed structural control methods. For the passive case, design optimization process is performed on a TMD installed in the spar platform, and the obtained numerical simulation results have indicated the both their effectiveness and limits regarding different system parameters and installations. Regarding the active case, a gain scheduling H_2/H_∞ active structural full-state feedback controller is designed for an HMD installed at the tower top of a spar-type FOWT, aiming at both reducing tower bottom load and mitigating the aerodynamic disturbance. The results demonstrate that more load reduction could be achieved at the expense of more energy consumption. At the same time, this will bring the risk of instability. Moreover, the full-state feedback controller is not very practical from a technical point of view due to the lack of sensors and measurement inaccuracy.

A number of research directions can be proposed to further investigate different structural control designs for load reduction of OWTs. First, it will be interesting to see the effectiveness of structural control ideas on other types of OWT models, such as monopile, barge, TLP, and semisubmersible. Second, a TLCD also offers the possibility of the semiactive control design by using the tunable orifice, which leaves the space for further load reduction without too much energy consumption. Third, due to lack of measured data, state-feedback is almost impossible for control design in practice. Alternatively, output feedback theory provides a possible method to deal with this problem, which only require a few types of measurable data, such as the acceleration on tower top and platform tilt angle. Last but not the least, despite extensive numerical simulations, the more convincing measure is to perform the field test to demonstrate the effectiveness of structural control. The test can be first performed on a scaled wind turbine model with structural vibration control devices mounted.

References

[1] C. L. Archer and M. Z. Jacobson. Evaluation of global wind power. *Journal of Geophysical Research*, 110:D12110, 2005.

[2] J. K. Kaldellis and M. Kapsali. Shifting towards offshore wind energy – recent activity and future development. *Energy Policy*, 53:136–148, 2013.

[3] Global Wind Energy Council. Global wind report. *GWEC, Brussels, Belgium, Technical Report*, 2016.

[4] W. Musial, S. Butterfield, and B. Ram. Energy from offshore wind. In *Offshore Technology Conference*, pages 1888–1898, 2006.

[5] J. Jonkman. Overview of the HydroDyn hydrodynamics module. *Presentation for the CREW/NREL Wind Turbine Design Code Workshop*, 2011.

[6] DNV. Offshore standard: design of offshore wind turbine structures. *Technical Report, DET NORSKE VERITAS*, 2011.

[7] J. Jonkman and D. Matha. A quantitative comparison of the responses of three floating platform concepts. *European Offshore Wind 2009 Conference & Exhibition*, 14–16, 2009.

[8] D. Matha. Model development and loads analysis of an offshore wind turbine on a tension leg platform with a comparison to other floating turbine concepts. *Technical Report, NREL/SR-500-45891*. Golden, CO: *National Renewable Energy Laboratory (NREL)*, 2010.

[9] T. Burton, N. Jenkins, D. Sharpe, and E. Bossanyi. *Wind energy handbook*. Chichester: John Wiley & Sons, 2011.

[10] S. D. Downing and D. F. Socie. Simple rainflow counting algorithms. *International Journal of Fatigue*, 4(1):31–40, 1982.

[11] T. J. Larsen and T. D. Hanson. A method to avoid negative damped low frequent tower vibrations for a floating, pitch controlled wind turbine. *Journal of Physics: Conference Series*, 75(1):1–11, 2007.

[12] J. M. Jonkman. Influence of control on the pitch damping of a floating wind turbine. In *2008 ASME Wind Energy Symposium. Reno, Nevada. January 7–10*. National Renewable Energy Laboratory, 2008.

[13] M. A. Lackner. An investigation of variable power collective pitch control for load mitigation of floating offshore wind turbines. *Wind Energy*, 16(3):435–444, 2013.

[14] H. Namik and K. Stol. Individual blade pitch control of floating offshore wind turbines. *Wind Energy*, 13(1):74–85, 2010.

[15] S. Korkmaz. A review of active structural control: challenges for engineering informatics. *Computers & Structures*, 89:2113–2132, 2011.

[16] P. J. Murtagh, A. Ghosh, B. Basu, and B. M. Broderick. Passive control of wind turbine vibrations including blade/tower interaction and rotationally sampled turbulence. *Wind Energy*, 11(4):305–317, 2008.

[17] S. Colwell and B. Basu. Tuned liquid column dampers in offshore wind turbines for structural control. *Engineering Structures*, 31(2):358–368, 2009.

[18] A. F. Mensah and L. Dueñas-Osorio. Reliability analysis of wind turbines equipped with tuned liquid column dampers (TLCD). In *Structures Congress*, pages 1190–1200, 2012.

[19] J. Li, Z. Zhang, and J. Chen. Experimental study on vibration control of offshore wind turbines using a ball vibration absorber. *Energy and Power Engineering*, 4(3):153–157, 2012.

[20] J. M. Jonkman and M. L. Buhl Jr. FAST users guide. *Technical Report, NREL/EL-500-29798*. Golden, CO: *National Renewable Energy Laboratory*, 2005.

[21] M. A. Lackner and M. A. Rotea. Passive structural control of offshore wind turbines. *Wind Energy*, 14(3):373–388, 2011.

[22] G. M. Stewart. *Load reduction of floating wind turbines using tuned mass dampers*. PhD thesis, University of Massachusetts Amherst, 2012.

[23] G. Stewart and M. Lackner. Offshore wind turbine load reduction employing optimal passive tuned mass damping systems. *IEEE Transactions on Control Systems Technology*, 21(4):1090–1104, 2013.

[24] H. Namik and K. Stol. Performance analysis of individual blade pitch control of offshore wind turbines on two floating platforms. *Mechatronics*, 21(4):691–703, 2011.

[25] J. Jonkman. Definition of the floating system for phase IV of OC3. *NREL/TP-500-38060*. Golden, CO: *National Renewable Energy Laboratory*, 2010.

[26] C. Roderick. Vibration reduction of offshore wind turbines using tuned liquid column dampers. Master's thesis, University of Massachusetts Amherst, 2012.

[27] M. A. Lackner and M. A. Rotea. Structural control of floating wind turbines. *Mechatronics*, 21(4):704–719, 2011.

[28] G. M. Stewart and M. A. Lackner. The effect of actuator dynamics on active structural control of offshore wind turbines. *Engineering Structures*, 33(5):1807–1816, 2011.

[29] Y. Si, H. R. Karimi, and H. Gao. Modelling and optimization of a passive structural control design for a spar-type floating wind turbine. *Engineering Structures*, 69:168–182, 2014.

[30] Y. Si and H. R. Karimi. Gain scheduling H_2/H_∞ structural control of a floating wind turbine, IFAC Proceedings Volumes, 47(3):6788–6793, 2014.

[31] J. J. Connor. *Introduction to structural motion control*. Upper Saddle River, NJ: Pearson Education, Inc., 2003.

[32] J. P. Den Hartog. *Mechanical vibrations*. Chelmsford, MA: Courier Corporation, 1985.

[33] H.-C. Tsai and G.-C. Lin. Optimum tuned-mass dampers for minimizing steady-state response of support-excited and damped systems. *Earthquake Engineering & Structural Dynamics*, 22(11):957–973, 1993.

[34] H. Wang and H. Ohmori. Elasto-plastic analysis based truss optimization using genetic algorithm. *Engineering Structures*, 50:1–12, 2013.

[35] A. R. Hedar and M. Fukushima. Minimizing multimodal functions by simplex coding genetic algorithm. *Optimization Methods and Software*, 18(3):265–282, 2003.

[36] C. Kelley. *Iterative methods for optimization*. Frontiers in Applied Mathematics. Philadelphia, PA: Society for Industrial and Applied Mathematics, 1999.

[37] B. Jonkman and M.L. Buhl. Turbsim users guide. *Technical Report, NREL/EL-500-36970*. Golden, CO: *National Renewable Energy Laboratory*, 2005.

[38] C. W. Scherer. Multiobjective H_2/H_∞ control. *IEEE Transactions on Automatic Control*, 40(6):1054–1062, 1995.

[39] J. Doyle, K. Zhou, K. Glover, and B. Bodenheimer. Multiobjective H_2/H_∞ control. *IEEE Transactions on Automatic Control*, 39(8):1575–1587, 1994.

Chapter 5

Advanced control of wind turbine system

Fanzhong Meng[1], *Jan Wenske*[1], *Mohsen Neshati*[1], *and Arne Bartschat*[1]

5.1 A state-of-the-art wind turbine controller

Wind turbines are complex and nonlinear dynamic systems driven by stochastic wind disturbances together with gravitational, centrifugal and gyroscopic forces. The aerodynamic of wind turbines is nonlinear, unsteady and complex. The rotor of a wind turbine is subjected to complicated turbulent wind inflow and periodic gravity loading that drives fatigue loads. The rotation of rotor adds even more complexity to the dynamic model. Hence, wind turbine modeling is challenging and complex. The design of control algorithms for wind turbines must account for these complexities. The models used for control design purposes must contain enough degrees of freedom (DOFs) to capture the most important turbine dynamics without being too complex.

The state-of-the-art controllers widely used in large commercial pitch-regulated, variable-speed wind turbines are typically designed using classical control theory such as proportional–integral–derivative (PID) control for wind turbine pitch and torque regulation [1]. In partial-load region (sometimes called Region 2), the generator torque (T_g) is regulated to balance the aerodynamic torque while keeping optimum tip–speed ratio (λ_{opt}) to obtain the maximum power coefficient (C_p). The generator torque demand is calculated as

$$T_g = K_{opt}\Omega_g^2 \tag{5.1}$$

where Ω_g is the measured generator speed in rad/s, K_{opt} is the optimal generator torque-speed gain calculated as

$$K_{opt} = \frac{\pi\rho R^5 C_p}{2\lambda_{opt}^3 N^3} \tag{5.2}$$

where ρ is the air density, R is the rotor radius and N is the gearbox ratio. In the full-load region (Region 3), the blade pitch angle is controlled to maintain constant rotor speed, while the generator torque is held constant.

[1]Division Wind Turbine and System Technology, Fraunhofer Institute for Wind Energy Systems IWES, Bremerhaven, Germany

Classical PID control methods are used to design the blade pitch controller. Furthermore, a refined switch mechanism is incorporated, resulting in smooth switching between partial and full-load operation. Aerodynamic gain scheduling is applied due to the fact that the sensitivity of aerodynamic torque with respect to the pitch angle varies at different operating points above the rated wind speed. One way to implement aerodynamic gain scheduling is to calculate the aerodynamic gain as illustrated in the following equation [2]:

$$GK(\theta) = \frac{1}{1 + (\theta/\theta_K)} \tag{5.3}$$

where θ is the current pitch angle and θ_K is the pitch angle chosen at which, the partial derivative of aerodynamic torque with respect to pitch angle is doubled in comparison with that value at rated operating point.

The antiwindup is required by a state-of-the-art wind turbine pitch controller in order to react quickly during a sudden change of operating point from Region 2 to Region 3 caused by gust. One way to implement the antiwindup is using a minimum pitch angle to limit the summation of proportional, differential and integral terms, which results in the demanded collective pitch angle in current time step expressed as in the following equation [3]:

$$\theta_{ref,k} = max(\theta_{min,k}, (\theta_{P,k} + \theta_{I,k} + \theta_{D,k})) \tag{5.4}$$

The integral term to be used for the integration in the further time step is recalculated as $\theta_{I,k+1} = \theta_{ref,k} - \theta_{P,k} - \theta_{D,k}$, which only makes a change to the integral term if the demanded collective pitch angle is equal to the minimum pitch angle limit. When the wind turbine is operating in Region 2, where the proportional term ($\theta_{P,k}$) is negative, in order to keep the rotor speed error close to zero by the PID pitch controller, the integral term ($\theta_{I,k}$) will be positive. If the aerodynamic power is increased and becomes close to rated power due to a gust wind, then the proportional term will be close to zero, whereas the integral part remains positive. The demanded collective pitch angle will be positive, whereby large power and speed variations are avoided. The same antiwindup scheme can be also used in the generator torque PID controller.

Another common feature of a state-of-the-art controller is to mitigate structural dynamic loads. In commercial wind turbines, an additional generator torque control loop in Regions 2 and 3 is used to actively damp the drivetrain torsional vibration as well as the tower side-to-side vibrations [4]. In Region 3, an additional pitch control loop is added to damp the tower fore–aft vibration. An additional pitch action to actively damp tower fore–aft vibration is usually implemented as a single-input–single-output control loop on top of the basic collective pitch control [4].

In the previous paragraphs, the classical PID blade pitch and generator torque controller used in a state-of-the-art turbine controller have been described. The measured rotor or generator speed is the input signal to the blade pitch and generator torque controller. It will contain oscillations at the first drivetrain torsional frequency due to the aerodynamic torque fluctuation created by the interaction of the rotor with a stochastic wind field. When the PID blade pitch controller tries to regulate the speed, it will attempt to respond at the first drivetrain torsional frequency. Since the

PID controller does not have the needed information to provide a stable control on this mode, it will cause instability at this mode. A solution would be to implement a low-pass filter. The oscillations at the first drivetrain torsional frequency are filtered out when a low-pass filter is applied on the measured rotor or generator speed signal. Usually, a first-order low-pass filter with a transfer function as expressed in (5.5) is selected.

$$G(s) = \frac{1}{1 + \tau s} \tag{5.5}$$

The time constant (τ) is selected based on the first drivetrain torsional frequency by examining the bode plot of the transfer function.

5.2 Design of controllers for load reduction in wind turbines

Nowadays, the size of wind turbines is becoming larger and larger due to the high demand of power production from the renewable energy. This has led to new design issues for designing large and flexible wind turbines. One important issue for large and flexible wind turbines is to mitigate fatigue and extreme loads due to the stochastic wind field on the structure components of wind turbines to increase their life time. This section describes different control methods to be added in the state-of-the-art collective blade pitch controller including individual pitch controller (IPC) for reducing the loads on wind turbines.

This section is divided into three subsections. The first subsection is the design of IPC to mitigate loads on blade root and hub center. The second subsection presents a method for estimating the effective wind speed based on the work presented in [5,6] together with some simulation results. The third subsection shows the development and implementation of a feedforward feedback (FF-FB) collective pitch controller based on the effective wind speed estimation.

5.2.1 Individual pitch controller

The concept of using IPC for load alleviation has been proposed many times over the years, and the most important contribution to the development of IPC was made by Bossanyi [7]. The IPC creates an additional zero-mean demanded pitch angle for each blade to be superimposed to the collective demanded pitch angle. The individual pitch action requires a multivariable controller with at least two inputs (measurements) and two outputs. Three individual pitch angles are calculated based on measured blade root out-of-plane bending moments in such a way to minimize asymmetric loads (yaw moment and tilt moment) on the rotor, produced by inhomogeneity of turbulence, wind shear, yaw misalignment, tower shadow and aerodynamic asymmetry of the rotor. The approach is to transform the three measured blade root out-of-plan bending moments from the rotating reference frame into two orthogonal components in the nonrotating reference frame. The same method was used in the helicopter control, where it is known as the Coleman transformation [8].

The transformed two load components are representing the asymmetrical loads in tilt and yaw directions. It is possible to treat the two directions as being almost independent. This means that two classical PI controllers can be applied to the two orthogonal components separately, in order to calculate the two additional pitch angles. Usually, the two controllers are designed to be the same, but they can be different to account for the differences due to yaw dynamics or the tower dynamics. If it is required to create yawing moment or nodding moment for other control purposes, nonzero set-point can be used, but for common application, zero set-points are used. Afterwards, an inverse Coleman transformation calculates the appropriate individual pitch angles for the three blades in the rotating reference frame, which can be superimposed to the collective demanded pitch angle.

Coleman transformation

The Coleman transformation applied on three blade root out-of-plane bending moments can be expressed as

$$\begin{bmatrix} M_y \\ M_t \end{bmatrix} = \frac{2}{3} \begin{bmatrix} \cos(\phi) & \cos\left(\phi + \frac{2\pi}{3}\right) & \cos\left(\phi + \frac{4\pi}{3}\right) \\ \sin(\phi) & \sin\left(\phi + \frac{2\pi}{3}\right) & \sin\left(\phi + \frac{4\pi}{3}\right) \end{bmatrix} \begin{bmatrix} M_1 \\ M_2 \\ M_3 \end{bmatrix} \tag{5.6}$$

The corresponding inverse Coleman transformation is described as

$$\begin{bmatrix} M_1 \\ M_2 \\ M_3 \end{bmatrix} = \begin{bmatrix} \cos(\phi) & \sin(\phi) \\ \cos\left(\phi + \frac{2\pi}{3}\right) & \sin\left(\phi + \frac{2\pi}{3}\right) \\ \cos\left(\phi + \frac{4\pi}{3}\right) & \sin\left(\phi + \frac{4\pi}{3}\right) \end{bmatrix} \begin{bmatrix} M_y \\ M_t \end{bmatrix} \tag{5.7}$$

where, M_i $(i = 1, 2, 3)$ are the three blade loads in the rotating reference frame, M_y and M_t are the transformed two load components which are orthogonal to each other in the nonrotating reference frame and ϕ is then the rotor azimuth angle. It is useful to select the vertical and lateral directions which are corresponding to yawing and nodding direction as two orthogonal axes in Coleman transformation, since this gives an axis system which is fixed in space. Wind speed variations in this coordinate system are unaffected by rotational sampling.

If the blade loads are measured, the Coleman transformation in (5.6) is used to convert the measured loads into nonrotating reference frame. If rotating hub or shaft loads are used, a simple rotational transformation through the azimuth angle is required. Loads measured on a stationary part of the wind turbine, such as the main bearing housing or the yaw bearing, is considered to be in the nonrotating coordinate system already. The inverse Coleman transformation in (5.7) is used to calculate the individual pitch angle for the three blades from the two pitch angles calculated by two classical PI controllers.

Sample results: with 1P individual pitch control

This section presents sample results to illustrate the performance of the IPC. The effect of individual pitch control is demonstrated by comparing the results with collective pitch control and with individual pitch control. The test case shown here is simulated using DNV-GL Bladed with a 1P (first harmonic) IPC on a 5-MW reference wind turbine [9]. A turbulent wind field with 18 m/s mean wind speed is used. The turbulence intensities are 17% in longitudinal direction, 13.6% in lateral direction and 8.5% in vertical direction. The results are very similar at other wind speeds across the above-rated range. The effect of IPC is explained by investigating the blade root out-of-plane (M_y) bending moment, the hub nodding moment M_y or yawing moment M_z (the effect on these two loads is very similar) in rotating reference frame and the nodding moment M_y or yawing moment M_z at the yaw bearing in the fixed (nonrotating) reference frame.

The power spectrum density of the above-mentioned loads are shown in Figure 5.1. Figure 5.1 (a) and (b) illustrates the loads on the rotating components which are the blade root out-of-plane bending moment and the hub nodding moment. It can be seen that the large dominating pack at 1P (around 0.2 Hz) is removed by the individual pitch control. The smaller peak at 2P, which is not affected by the 1P IPC, could be removed by a 2P IPC. But the implementation of a 2P IPC is not presented here. Figure 5.1 (c) represents the nonrotating load which is the yaw bearing yawing moment dominated by a low frequency peak representing the asymmetry in the wind field. In this case, the effect of the individual pitch control reduces the 0P (low frequency) variations.

5.2.2 Effective wind speed estimation

The wind speed is a very important information for the wind turbine controller. If the wind speed is known, the control algorithms can be further improved. One way to obtain the wind speed at rotor plane is using light detection and ranging (LIDAR) to measure the wind speed at a certain distance in front and predict the wind speed in the rotor. Early works on this LIDAR-assisted control method are reported [10]. The LIDAR measurement suffers from measurement and the prediction errors. The prediction is usually based on the assumption that the same wind speed which is measured at a distance in front of the wind turbine reaches the wind turbine rotor. This assumption is often not valid in practice. Moreover, a classical frozen turbulence assumption used for wind speed prediction on the rotor introduces additional errors. Furthermore, wind speed measurement requires installation of the LIDAR system, which is expensive.

In this section, a method presented by Meng *et al.* is shown [6]. The state and input estimation problem based on a state-space model are split into two separate parts. First, a state observer based on the Kalman filter approach is designed for the state estimation, which is rotor speed in this case. Second, an input observer based on the ideas from optimal tracking control strategy is designed for estimating the rotor aerodynamic torque. By using this approach, it is expected that the tracking

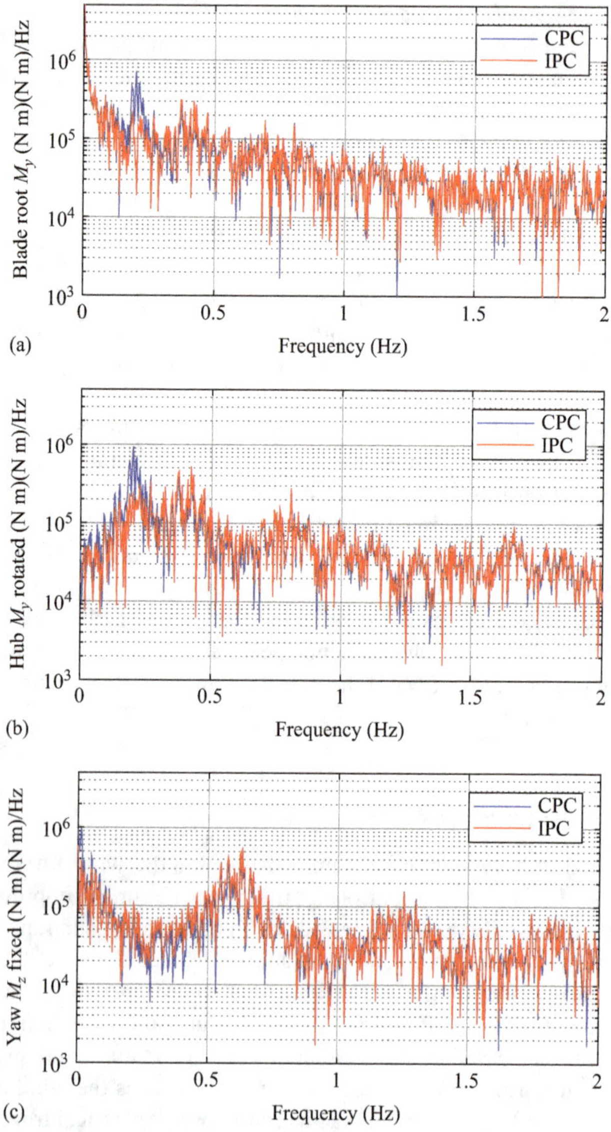

Figure 5.1 Power spectrum density of the key loads on the wind turbine

performance will be significantly better than the state estimation method. Furthermore, the problem is simplified as it will be illustrated in the following sections, and the difficulties of modeling the wind and rotor aerodynamics in the state-space model of the wind turbine are avoided by changing it into the design of an optimal tracking controller, which is efficiently solved by a PI controller.

5.2.2.1 Estimation of rotor speed and aerodynamic torque

The drivetrain of the wind turbine is simplified into a two DOF system with two inertias interconnected by a spring and damper to keep a sufficient level of accuracy. To keep the state-space model simple, the tower and rotor blades are assumed to be rigid, which are ignored in the equations of motion of the wind turbine. The external forces are the aerodynamic torque, T_a, on the low speed shaft and the generator torque, T_g, on the high speed shaft. This results in the equations of motion shown below:

$$\begin{aligned} J_r\dot{\Omega}_r &= T_a - D_d\Omega_r + \frac{D_d}{N}\Omega_g - K_d\delta\varphi \\ J_g\dot{\Omega}_g &= -T_g + \frac{D_d}{N}\Omega_r - \frac{D_d}{N^2}\Omega_g + \frac{K_d}{N}\delta\varphi \\ \delta\dot{\varphi} &= \Omega_r - \frac{1}{N}\Omega_g \end{aligned} \tag{5.8}$$

where J_r and J_g are the rotor and generator inertia, respectively. Ω_r and Ω_g are the rotor and generator speed, respectively. $\delta\varphi$ is the equivalent torsional deformation of the shaft on the low speed side. K_d and D_d are the equivalent spring stiffness and damping factor of the drivetrain, respectively. N is the gearbox ratio. For the state and input estimation, equation (5.8) is rewritten into the state-space form shown in (5.9) and (5.10) with $\mathbf{Z} = [\ \Omega_r,\ \Omega_g,\ \delta\varphi\ ]^T$ as the state vector and $\mathbf{u} = [\ T_a,\ T_g\ ]^T$ as the input vector.

$$\begin{cases} \dot{\mathbf{Z}} = \underline{\mathbf{A}}\,\mathbf{Z} + \underline{\mathbf{B}}\,\mathbf{u} \\ y = \mathbf{C}\,\mathbf{Z} + D\,\mathbf{u} \end{cases} \tag{5.9}$$

$$\underline{\mathbf{A}} = \begin{bmatrix} -\frac{D_d}{J_r} & \frac{D_d}{J_r N} & -\frac{K_d}{J_r} \\ \frac{D_d}{J_g N} & -\frac{D_d}{J_g N^2} & \frac{K_d}{J_g N} \\ 1 & -\frac{1}{N} & 0 \end{bmatrix} \tag{5.10a}$$

(5.10b)

$$\underline{\mathbf{B}} = \begin{bmatrix} \frac{1}{J_r} & 0 \\ 0 & -\frac{1}{J_g} \\ 0 & 0 \end{bmatrix} \tag{5.10c}$$

(5.10d)

$$\mathbf{C} = \begin{bmatrix} 0 & 1 & 0 \end{bmatrix} \tag{5.10e}$$

(5.10f)

$$D = 0 \tag{5.10g}$$

For the state estimation, the rotor speed, generator speed and generator torque are assumed to be available from the measurements, which is true in practice. At this

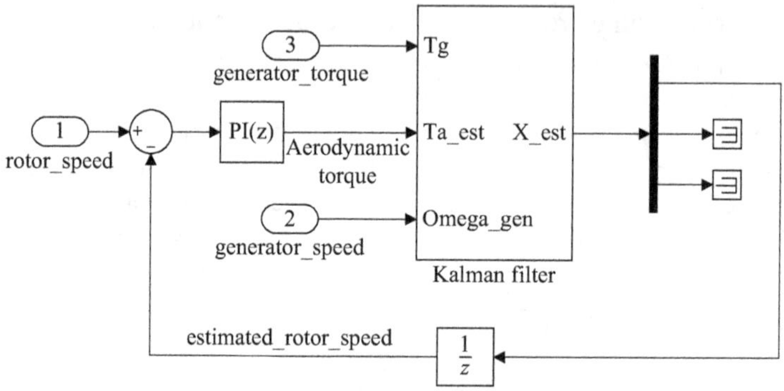

Figure 5.2 Block diagram of the wind speed estimator

moment, the rotor aerodynamic torque is assumed to be known. The Kalman gain, **L**, can be calculated using the Kalman filter approach, which is a well-known method for estimating the states. The main idea of this method is to minimize the estimate error covariance. The detailed description of Kalman filter approach is discussed in [11]. The complete observer structure is as shown in Figure 5.2. As it is mentioned before, the observer design is split into two parts operating in a coupled loop. The inner part is a Kalman filter designed on the basis of (5.9) and (5.10) under the assumption that T_a is available. The outer part is a tracking loop with the measured Ω_r as the tracking variable and T_a as the output of the tracking controller. The outer loop is chosen to be of the PI structure in order to have an integral term taking care of the asymptotic tracking and a direct proportional gain handling the faster response to the variations. During the design of the Kalman filter (the inner loop), it is assumed that the rotor aerodynamic torque is known, and it is the input of the Kalman filter. Actually, it is the output of the tracking controller (see Figure 5.2). This means that at the first time step, the rotor aerodynamic torque is an unknown input of the Kalman filter; therefore, an initial value of rotor aerodynamic torque should be tuned and given to the PI controller. Increasing the bandwidth of the outer loop to the value close to the bandwidth of the inner loop can potentially lead to instability; therefore, the stability of the coupled loop must be checked during the observer design. The main advantage of this method is that the tracking controller plays the role to observe the rotor aerodynamic torque instead of using the simplified rotor aerodynamic model, which contains more uncertainty.

5.2.2.2 Calculation of effective wind speed based on the Newton–Raphson method

In the previous section, an observer was presented for estimating the rotor speed and aerodynamic torque. These two variables together with the measured pitch angle will be used in this section to calculate the effective wind speed as shown below:

$$T_a = \frac{1}{2}\rho\pi R^2 \frac{v^3}{\Omega_r} C_p(\lambda, \theta) \tag{5.11}$$

where λ is the tip speed ratio defined as $\lambda = \Omega_r R/v$, θ is the measured blade pitch angle, $C_p(\lambda, \theta)$ is the wind turbine power coefficient function in steady-state and v is the effective wind speed defined as the spatial average of the wind field over the rotor plane that needs to be estimated. For the calculation of the effective wind speed, equation (5.11) is rewritten into the following form:

$$\frac{2T_a}{\rho\pi R^5\Omega_r^2} = \frac{C_p(\lambda, \theta)}{\lambda^3} \tag{5.12}$$

On the left-hand side of (5.12), the air density, rotor radius, rotor aerodynamic torque and rotor speed are all available, while on the right-hand side, the power coefficient function $C_p(\lambda, \theta)$ can be calculated by steady-state calculation and stored in a look-up table as function of λ and θ. The blade pitch angle (θ) is also available from the measurement. For a given pitch angle, the power coefficient function is expressed as $C_{p,\theta}(\lambda)$. An auxiliary function $f(\lambda)$ is defined as

$$f(\lambda) = \frac{2T_a}{\rho\pi R^5\Omega_r^2} - \frac{C_{p,\theta}(\lambda)}{\lambda^3} = 0 \tag{5.13}$$

The effective wind speed is then calculated by solving the nonlinear equation (5.13) for λ, and then the effective wind speed is calculated by $v = \Omega_r R/\lambda$. Because $C_{p,\theta}(\lambda)/\lambda^3$ is a nonmonotonic function for some values of blade pitch angle [5], this function is invertible within this certain range of pitch angle. The main idea of the Newton–Raphson method is finding the root of an equation when it locally has an unique solution. Therefore, in this case, the Newton–Raphson method is suitable for solving such a nonlinear equation (5.13), when an initial value of λ is given. In order to solve this nonlinear equation, the derivative of $f(\lambda)$ with respect to λ is obtained as

$$f'(\lambda) = \frac{dC_{p,\theta}(\lambda)}{d\lambda}\lambda^{-3} - \frac{3C_{p,\theta}(\lambda)}{\lambda^4} \tag{5.14}$$

where $dC_{p,\theta}(\lambda)/d\lambda$ can be calculated numerically by center difference method based on the power coefficient look-up table as

$$\frac{dC_{p,\theta}(\lambda)}{d\lambda} = \frac{dC_{p,\theta}(\lambda + h) - dC_{p,\theta}(\lambda - h)}{2h} \tag{5.15}$$

where h denotes $\Delta\lambda$.

The Newton–Raphson iteration algorithm is explained in the following:

1. The measured pitch angle θ and the initial value of λ at $i = 0$ as $\lambda^{(0)}$ are given;
2. Calculate $\lambda^{(i)} = \lambda^{(i-1)} - (f(\lambda^{(i-1)})/f'(\lambda^{(i-1)}))$ using (5.13)–(5.15) for $i = 1 \ldots n$;
3. Check the solution convergence by evaluating $\left|(f(\lambda^{(i-1)})/f'(\lambda^{(i-1)})\right| < \varepsilon$. If it is true, the iteration is completed and the solution of λ is found;
4. If the required accuracy ε is not satisfied, increase i and repeat steps 2 and 3.

Results: wind speed estimation

All test cases shown in this subsection are simulated using DNV-GL Bladed based on a 5-MW reference wind turbine [9] designed by the National Renewable Energy Laboratory (NREL) in order to demonstrate the performance of the presented effective wind speed estimator.

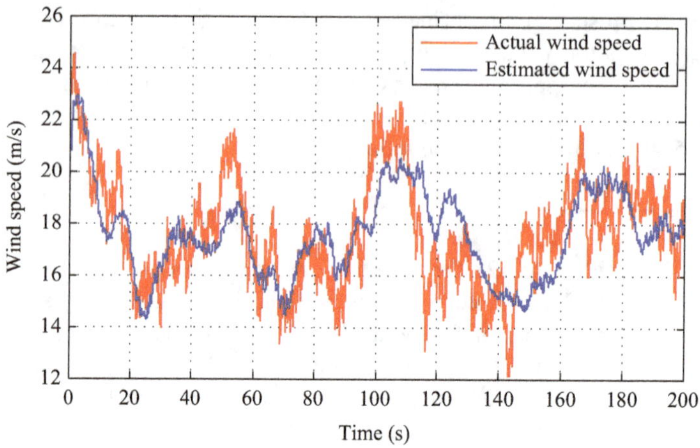

Figure 5.3 Comparison of estimated wind speed with the actual wind speed for normal turbulent wind

Test case 1: stochastic wind

The first test is performed by using a normal turbulence model (NTM) defined according to the IEC-61400 standard [12] at an 18-m/s mean wind speed. Figure 5.3 shows that on average, the estimated wind speed agrees well with the actual wind speed (generated by Bladed), while the very fast changes and peaks are not captured in the estimation. As it is expected, a time delay in the estimated wind speed can be found, which needs to be compensated by an extrapolation of a wind speed to be mentioned in Section 5.2.3.

Test case 2: deterministic wind

The second test is performed using the extreme operating gust (EOG) with a hub height wind speed equal to 18 m/s in order to demonstrate the performance of the presented wind speed estimation method in reaction to the gust. The result is shown in Figure 5.4.

It can be found out that the wind speed estimator converges in about 5 s and there is a very small difference between the estimated wind speed and the actual wind speed before and after the gust. The time delay is about 0.4 s, and there is about 3.5% over prediction at the location where the wind speed reaches its maximum value. Both of these results show that the presented effective wind speed estimator works correctly.

5.2.3 Feedforward feedback controller based on the effective wind speed estimation

In this section, the FF-FB controller and the preview controller using the estimated wind speed information are described. For the purpose of comparison, an existing baseline controller (feedback only) as described in [13] is extended by adding a feedforward loop. It combines a variable speed generator torque controller as well

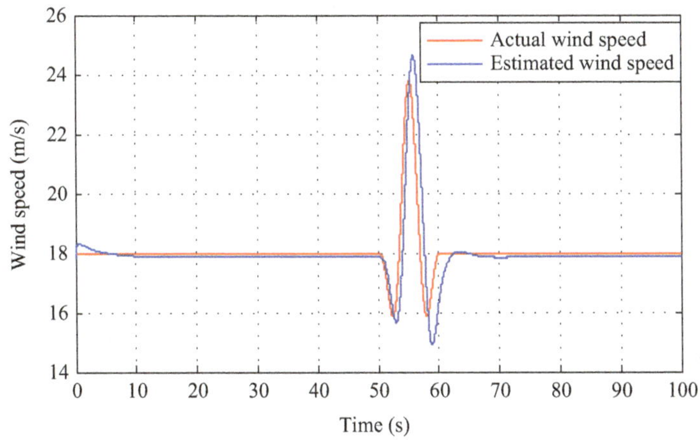

Figure 5.4 Comparison of estimated wind speed with the actual wind speed for extreme operating gust

as a collective pitch controller with a feedforward loop. A simple linear piecewise extrapolation algorithm is developed to estimate wind speed on the rotor plane several time steps further in time, resulting in an improved performance of the FF-FB pitch controller in terms of the reaction to the gust.

5.2.3.1 Control structure and wind speed extrapolation

Previous research works focus on using LIDAR assist measurements of wind speed in order to improve the collective pitch control action by adding a feedforward loop on the existing feedback pitch control loop. Several FF-FB controllers described in [14–17] have been proposed for load reduction or increasing the annual energy production. The present work illustrates that the estimated effective wind speed instead of the measured wind speed can be used in the feedforward loop for improving the control quality in terms of loads reduction. The proposed feedforward loop, in which an additional collective pitch rate is calculated based on the steady-state pitch angle corresponding to the wind speed, is combined with the conventional feedback collective pitch control loop. The complete control diagram is shown in Figure 5.5. As is shown, the feedforward pitch angle θ_{ff} is calculated using the estimated and filtered effective wind speed V_{ef} and the steady-state look-up table that contains the steady pitch angles versus the wind speeds. The additional feedforward pitch rate $\dot{\theta}_{ff}$ adding to the pitch rate of the feedback PID loop is determined by feeding back the measured actual pitch angle θ to the feedforward loop. The total FF-FB collective pitch angle θ_{ff_fb} is then fed into the wind turbine system or in this case the aeroelastic simulation tool.

However, the estimated effective wind speed on the rotor plane as an input signal of the feedforward loop contains high frequencies which either the pitch system cannot respond or introduces additional high-frequency dynamics to the control system. This eventually results in higher fatigue loads on the wind turbine. Therefore, a low-pass

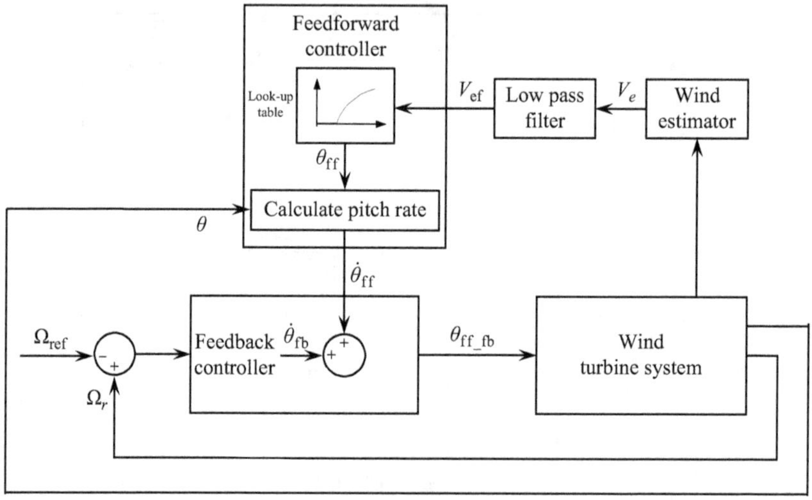

Figure 5.5 Control diagram of the feedforward feedback collective pitch controller using estimated wind speed

filter as described in (5.16) is designed to remove the high-frequency contents in the estimated effective wind speed, but this will introduce more delay. This delay can be compensated by extrapolating the estimated effective wind speed several time steps future before passing it to the low-pass filter. The transfer function of the second-order low-pass filter is expressed in the following equation, where ω_0 is the cut-off frequency and ζ_0 is the damping ratio:

$$G(s) = \frac{\omega_0^2}{s^2 + 2\zeta_0\omega_0 s + \omega_0^2} \tag{5.16}$$

The extrapolation of the estimated effective wind speed is calculated in the following equation:

$$\begin{aligned} V_e(t_i + \Delta T) &= V_e(t_i) + \Delta T V_e'(t_{i-1}) \\ V_e'(t_{i-1}) &= \frac{V_e(t_i) - V_e(t_{i-1})}{\delta t} \end{aligned} \tag{5.17}$$

where ΔT is the preview time and δt is the simulation time step. The preview time affects the stability of the controller; therefore, it has to be selected carefully. It is suggested that the value of 0.5 s is a good choice. Finally, the calculation of the feedforward pitch rate $\dot{\theta}_{ff}$ can be done as in the following equation:

$$\dot{\theta}_{ff} = \frac{\theta_{ff}(V_{ef}) - \theta}{\delta t} \tag{5.18}$$

where V_{ef} is the low-pass filtered and extrapolated effective wind speed and $\theta_{ff}(V_{ef})$ is a function of steady-state pitch angle with respect to the wind speed. θ is the current measured pitch angle.

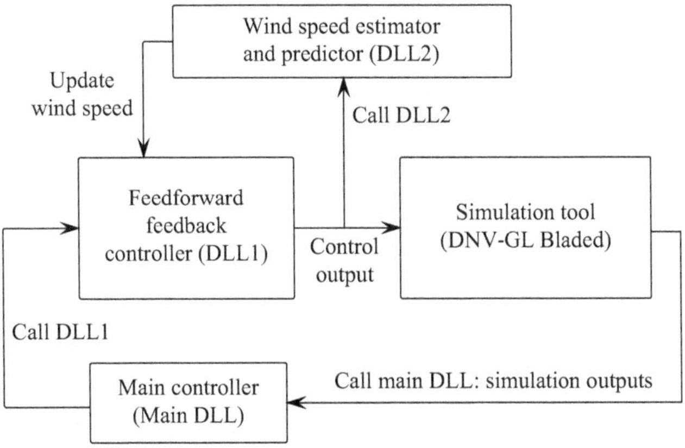

Figure 5.6 Implementation diagram of the feedforward feedback collective pitch controller using estimated wind speed

It is known that the collective pitch controller design involves a compromise between accurate speed regulation and turbine load reduction. With the wind speed information, the feedforward controller takes over some of the low-frequency speed regulation duty, so the feedback PID controller can be optimized by reducing the gains. This is to ensure that the quality of the speed controller remains acceptable, but a reduction of thrust related loads can be achieved.

5.2.3.2 Implementation example

One application example is that the presented FF-FB controller with the effective wind speed estimation and prediction is implemented as external dynamic link libraries (DLLs) which can be called from DNV-GL Bladed. The implementation diagram is shown in Figure 5.6. As is shown, three DLLs are developed. The first one is the main DLL, which handles the communication between the aeroelastic simulation tool (DNV-GL Bladed) and the controller, implemented as Bladed style external DLL in Fortran programming language [18]. The second DLL is the FF-FB controller (DLL1) that includes the generator torque controller, drivetrain damping and tower top vibration damping. The last one is the wind speed estimator and predictor (DLL2), which is designed and implemented in MATLAB®/Simulink™ and compiled into a DLL using the MATLAB Coder. The main advantage of this implementation scheme is that the design and optimization of the wind speed estimator can be easily achieved in MATLAB/Simulink, while the existing baseline feedback collective pitch controller can still be reused with small changes. For the real application on the wind turbine, the presented FF-FB controller has the advantage that it can be easily integrated in the state-of-the-art wind turbine control system without the installation of new hardware, for example, the LIDAR system. Only few changes on the software part of the current control system are needed.

Results: feedforward feedback controller based on the effective wind speed estimation

The fatigue and extreme loads on some of the wind turbine components based on the simulated load cases are presented here.

Test case 1: fatigue loads

As it is known, the collective pitch controller only controls the wind turbine when it operates at over-rated wind speeds. To evaluate the effect of the presented FF-FB collective pitch controller in terms of fatigue loads, a set of test load cases are defined by focusing on the wind speeds above rated. A group of stochastic wind fields including mean wind speed of 14, 16 and 18 m/s are generated using NTM with 6 random seeds for each wind speed. Simulation experiments are carried out with and without the feedforward loop for a duration of 650 s. The first 50 s are removed in order to take out the transient period. Damage equivalent loads (DELs) on the key components (blades, hub and tower) are calculated for each of the wind speeds mentioned above by assuming that the appearance of each wind speed is 20 years. This is an important criterion to judge the control performance of the FF-FB pitch controller on the fatigue loads on some of the wind turbine components.

In this investigation, the thrust at the hub center, the blade flap-wise root bending moment and the tower bottom fore–aft bending moment, which are assumed to be affected by the pitch controller mostly, are selected to calculate the DELs. All the DELs are normalized with respect to the results of the baseline controller. The normalized standard deviation of rotor speed is also calculated for each of the three wind speeds. Figure 5.7 shows the comparison of the DELs and the standard deviation of rotor speed simulated with and without the feedforward loop.

The standard deviation of rotor speeds are reduced for all three wind speeds when using the presented FF-FB controller. Specially, at wind speeds of 16 and 18 m/s, the reduction is about 20%. This shows that the speed regulation is improved with the FF-FB controller. The DELs on hub center which are affected mainly by the pitch controller are reduced clearly for all the three wind speeds. The DELs based on the blade root flap-wise bending moments are only slightly reduced. This means the pitch activity can be further reduced by optimizing the feedback control parameters. This will result in further load reduction on the blade, while the standard deviation of the rotor speed will be increased. The detailed difference of DELs in percentage compared with the baseline controller is listed in Table 5.1.

Test case 2: extreme loads

To investigate the effect of the presented FF-FB collective pitch controller in terms of extreme loads, several simulations are performed using the EOG wind model with the hub height wind speed of 18 m/s. Four different rotor initial azimuth positions (0°, 30°, 60° and 90°) are simulated. The extreme load results are shown in Figure 5.8. Here, Hub_Fx is the thrust at the hub center, Blade1_My is the blade flapwise root bending moment and Tower_My is the tower bottom fore–aft bending moment. These three output sensors are selected for analyzing the extreme loads. All the extreme

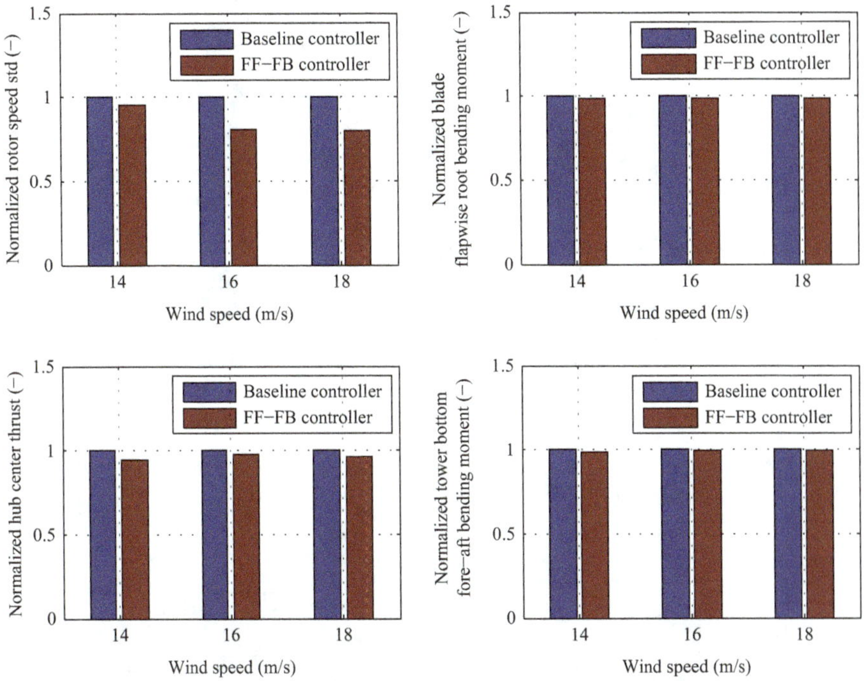

Figure 5.7 Comparison of the DELs with and without feedforward controller

Table 5.1 The difference of the damage equivalent loads

	14 m/s	**16 m/s**	**18 m/s**
Sensors	Δ	Δ	Δ
Blade M_y(%)	−2.0	−1.5	−1.6
Tower M_y(%)	−1.4	−0.6	−1.0
Hub center F_x(%)	−5.7	−2.0	−3.6

loads are normalized with respect to the results of the baseline controller. It can be found out that the extreme loads at these sensors are reduced about 10%.

Figure 5.9 shows the change of the rotor speed and the pitch angle with respect to the gust. The upper part of the figure shows that the fluctuation of the rotor speed during the gust is reduced significantly with the presented control algorithm using the preview time of 0.5 s. The lower plot shows that the blades pitch earlier and faster in reaction to the gust, which results in the lower extreme loads shown in Figure 5.8. An increased pitch action after the gust is detected, which might increase the fatigue load on the pitch drive system. But it is not relevant regarding to the extreme loads reduction.

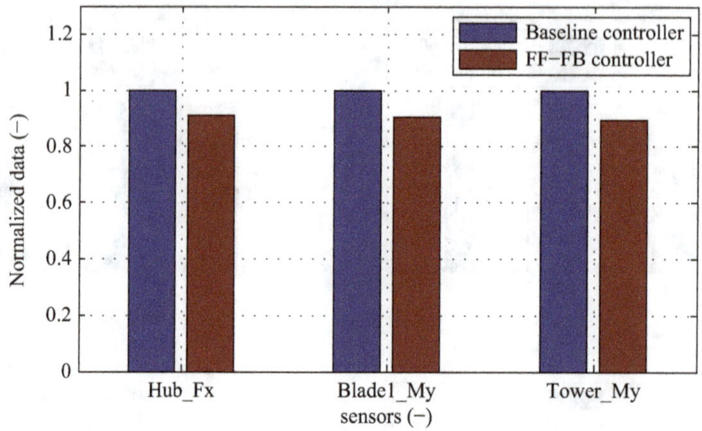

Figure 5.8 Comparison of extreme loads with and without feedforward loop at 18 m/s EOG

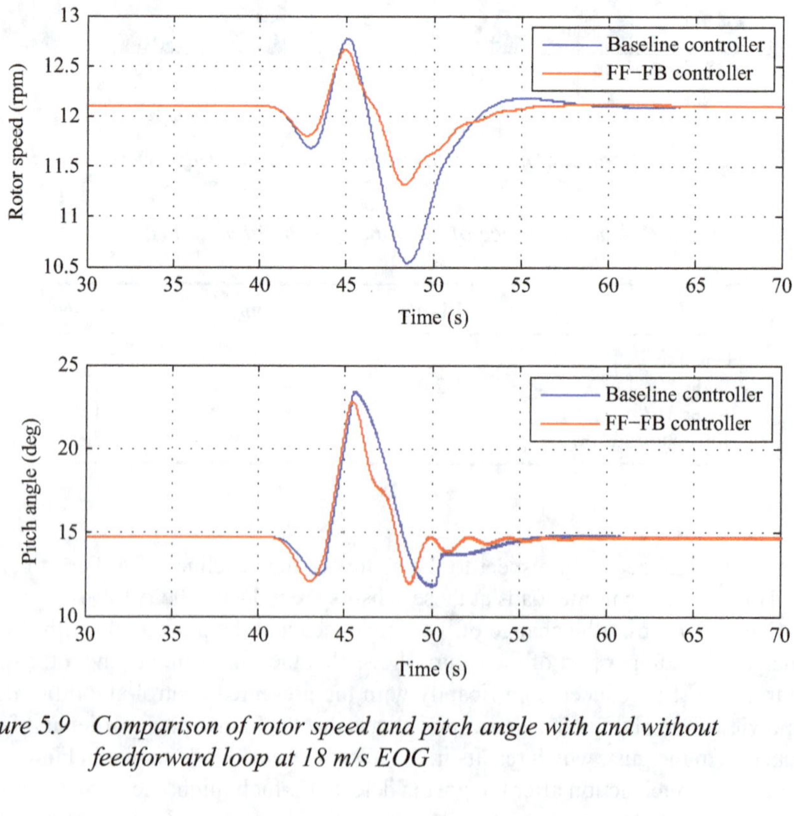

Figure 5.9 Comparison of rotor speed and pitch angle with and without feedforward loop at 18 m/s EOG

5.3 Drivetrain damping

State-of-the-art controllers which are widely used in typical pitch-regulated and variable speed wind turbines are controlling the generator torque to keep the turbine operating at the optimum tip–speed ratio and the maximum power point. Therefore, the generator torque is defined by a function of the rotor speed [see (5.19)] and an optimal generator torque-speed gain in the partial-load region. When pitch control is activated to maintain rated rotor speed above rated wind speed, the generator torque can be set to constant to its nominal value but can also be partially controlled to limit harmonics of the produced power due to the slow pitch control dynamics.

The drivetrain of wind turbines usually can be considered as a linkage of inertias coupled through springs representing the soft mountings, the twisting of shafts and the interaction of gear tooth in gearboxes. Therefore, the drivetrain of a wind turbine generally has a wide variety of torsional modes. Due to effects like tower damping influence, variable meshing stiffness as well as nonideal back electromagnetic force shape of the generator, the probability to induce torsional vibrations even during normal operation of a wind turbine drivetrain is very high. In transient operation, the torque control of the generator adds some damping to the torsional excitations. But in case of constant generator torque in operation at rated rotor speed, very little damping is provided by the generator. In order to prevent the drivetrain from excessive loads due to torsional vibrations, different control approaches are possible.

Today, there are two common ways to consider dynamics in wind turbine simulation. Mostly, all of the established simulation software for the entire wind turbine (DNV-GL Bladed, Flex5, Hawk2, FAST, etc.) use very simple low order, 2-mass-oscillator models, to represent the drivetrain behavior. In case of direct drive concept, this system degrades to a single mass model for the inertia of the generator. Generally, these software tools allow the implementation of more sophisticated models by use of DLL-interfaces, as for implementing the wind energy converter (WEC) control already described in Section 5.2.3.2. Furthermore, other dedicated software tools have been raised for wind turbine applications and designed for modeling, simulation and analysis of complex multibody systems, such as Simpack, Alaska and Sam-Cef4Windturbine. Originally coming from drivetrain applications in general, they were adapted and extended by considering aerodynamic and aeroelastic effects. The usage of such software results in a very complex and detailed mechanical drivetrain model providing high-order, multi-DOF and crosscoupled system descriptions. But both classes of models lack a representation of the strong electromechanical coupling between generator/converter and the pure mechanical drivetrain system of a modern multi-MW WEC. Only a mechatronic approach, taking into account the entire system behavior, leads to sufficient solutions for even more powerful drivetrains with rising challenges due to cost-driven development with its inherent lower stiffness to inertia ratios within the new designs.

5.3.1 Traditional drivetrain damping

Traditional approaches for the damping of torsional vibrations of WEC drivetrains are based on passive load reduction. Therefore, additional mechanical damping is

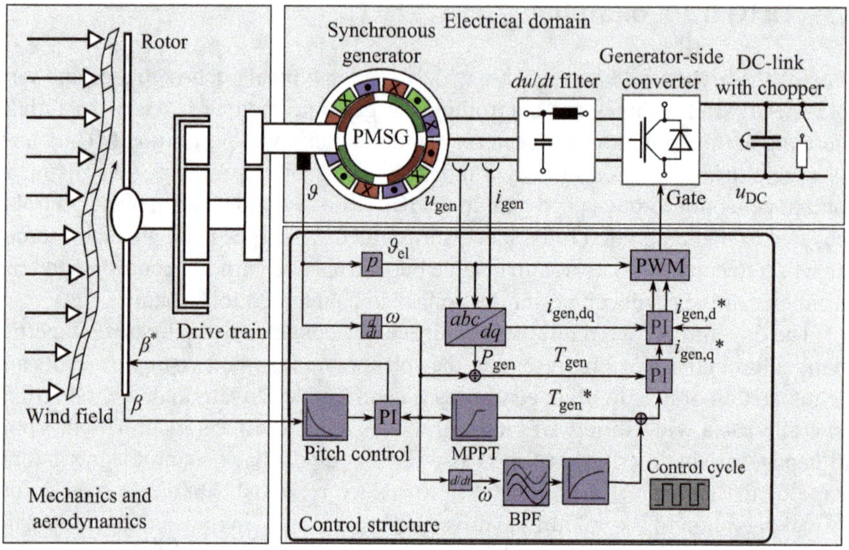

Figure 5.10 Control diagram of a wind turbine with active torsional damping provided by the torque control of the generator [21]

added to the drivetrain, by using rubber mounts in case of the torque support of the gearbox. In order to mitigate the effects of misalignment of the generator and the gearbox, clutches are used to connect these parts of the drivetrain. In addition, these clutches provide a reduction of the stiffness of the connecting shafts and add damping to the torsional modes. While capabilities of this solution are somehow limited and more advanced mechanical solutions also affect the overall cost and reliability of the drivetrain, the active torque control by the generator converter system provides another possibility for lowering loads based on torsional vibrations in the drivetrain [4,19]. This approach greatly benefits from the generally fast torque control (FOC, field oriented control) of modern converter driven generators. Unlike the pitch control of the wind turbine, which is highly dependent on the high inertia of the blades and the comparatively slow dynamics of the wind loads, the generator is able to ensure fast and effective reaction on oscillations in the drivetrain. Commonly, the control of the generator torque is slightly adapted to provide a more dynamic and oscillating set-point torque component with small amplitude superposed to the basic torque value fed into the FOC. Figure 5.10 shows a simple control diagram which takes active drivetrain damping into account. In order to detect vibrations near the modes of the drivetrain which are characterized with very little damping, a band-pass filter described by (5.19) is used [20]. Another option for this control structure is the implementation of a notch-filter cascaded with the band-pass filter to improve the damping for specific excitations like the blade-passing frequencies at three or six times the rotor rotational speed.

$$G_{BPF}(s) = \frac{\beta_i s}{s^2 + \alpha_i s \omega_i + \omega_i^2} \tag{5.19}$$

This band-pass filter is used to filter the speed or acceleration from the generator speed encoder signal, and it is tuned for one or several resonant frequencies (ω_i) of the drivetrain. The parameters β_i and α_i are used to tune gain and damping of each individual band-pass filter. The band-pass filter is followed by an inverter and a simple low-pass filter which adds the desired phase lag and filter gain to provide active torsional damping to the drivetrain in terms of adjusting the set-point of the demanded generator torque dynamically. The transfer function of the low-pass filter can be described by

$$G_{LPF}(s) = \frac{G_i}{s + \omega_{comp,i}} \tag{5.20}$$

where the parameter G_i tunes the gain of the low-pass filter and therefore the amplitude of the additional oscillating torque signal. Usually, the dynamic of the torque which can be provided by the generator is limited as modern FOCs reach a theoretical bandwidth for generator airgap torque between 100 and 200 Hz.

Hence, the parameter G_i has to be chosen carefully in terms of the limitations of the system input. The parameter $\omega_{comp,i}$ describes the phase lag of the torque signal and should be tuned regarding the individual dynamics of the generator and converter system in order to provide a torque signal which is able to extinguish torsional vibrations within the mechanical drivetrain. The sum of basic generator torque (for MPPT purpose) and dynamic generator torque describes the control input for the FOC of the generator–converter system. This approach for adding damping to different torsional modes of the drivetrain is particularly simple to implement but has also some limitations. Because the speed or acceleration signal of the generator is used in this method, the resolution and accuracy of the sensor has to be considered. Therefore, the possibility to detect torsional vibrations somewhere else within the drivetrain is limited without the implementation of further sensors which will increase the costs or is rather not possible.

In order to show the effect of the presented drivetrain damping approach, a model based analysis is carried out. An effective two-mass-model (2-DOF) of the drivetrain of a wind turbine is able to consider the overall dynamics of the drivetrain based on the first torsional natural frequency like in DNV-GL Bladed (see Section 5.2.2.2). However, the representation of higher torsional modes requires a more detailed model. This can be achieved by considering a higher quantity of equations of motion to represent the torsional dynamics. In addition, a more detailed representation of the drivetrain is able to allow the implementation of gearbox properties like backlash and dynamic meshing force of the gears. This adds further possibilities for the excitation of torsional modes of the drivetrain.

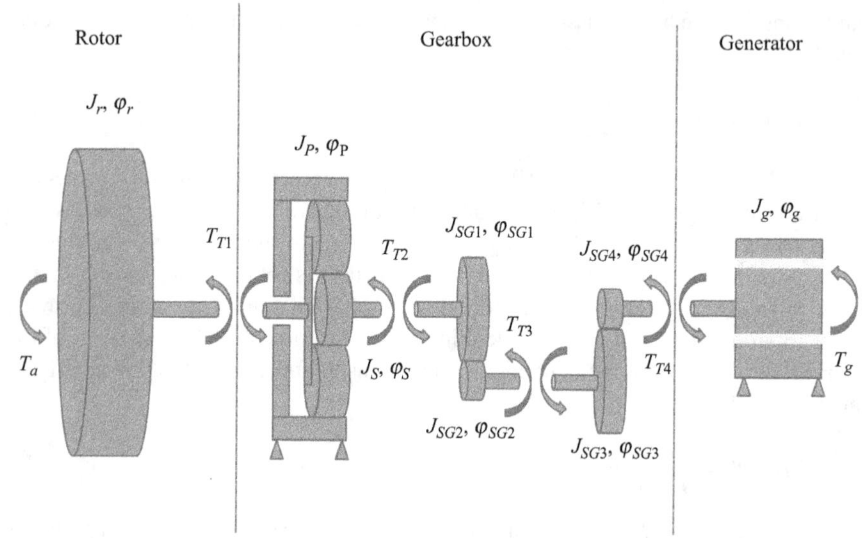

Figure 5.11 Torsional elastic drivetrain model with eight inertias J and the corresponding angular positions φ

The following model-based example of the presented drivetrain damping approach is based on a drivetrain model with 8-DOF (see Figure 5.11). The equations of motion for a drivetrain with a three-stage gearbox can be described as follows:

$$\begin{aligned}
J_r\dot{\Omega}_r &= T_a - T_{T1} \\
J_P\dot{\Omega}_P &= T_{T1} - M_{P,drive} \\
J_S\dot{\Omega}_S &= M_{S,react} - T_{T2} \\
J_{SG1}\dot{\Omega}_{SG1} &= T_{T2} - M_{SG1,drive} \\
J_{SG2}\dot{\Omega}_{SG2} &= M_{SG2,react} - T_{T3} \\
J_{SG3}\dot{\Omega}_{SG3} &= T_{T3} - M_{SG3,drive} \\
J_{SG4}\dot{\Omega}_{SG4} &= M_{SG4,react} - T_{T4} \\
J_g\dot{\Omega}_g &= T_{T4} - T_g
\end{aligned} \tag{5.21}$$

Inertias in the equations of motion refer to the rotor J_r, the planetary gear J_P, the sun gear J_S, the spur gears J_{SGi} and the generator J_g. External forces to this 8-DOF system are the aerodynamic force T_a and the generator torque T_g. Furthermore, torsional moments of the low speed shaft, the shafts between the stages of the gear box and the high speed shaft are defined by

$$T_{Ti} = D_{Ti}\Delta\Omega_i + K_{Ti}\delta\varphi_i \tag{5.22}$$

where K_{Ti} and D_{Ti} describe the specific stiffness and damping values for different shafts, respectively. $\Delta\Omega_i$ and $\delta\varphi_i$ are the differences in speed and angular position of the specific inertias connected by the shafts. The driving and reaction moments of the planetary gear and sun gear of the spur gears can be expressed as functions of varying meshing stiffness ($K_{SG1,SG2}(\varphi_{SG1})$) and the deformations of the gear tooth based on the radius (r_{SG1}, r_{SG2}) and the angular positions of the interacting gears ($\varphi_{SG1}, \varphi_{SG2}$). The following equations show an example for spur gear 1 and spur gear 2:

$$\begin{aligned} M_{SG1,drive} &= r_{SG1}F_{SG1,SG2} = r_{SG1}[K_{SG1,SG2}(\varphi_{SG1})(\varphi_{SG1}r_{SG1} - \varphi_{SG2}r_{SG2})] \\ M_{SG2,drive} &= r_{SG2}F_{SG1,SG2} = r_{SG2}[K_{SG1,SG2}(\varphi_{SG1})(\varphi_{SG1}r_{SG1} - \varphi_{SG2}r_{SG2})] \end{aligned} \tag{5.23}$$

$$K_{SG1,SG2}(\varphi_{SG1}) = K_{static} + K_{var}(z, \Omega_{SG1}t) \tag{5.24}$$

The varying meshing stiffness of the contact between the individual spur gears considers a static value (K_{static}) and a fluctuating part ($K_{var}(z, \Omega_{SG1}t)$) as a function of the rotational speed, the number of teeth (z) and other tooth parameters [22]. These representations of the drivetrain give a detailed insight in the torsional dynamics during normal operation and under the influence of transient excitations while being coupled with aeroelastic and electrical parts of a wind turbine model (see Figure 5.10). In order to show the functionality of the described control structure, simulations were carried out. The simulation is based on a dynamic model of a variable speed wind turbine considering the described representation of a drivetrain with 8-DOF. The stiffness and inertia values can be derived from CAD-drawings and finite element models. Furthermore, they can be chosen empirically to represent specific target values regarding the torsional dynamics.

In this example, the model [21] was torsional excited by a symmetric three-phase voltage drop by 80% at $t = 20$ s which causes a transient excitation of the drivetrain due to the transient disturbance of the generator torque during the grid fault [23,24]. Three band-pass filters were used in the example to add damping to the first three torsional resonant frequencies of the drivetrain at 2.55, 57.73 and 330.79 Hz. Figure 5.12 shows the simulated speeds of the inertias of the generator and of one spur gear inside the gear box in addition to the FFT-results of the corresponding acceleration signals after the grid fault. It can be seen that the described-active torsional damping approach is able to lower the amplitudes of the torsional vibrations within the drivetrain. The amplitudes of the vibrations with respect to the first and the second natural frequency (2.55 and 57.73 Hz) are lowered by approximately 50%. Higher frequencies, which can be excited by the varying meshing stiffness, seem to be less affected by the active torsional damping approach. Therefore, it has to be mentioned that this approach is somehow inherently limited by the bandwidth of sensors and the requirements on an overall robust generator control to avoid transient instabilities.

Sensors for mechanical quantities are usually less precise and inaccurate compared to sensors for electrical measurements and are an essential bottleneck in wind applications. It can be showed that the utilization of high-resolution (regarding time) measurements in the mechanical system as well as the electrical system adds a strong benefit to modern wind turbine design and operation. Therefore, more advanced

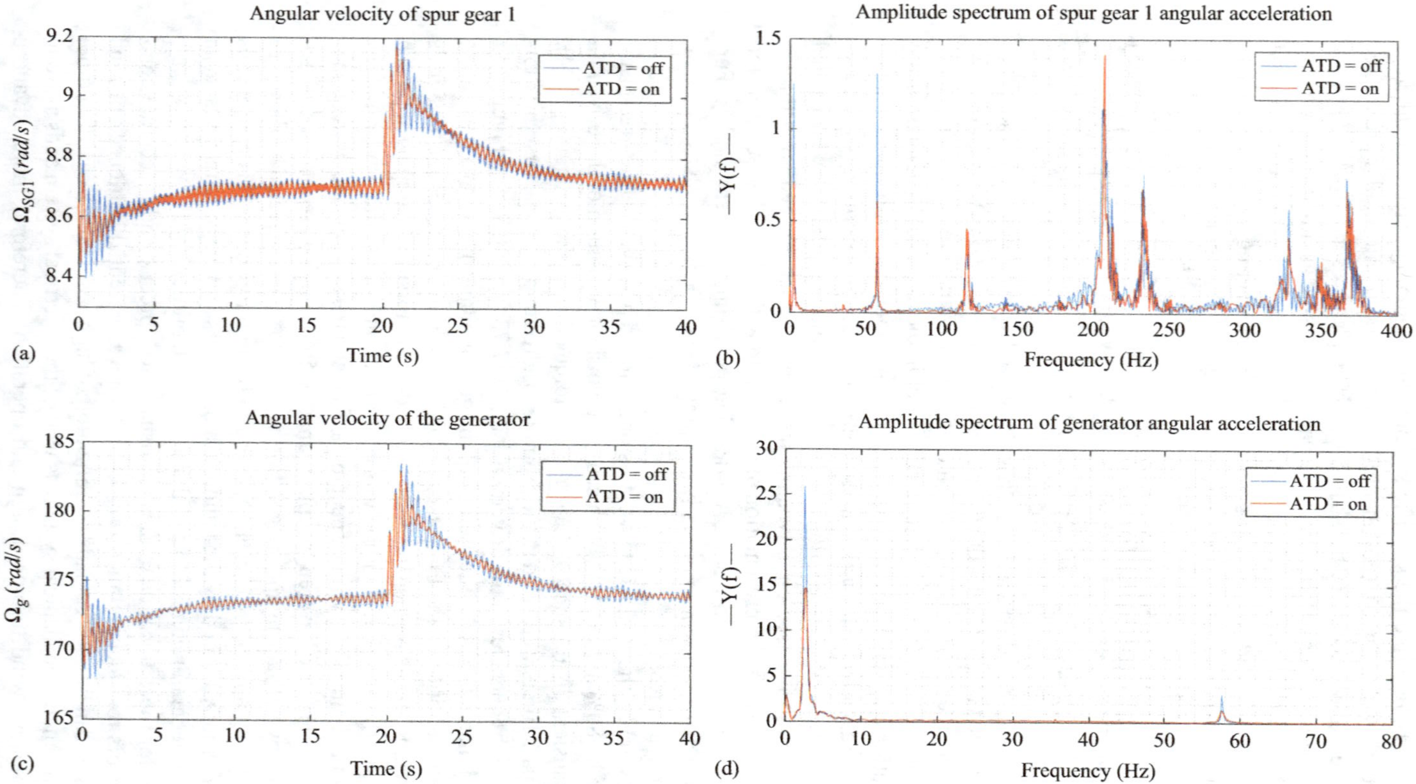

Figure 5.12 Active drivetrain torsional damping performance; torsional damping active (red); torsional damping deactivated (blue) [21]. (a) Spur gear 1 angular velocity, (b) Amplitude spectrum of spur gear 1 angular acceleration, (c) Generator angular velociy, (d) Amplitude spectrum of generator angular acceleration.

control strategies could be the solution for providing additional damping to the drive-train, but they require real-time information about internal variables and states. An alternative approach described in the next section helps to overcome the previously mentioned restrictions based on a model-based approach. The goal is to utilize minimum measurements of the electrical system (voltage and currents) and incorporate model-based estimation algorithms in order to obtain the unmeasured variables.

5.3.2 Model-based drivetrain active vibration damping

Active damping of torsional oscillations is an essential requirement in any industrial applications incorporating electromechanical systems with elastic couplings [20,25,26]. The excitation of drivetrain torsional oscillations in WECs typically occur due to the stochastic wind loads and during grid side transient events, due to the discrete mass structure of the drivetrain and finite stiffness of the intermediate couplings. Generally, active vibration suppression can be fulfilled by a suitable motion control technique that takes this torsional behavior into account. In this content, incorporating modern control algorithms based on state-space models provides the chance of an optimal solution to the control objective by utilizing complete information on system dynamics, in the case of minimum available measurement and presence of measurement noise.

Like described in the previous section, active damping of WEC drivetrain torsional oscillation is usually accomplished in full-load operation, by calculating an additional torque set-point for the generator. This will lead to small variations in the generator torque on top of the nominal value. The control objective is to modify eigenfrequencies of the system in order to reach desired response in and around the torsional mode and enhance the corresponding damping. When incorporating modern control algorithms, a full-state feedback control will be implemented based on direct knowledge of system dynamics using a state-space model. The state-space model in this case can be a reduced-order model containing only the states describing the drivetrain torsional mode at full-load operation. In fact, this model does not have nonlinear dependencies for the case of generator torque control in full-load operation. However, for the case of pitch control, system input gains corresponding to the pitch actuator are sensitive to changes in wind speed, and therefore, control design at a single operating point would fail to fulfill the control objective.

Although complete WEC simulation tools incorporate well developed and detailed nonlinear dynamic models of the overall system for simulation purposes, a reduced order model containing the important dynamics in accordance with the control objective is of interest for purpose of control design. Thanks to the linearization tool of the high fidelity FAST dynamics code, such a linear model containing a subset or all of the DOFs modeled can be extracted [27]. This tool, however, exports periodic state matrices as a result of numerical linearization of the FAST model about a previously calculated periodic steady-state operating point for the selected DOFs. The output state matrices have to be then azimuth-averaged for nonperiodic or time-invariant control design. For the introduced control objective here, only the two drivetrain rotational flexibility and the generator DOF have to be activated and the linearization has to trim on generator torque as the control input [2]. Although the linear state-space matrices are provided by FAST in a numerical form, knowledge on the state-space variables is required for the control design. Consider the following system equations based on the previously introduced active DOFs:

$$J_r\ddot{\varphi}_r = T_a(v_w, \dot{\varphi}_r, \theta) - T_t \tag{5.25}$$

$$T_t = K(\varphi_r - \varphi_g) + D(\dot{\varphi}_r - \dot{\varphi}_g) \tag{5.26}$$

$$J_g\ddot{\varphi}_g = T_t - T_g \tag{5.27}$$

where J_r and J_g correspond to the rotor and generator moment of inertia, and the rotational position of those are defined by φ_r and φ_g, respectively. Furthermore, T_a is the aerodynamic torque of the rotor, T_t is the torsional torque of the shaft and T_g defines the generator torque. The shaft torsional stiffness coefficient presented by K and D is the corresponding damping. Since the aerodynamic torque in (5.25) is a continuous function in terms of wind speed v_w, rotor speed $\dot{\varphi}_r$ and pitch angle θ, it can expanded by the following Taylor series as a good approximation:

$$T_a(v_w, \dot{\varphi}_r, \theta) \backsimeq T_a(v_{w0}, \dot{\varphi}_{r0}, \theta_0) + \alpha\,\delta v_w + \gamma\,\delta\dot{\varphi}_r + \xi\,\delta\theta \tag{5.28}$$

where $(v_{w0}, \dot{\varphi}_{r0}, \theta_0)$ determine the nominal values at equilibrium and δv_w, $\delta\dot{\varphi}_r$ and $\delta\theta$ represent the corresponding perturbations of the variables. Furthermore, α, γ and ξ are the partial differentiations of T_a with respect to v_w, $\dot{\varphi}_r$ and θ, respectively. These terms contribute to the perturbation of the aerodynamic torque to be represented by δT_a.

$$T_t = K(\varphi_{r0} - \varphi_{g0}) + D(\dot{\varphi}_{r0} - \dot{\varphi}_{g0}) + K(\delta\varphi_r - \delta\varphi_g) + D(\delta\dot{\varphi}_r - \delta\dot{\varphi}_g) \tag{5.29}$$

Moreover, the reaction torque of the shaft in (5.26) could also be rewritten in terms of the equilibrium operating point, as well as the corresponding perturbation δT_t, as given in (5.29) where $\delta\varphi_r$ and $\delta\varphi_g$ correspond to the perturbation in the rotor and generator speed, respectively. Overall, by combining (5.25)–(5.29), the new system equations can be obtained as provided in (5.30) in terms of the perturbed variables, by considering the fact that T_a and T_t are equal at equilibrium.

$$\begin{aligned} J_r\delta\ddot{\varphi}_r &= \delta T_a - \delta T_t \\ \delta T_t &= K(\delta\varphi_r - \delta\varphi_g) + D(\delta\dot{\varphi}_r - \delta\dot{\varphi}_g) \\ J_g\delta\ddot{\varphi}_g &= \delta T_t - \delta T_g \end{aligned} \tag{5.30}$$

Finally, the provided state-space matrices by the FAST linearization tool form a linear and nonperiodic state-space model in the following form, where $\underline{\mathbf{A}}$ is the system matrix and $\mathbf{Z}$ defines the state variable vector. Furthermore, $\mathbf{B}$, $\mathbf{B}_d$ and $\mathbf{E}$ are input vectors and $\mathbf{C}$ is the output vector corresponding to the measured variable $y(t)$.

$$\begin{cases} \dot{\mathbf{Z}}(t) = \underline{\mathbf{A}}\,\mathbf{Z}(t) + \mathbf{B}\,\delta T_g(t) + \mathbf{B}_d\,\delta v_w(t) + \mathbf{E}\,\delta\theta(t) \\ y(t) = \mathbf{C}\,\mathbf{Z}(t) \end{cases} \tag{5.31}$$

The delivered state-space system determines the perturbed state-variables as introduced below. This is important to be considered, for the control design in accordance with the objective.

$$
\begin{aligned}
z_1 &= \delta\varphi_r - \delta\varphi_g && \text{Perturbed torsional deflection of the shaft} \\
z_2 &= \delta\dot{\varphi}_g && \text{Perturbed generator rotational speed} \\
z_3 &= \delta\dot{\varphi}_r - \delta\dot{\varphi}_g && \text{Perturbed torsional velocity of the shaft}
\end{aligned}
\tag{5.32}
$$

Generally, suppression of torsional oscillations introduces a minimization problem in terms of the introduced state-variables. This can be well handled by a feedback controller implementation, which leads state-variables of the system to reach the equilibrium point. This can be implemented by formulating the control law as a linear combination of system states as illustrated in the following equation:

$$\delta T_g(t) = \mathbf{G}\,\mathbf{Z}(t) \tag{5.33}$$

where **G** corresponds to the feedback gain vector, being a solution to an optimization problem in terms of the system states and the input. Implementation of the feedback law leads to a new system matrix ($\underline{\mathbf{A}} - \mathbf{B}\,\mathbf{G}$) with manipulated eigenfrequencies that help to suppress the perturbations in system dynamics, whereas rejection of wind disturbances in full-load operation mode is going to be realized by the pitch controller.

$$
\begin{cases}
\dot{\widehat{\mathbf{Z}}}(t) = (\underline{\mathbf{A}} - \mathbf{L}\mathbf{C})\,\widehat{\mathbf{Z}}(t) + \begin{pmatrix} \mathbf{B} & \mathbf{L} \end{pmatrix} \begin{pmatrix} \delta T_g(t) \\ \delta\dot{\varphi}_g(t) \end{pmatrix} \\
\widehat{y}(t) = \mathbf{C}\,\widehat{\mathbf{Z}}(t)
\end{cases}
\tag{5.34}
$$

For a realizable controller, an estimation of the perturbed system state-variables can be obtained by a full-state observer as defined in (5.34) in a closed-loop form. Here, $\widehat{\mathbf{Z}}$ and $\widehat{y}$ represent the estimation of state vector and the measurement, respectively, and **L** represents the observer gain vector. The only measurements to be provided for the observer are the δT_g being the actuating variable as well as $\delta\dot{\varphi}_g$ which can be extracted from the $\dot{\varphi}_g$ measurement provided by the FAST code. For a justification of the introduced procedure in this section, control design, implementation and simulation runs have been performed in the MATLAB Simulink environment for the NREL 5-MW reference turbine using the nonlinear FAST code and the corresponding interface, providing a valuable chance for control design evaluation [9,27]. For the introduced control objective, a linear-quadratic regulator has been implemented, providing a feedback vector **G** as a solution to an optimization problem based on a cost function in terms of the system states and the actuating variable. In addition, the observer gain vector **L** is calculated using the pole placement procedure.

Simulations results provided here have been obtained considering a turbulent wind inflow with an average wind speed of 12 m/s, while the baseline pitch controller has been activated using the available DLL. Furthermore, for an illustration of the control performance, results from the implemented controller is compared with the case of a similar controller without any vibration damping functionality, but also with the baseline torque controller available as DLL. Figure 5.13 illustrates the power

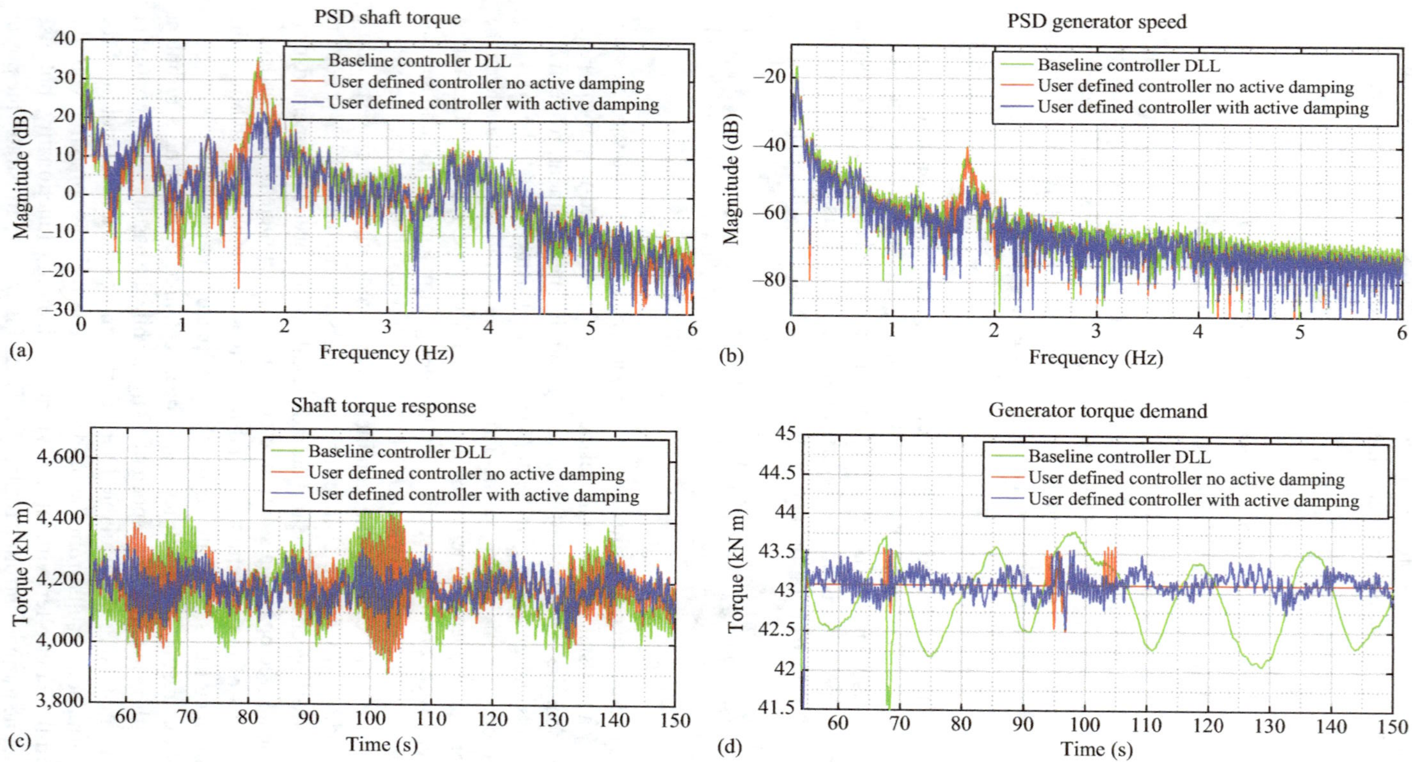

Figure 5.13 *(a) Power spectral density of the low sped side shaft torque measurement, (b) power spectral density of the generator speed measurement, (c) low speed side shaft torque measurement, (d) generator torque demand (high-speed side)*

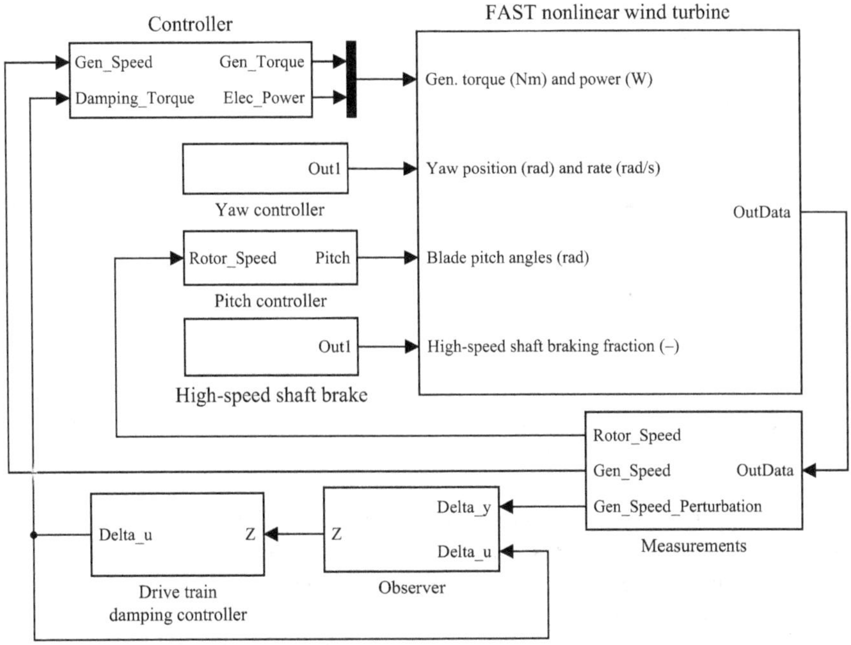

Figure 5.14 Control block diagram in MATLAB Simulink environment incorporating the FAST s-function

spectral density of the shaft torque and the generator speed measurements in full-load operation region.

It can be clearly seen that the first drivetrain torsional mode at 1.7 Hz is damped by 15 dB in the magnitude of both the torque and speed measurements using the presented control design. This can also be clearly seen in the shaft torque measurement in time domain illustrated in Figure 5.13(c), demonstrating lower amplitudes of oscillations. This is as a result of the generator torque demand illustrated in Figure 5.13(d), allowing small variations in the generator torque on top of the nominal value. Figure 5.14 illustrates the overall control block diagram in MATLAB Simulink environment.

5.3.3 Sensorless generator control techniques for drivetrain

In the past and still today in price sensible WEC market segments, the most common drivetrain concept for multimegawatt wind turbines is still the classic geared design with multistage gearbox and high speed (1,000–1,800 rpm) doubly fed induction generator (DFIG). However, a clear trend toward synchronous generator types also exists, whether electrical or permanent magnet excited and combined with hybrid (1–2 gear stages) or gearless drivetrain concepts. These gain the market share in new installations, especially for offshore and grid requirement driven market segments [23,24].

As generally discussed in the previous sections, control strategies for the damping of drivetrain torque oscillations already exist, but they are usually based on measurements from a speed or incremental encoder at the nondrive end of the generator (see Section 5.3.1), which can fail or distract within the harsh environment of a wind turbine. To achieve the highest possible system reliability and realize an optimal drivetrain system behavior regarding its damping characteristic, it is advantageous to implement a model-based control approach and to minimize the number of required sensor installations at the same time. In this content, the outstanding feature of the so-called sensorless (encoderless) generator control is the fact that no sensor for the rotational angle, speed or shaft torque is required. Instead, only measurements of the electrical quantities such as generator phase voltages and currents are needed. These are already provided by the converter with high resolution due to the internal FOC structure for dynamic magnetic flux and torque control of the generator, required for active and reactive power output regulation.

There are several sensorless control techniques available and in operation for the both asynchronous and synchronous generator classes [28–30]. However, this is not actually the case in wind energy applications due to the great variety of design variations in drivetrain concepts. Furthermore, the up-to-date valid requirements for wind turbine certification requires installation of at least one speed encoder at the drivetrain due to safety issue argumentation.

Figure 5.15 illustrates an overview of the various WEC drivetrain concepts which are all are suitable for sensorless control techniques. Here, the converter is a key element having an actuating bandwidth as high as 100–200 Hz of generator air gap torque. As shown in Figure 5.16, the converter uses power semiconductors with switching frequencies in the range of 700 Hz to 3 kHz for single switches. Furthermore, high-performance synchronized measurement devices are incorporated as standard equipment with sample rates in the range of 10 kHz for currents and voltages. In wind power applications specifically, converters with DC-voltage link in a back-to-back configuration are a standard solution. The FOC usually runs on a CPU-board implemented within the converter. The set-point values for the air gap torque and magnetic flux can be commanded by the turbine controller by means of a digital communication line (field bus system).

Further in this section, the very robust principle of sensorless generator control will be explained for the commonly used DFIG drivetrain configuration and the control stability at parameter variations will be evaluated. This is similarly explained in detail for synchronous machines in [30,31].

Equations (5.35)–(5.37) demonstrate the well-known mathematical model of an induction machine including its fundamental components according to Figure 5.17 in the stator-related $(\alpha, j\beta)$ coordinate system. It is based on these equations that air gap torque and rotational speed estimation algorithms will be implemented for a DFIG.

Here, the stator and rotor related variables are distinguished with indices 1 and 2, respectively. Furthermore, current displacement (skin and proximity effects) and additional iron losses are not considered [32]. In fact, this system of equations is only

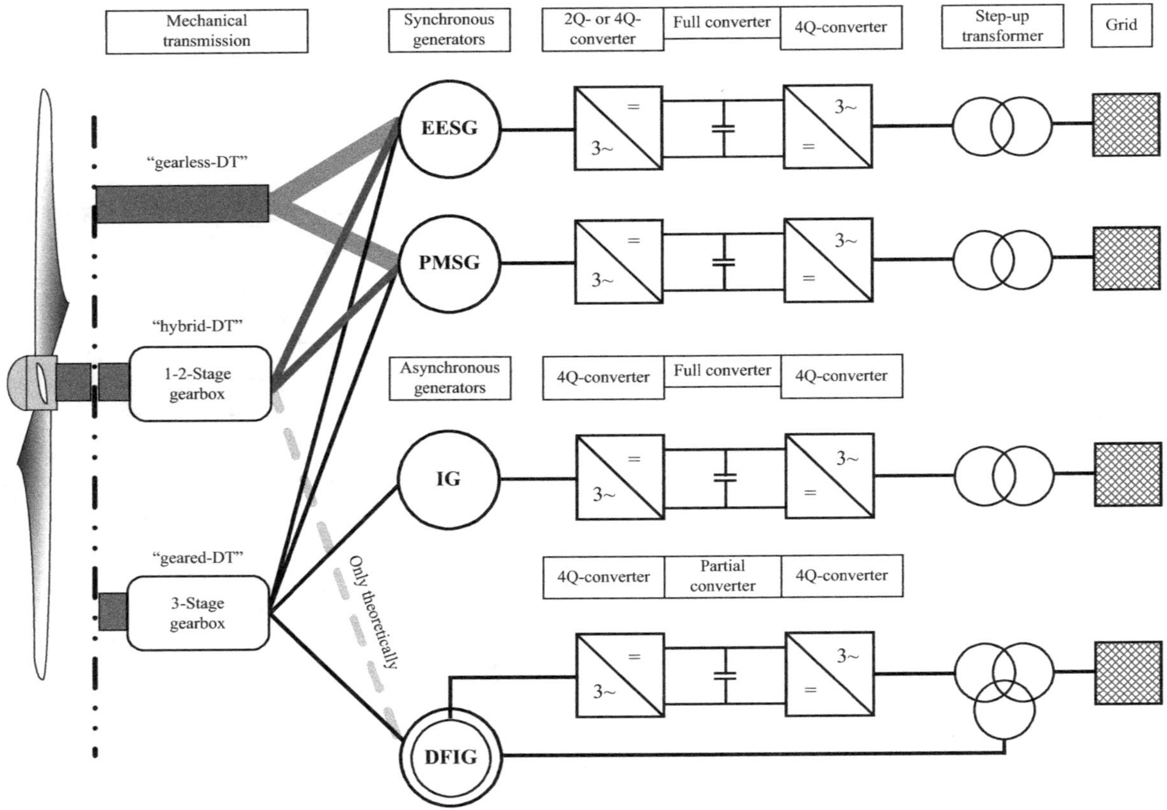

Figure 5.15 An overview of the currently implemented WEC drivetrain concepts

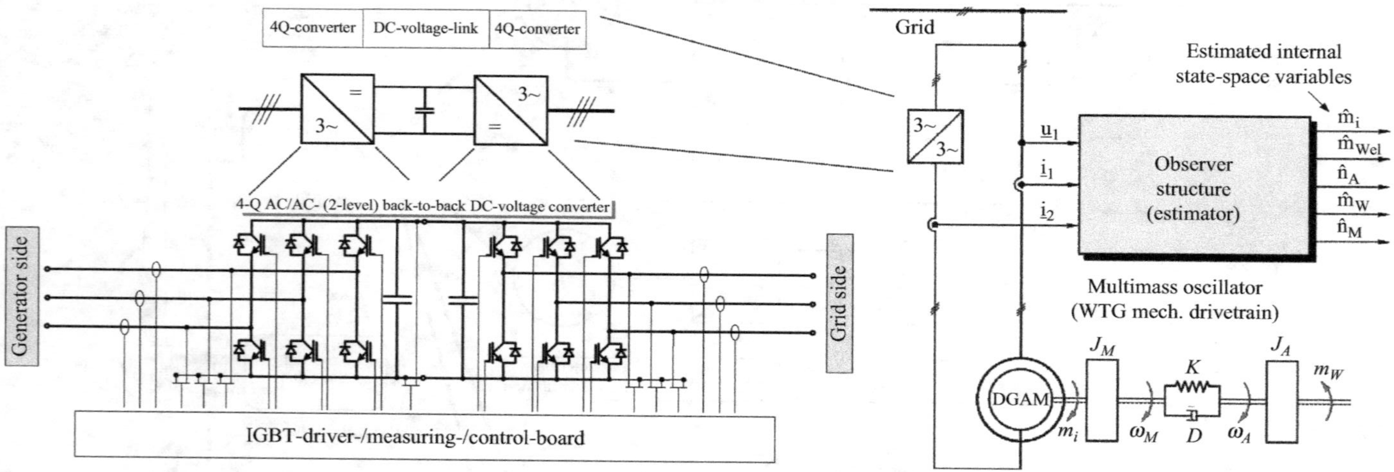

Figure 5.16 (Left) AC/DC-converter in back-to-back configuration as generalized actuator within the generator/converter systems of WEC (for asynchronous and synchronous generators) [29], (right) observer for drivetrains with DFIG

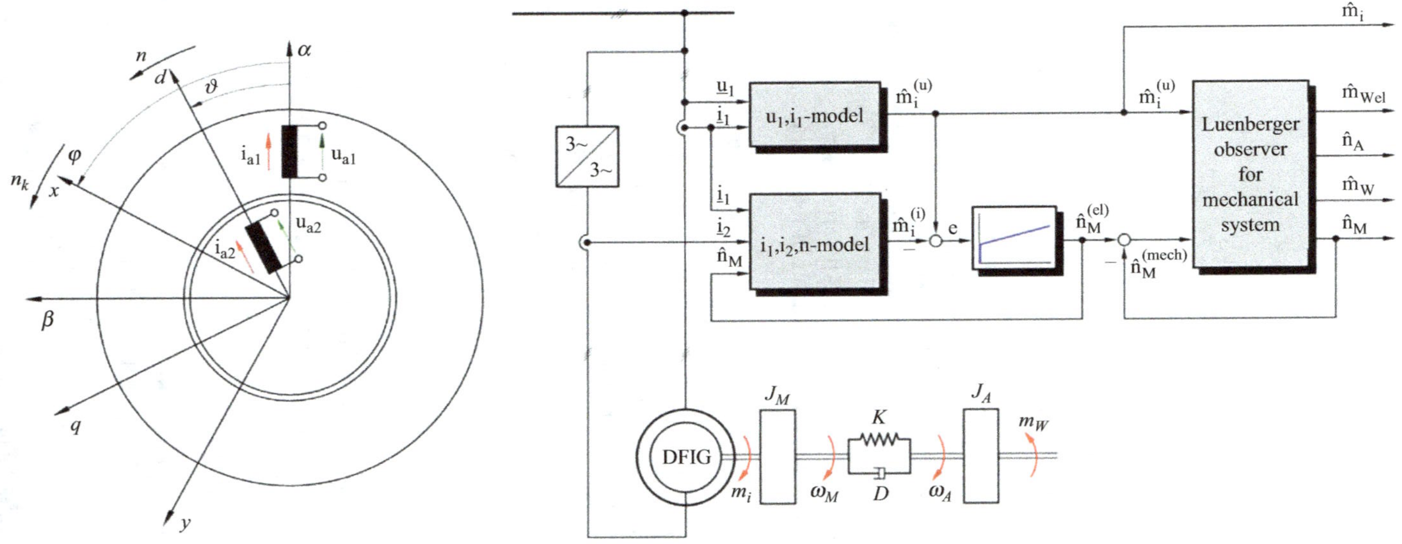

Figure 5.17 (Left) Equivalent circuit one phase and corresponding coordinate systems, (right) complete state variable estimator, divided into electrical subsystem (MRAS) and mechanical subsystem (Luenberger observer) [29]

valid under ideal assumptions of a perfectly symmetrical machine setup, a sinusoidal air gap field and a constant saturation level.

$$\frac{d\underline{\psi}_1}{dt} = \underline{u}_1 - R_1\underline{i}_1$$
$$\frac{d\underline{\psi}_2}{dt} = \underline{u}_2 + jn\underline{\psi}_2 - R_2\underline{i}_2 \tag{5.35}$$

$$\underline{\psi}_1 = X_1\underline{i}_1 + X_h\underline{i}_2$$
$$\underline{\psi}_2 = X_h\underline{i}_1 + X_2\underline{i}_2 \tag{5.36}$$

$$m_i = \underline{\psi}_1 \times \underline{i}_1 = \frac{X_h}{X_2}\left|\underline{\psi}_2 \times \underline{i}_1\right| \tag{5.37}$$

Therein, $\underline{u}_1$ and $\underline{u}_2$ denote the space vectors of stator and rotor voltages, respectively. Space vectors of the stator and rotor flux are presented by $\underline{\psi}_1$ and $\underline{\psi}_2$. Similarly, $\underline{i}_1$ and $\underline{i}_2$ stand for the space vectors of stator and rotor currents. Furthermore, R_1 and R_2 represent the ohmic stator and rotor resistances. X_1, X_2 and X_h correspond to the stator, rotor and main reactance, respectively. The illustrated complex space vector definitions apply according to KOVACS [33]. All variables and parameters are normalized (per-unit), where usually nominal values of the machine are considered for scaling purposes.

In this section, incorporating the so-called model reference adaptive system (MRAS) is considered for the estimation of air gap torque and rotational speed of the DFIG [34]. A MRAS consists of a reference model which defines the behavior of the observed system as a reference. Furthermore, it includes an adaptive model which will adapt to the reference model by a continuous feedback of a derived error signal, as shown in Figure 5.17. For the DFIG application, the $\underline{u}_1$, $\underline{i}_1$ - model$^{(u)}$ of an induction generator in the stator related $(\alpha, j\beta)$ coordinate system serves as reference model. Equation (5.38) is used to calculate the space vector of the stator flux from the measured input variables, the stator voltages and the stator current.

$$\frac{d\hat{\underline{\psi}}_1^{(u)}}{dt} = \underline{u}_1 - R_1\underline{i}_1 \tag{5.38}$$

Subsequently, a first estimated value for the DFIG air gap torque can be calculated with (5.39) using the estimated stator flux vector $\hat{\underline{\psi}}_1^{(u)}$ from (5.38) and the measured stator current vector $\underline{i}_1$.

$$\hat{m}_i^{(u)} = \left|\hat{\underline{\psi}}_1^{(u)} \times \underline{i}_1\right| = \hat{\psi}_{1\alpha}^{(u)} i_{1\beta} - \hat{\psi}_{1\beta}^{(u)} i_{1\alpha} \tag{5.39}$$

In general, open integration of (5.38) leads to a numeric instability. This can be avoided by adding a feedback loop and thereby transforming the integrator into a first order delay function with a corresponding gain crossover frequency. This operation is then valid for the necessary range of bandwidth for state variables of the DFIG. The second, inherent drawback of the $\underline{u}_1$, $\underline{i}_1$ - model$^{(u)}$ is also well known. Large errors in the estimation of the stator flux and thus of the air gap torque may be caused by temperature-dependent variations of the stator resistance R_1, especially at low generator speed ($n < 0.05$).

But in wind turbine applications, this is not relevant, because the DFIG in a variable-speed wind turbine generator application only operates in a narrow speed range of $n \approx (0.7 \ldots 1.3)n_{sn}$ close to its synchronous operation. If there is an absolute position signal required from the drivetrain other than the speed and relative position for any other purposes, there are alternative observer or estimator algorithms available which are slightly more complex but also effective [30,31]. As an adaptive model within the MRAS structure, the $\underline{i}_1$, $\underline{i}_2$, n - model$^{(\mathrm{i})}$ of the DFIG in the stator related $(\alpha, j\beta)$ coordinate system is considered. The adaptable space vector for the stator flux $\underline{\hat{\psi}}_1^{(i)}$, out of the measured input values for stator $\underline{i}_1$ and rotor current vector $\underline{i}_2$, is calculated by the following equation:

$$\underline{\hat{\psi}}_1^{(i)} = X_1 \underline{i}_1 + X_h \underline{i}_2 \tag{5.40}$$

Subsequently, using the measured stator current space vector $\underline{i}_1$, a second adaptable estimated value for the air gap torque of the DFIG can be calculated according to the following equation:

$$\hat{m}_i^{(i)} = \left| \underline{\hat{\psi}}_1^{(i)} \times \underline{i}_1 \right| = \hat{\psi}_{1\alpha}^{(i)} i_{1\beta} - \hat{\psi}_{1\beta}^{(i)} i_{1\alpha} \tag{5.41}$$

If applying the $\underline{i}_1$, $\underline{i}_2$, n - model$^{(\mathrm{i})}$, it must be noted that the rotor current vector $\underline{i}_2$ is always measured within the specific (d, jq)-coordinate system and has to be transformed into the stator related $(\alpha, j\beta)$ coordinate before incorporating it within (5.40). The required relationships for the transformation and the control principle of MRAS are shown in Figure 5.17.

$$\underline{i}_2(\alpha, \beta) = \underline{i}_2(d, q)e^{j\vartheta} \tag{5.42}$$

$\vartheta = p\vartheta_m$ therein is the electrical and ϑ_m the mechanical rotation angle of the DFIG rotor relative to the stator position, respectively. p characterizes the number of pole pairs, and the rotation angle $\vartheta(t)$ is calculated by integrating the normalized rotational speed of the DFIG according to the following equation:

$$\frac{d\vartheta}{dt} = n = n_M \tag{5.43}$$

When applying the proposed solution to estimate the rotational speed of DFIG by means of the MRAS structure, the $\underline{i}_1$, $\underline{i}_2$, n - model$^{(\mathrm{i})}$ is provided with an estimated value for the speed $(\hat{n}_M, n_M)$ delivered by an error-control loop instead of a measured value out of a speed encoder.

$$\Delta m_i = m_i^{(u)} - m_i^{(i)} \tag{5.44}$$

The error value of the air gap torque is generated from the estimated values of air gap torque, therefore the corresponding output values of $\underline{u}_1$, $\underline{i}_1$ - and $\underline{i}_1$, $\underline{i}_2$, n - model are used according (5.44) as input signal for a PI controller. Then, its output signal $(\hat{n}_M, n_M)$ corresponds to an estimated and normalized value, which converge to the rotational rotor speed value of the DFIG.

Overall, the analysis and the tests demonstrate the fact that, in order to achieve a very accurate estimation of the air gap torque $m_i(t)$ as well as the rotational speed $n_M(t)$ of the DFIG, it is recommended to consider the saturation dependencies of the stator main inductance as well as the stator main reactance $X_h = f(\psi_h)$ accordingly during operation.

References

[1] E. A. Bossanyi. Developments in Closed Loop Controller Design for Wind Turbines. In *Proceeding of 19th ASME Wind Energy Conference*, pages 64–74, Reno, NV, USA, 2000.

[2] A. D. Wright and L. J. Fingersh. Advanced Control Design for Wind Turbines Part I: Control Design, Implementation, and Initial Tests. Technical Report, National Renewable Energy Laboratory, 2008.

[3] M. H. Hansen and L. C. Henriksen. Basic DTU Wind Energy Controller. Technical Report, Department of Wind Energy, DTU, 2013.

[4] T. Burton, D. Sharpe, N. Jenkins, and E. A. Bossanyi. *Wind Energy Handbook*. John Wiley & Sons, Ltd., New York, 2001.

[5] K. Z. Østergaard, P. Brath, and J. Stoustrup. Estimation of Effective Wind Speed. *Journal of Physics: Conference Series*, 75:012082, July 2007.

[6] F. Meng, J. Wenske, and A. Gambier. Wind Turbine Loads Reduction Using Feedforward Feedback Collective Pitch Control Based on the Estimated Effective Wind Speed. In *2016 American Control Conference (ACC)*, pages 2289–2294, July 2016.

[7] E. A. Bossanyi. Individual Blade Pitch Control for Load Reduction. *Wind Energy*, 6(2):119–128, April 2003.

[8] R. P. Coleman and A. M. Feingold. Theory of Self-excited Mechanical Oscillations of Helicopter Rotors with Hinged Blades. Technical Report, National Advisory Committee for Aeronautics, VA, United States, January 1958.

[9] J. Jonkman, S. Butterfield, W. Musial, and G. Scott. Definition of a 5MW Reference Wind Turbine for Offshore System Development. Technical Report NREL/TP-500-38060, National Renewable Energy Laboratory, Golden, CO 80401, 2009.

[10] E. A. Bossanyi, A. Kumar, and O. Hugues-Salas. Wind Turbine Control Applications of Turbine-mounted LIDAR. *Journal of Physics: Conference Series*, 555:012011, 2014.

[11] G. Welch and G. Bishop. An Introduction to the Kalman Filter. Technical Report TR95-041, University of North Carolina, Chapel Hill, NC, USA, 2006.

[12] Wind Turbines – Part1: Design Requirements, 2005. Third Edition IEC 61400-1.

[13] E. A. Bossanyi. Controller for 5MW Reference Turbine. Technical Report 11593/BR/04, Garrad Hassan and Partners Ltd., Bristol, UK, 2009.

[14] D. Schlipf, E. A. Bossanyi, C. E. Carcangiu, T. Fischer, T. Maul, and M. Rossetti. LIDAR Assisted Collective Pitch Control. Technical Report D5.1.3, University of Stuttgart, Allmandring 5B, 70569 Stuttgart, Germany, 2004.

[15] J. Laks, L. Pao, A. Wright, N. Kelley, and B. Jonkman. Blade Pitch Control with Preview Wind Measurements. In *Proceedings of 48th AIAA Aerospace Sciences Meeting Including the New Horizons Forum and Aerospace Exposition*, Orlando, FL, USA, January 2010.

[16] D. Schlipf, T. Fischer, C.E. Carcangiu, M. Rossetti, and E. A. Bossanyi. Load Analysis of Looking-ahead Collective Pitch Control Using LIDAR. In *Proceedings of DEWEK*, Bremen, Germany, 2010.

[17] W. V. Rietveld. Wind Speed Feedforward Control for a Wind Turbine Using LIDAR. Master thesis, Delft University of Technology, Delft, the Netherlands, 2013.

[18] Garrad Hassan and Partners Ltd. *GL Bladed User Manual version 4.3*, 2012.

[19] J. Licari, C. E. Ugalde-Loo, J. B. Ekanayake, and N. Jenkins. Damping of Torsional Vibrations in a Variable-Speed Wind Turbine. *IEEE Transactions on Energy Conversion*, 28(1):172–180, March 2013.

[20] G. Mandic, A. Nasiri, E. Muljadi, and F. Oyague. Active Torque Control for Gearbox Load Reduction in a Variable-Speed Wind Turbine. *IEEE Transactions on Industry Applications*, 48(6):2424–2432, November 2012.

[21] A. Bartschat, M. Morisse, A. Mertens, and J. Wenske. Analysis of Dynamic Interactions between Different Drivetrain Components with a Detailed Wind Turbine Model. *Journal of Physics: Conference Series*, 753(8), 082022, September 2016. doi:10.1088/1742-6596/753/8/082022.

[22] C. Mehler. Regelungstechnische Minimierung Dynamischer Belastung in Antriebssträngen von Windenergieanlagen. Number Nr. 413 in Fortschritt-Berichte VDI Reihe 21, Elektrotechnik. VDI-Verlag, Düsseldorf, 2015.

[23] J. Wenske and U. Beckert. Voltage-induced Stresses During Low Voltage Ride Through (LVRT) in the Drive Train of Wind Turbines with DFIG. In *Proceedings of ICREPQ*, Santiago de Compostela, Spain, 2012.

[24] H. Xu, L. Chen, and J. Wenske. Active Damping Control of DFIG Wind Turbines During Fault Ride Through. In *2013 3rd International Conference on Electric Power and Energy Conversion Systems*, pages 1–6, October 2013.

[25] C. Sourkounis. Active Dynamic Damping of Torsional Vibrations by h-∞ Control. In *12th International Conference on Optimization of Electrical and Electronic Equipment (OPTIM)*, 2010.

[26] M. Neshati, T. Jersch, and J. Wenske. Model Based Active Damping of Drive Train Torsional Oscillations for a Full-Scale Wind Turbine Nacelle Test Rig. In *2016 American Control Conference (ACC)*, July 2016.

[27] J. M. Jonkman and M. L. Buhl. Fast User's Guide. Technical Report NREL/EL-500-38230, National Renewable Energy Laboratory, Golden, CO 80401, 2005.

[28] C. Schauder. Adaptive Speed Identification for Vector Control of Induction Motors without Rotational Transducers. In *Conference Record of the IEEE Industry Applications Society Annual Meeting*, pages 493–499, vol. 1, October 1989.

[29] J. Wenske and U. Beckert. Sensorless Control for Active Damping of Torsional Vibrations in Wind Turbine Drivetrains with Doubly-fed Induction Generator. In *Proceedings of ICREPQ*, Madrid, Spain, 2016.

[30] M. Schrdl. Sensorless Control of A.C. Machines. Number Nr. 117 in Fortschritt-Berichte VDI Reihe 21, Elektrotechnik. VDI-Verlag, Dsseldorf, 1992.

[31] U. Beckert and J. Wenske. Verfahren und Vorrichtung zur Sensorlosen Erfassung der Rotorposition einer Synchronmaschine, April 2015. DE Patent 10 2014 209 603 B3.

[32] U. Beckert. Theorie des Dynamischen Verhaltens der Doppeltgespeisten Asynchronmaschine mit Berücksichtigung der Stromverdrängung im Läufer. Technical Report, TU Bergakademie Freiberg, 2013.

[33] K.P. Kovacs and I. Racz. *Transiente Vorgänge in Wechselstrommaschinen.* Verlag der Ungarischen Akademie der Wissenschaften, Budapest, 1959.

[34] R. Cardenas, R. Pena, J. Proboste, G. Asher, and J. Clare. MRAS Observer for Sensorless Control of Standalone Doubly Fed Induction Generators. *IEEE Transactions on Energy Conversion*, 20(4):710–718, December 2005.

Chapter 6

Toward farm-level health management of wind turbine systems: status and scope for improvements

Surya Teja Kandukuri[1], *Kjell G. Robbersmyr*[1], *and Hamid Reza Karimi*[2]

6.1 Introduction

Renewable energy sources are gaining importance due to depleting fossil fuel reserves and their adverse environmental impact. The European Union aims to shift 20% of its energy reliability to renewable energy resources by 2020 and the European Wind Energy Association (EWEA) estimates about 14% of this to be fulfilled by wind energy [1]. Higher wind energy production translates to larger wind farms and higher capacity turbines. Offshore wind farms (OWFs) have become popular because of abundance of wind source, savings on valuable real-estate and little impact of wind turbine (WT) noise.

Although OWFs are advantageous in many ways, their availability is still not comparable to their onshore counterparts. It was found that while the onshore availability is reaching 99%, the OWFs in United Kingdom at *Barrow North*, *Hoyle*, *Kentish Flats* have availability figures between 67% and 85% [2]. The downtimes in OWFs are longer because of the finite weather windows in which personnel can perform the required maintenance. Besides, the maintenance costs are also comparatively higher at the range of 18%–23% of the life cycle costs owing to their location and associated difficulty of maintenance [3].

It is now well accepted that condition monitoring (CM) is important in predicting failures in remotely located WTs both by the wind farm operators and insurance companies [4]. Health monitoring, condition-based maintenance (CBM) and reliability centered maintenance (RCM) are established areas in aerospace sectors and well-proven in energy, oil and gas sectors. Often, these techniques are borrowed in providing a solution in wind energy sector with necessary modifications. It is worthwhile

[1]Department of Engineering Sciences, University of Agder, Norway
[2]Department of Mechanical Engineering, Politecnico di Milano, Italy

to note that unlike other sectors, equipment in OWFs has to withstand a highly corrosive offshore environment, bad weather conditions like storms and non-stationary operating profiles.

Given the understanding about the importance of CM and need for effective maintenance strategies for wind farms, a substantial amount of research has been performed through collaboration between industry and academia. ECN's CONMOW [5], ReliaWind [6] and Supergen [7] are some of the notable efforts undertaken by European Research Consortia for WT CM. Some of the significant surveys on operation and maintenance (O&M) data of wind farms have appeared in Ribrant *et al.* [8] on Swedish wind farms, Hahn *et al.* [9] on German wind farms and ReliaWind [6] on turbines across EU, wherein the most frequent failures and their corresponding downtimes were identified. Besides, ReliaWind program made substantial contributions such as a WT taxonomy, which is now generally accepted for identification of high-priority failure modes for various WT subsystems [10]. More recently, an OWF database OWMEP was setup in Germany to gain insights into maintenance best practices and reliability analyses across multiple wind farm operators [11].

Due to the widespread acceptance of CM for wind farms, there are now a variety of commercial CM systems that are offered by original equipment manufacturers (OEMs), turbine operators and third parties. A recent research claims that there are as many as 36 different products on the market [12], but almost all of these target generator, drive train, main bearings and blades.

Van Bussel [13] proposes that in a typical onshore wind farm of 120 units of 2 MW turbines, an average of 600 maintenance visits per year is necessary. Even if one optimistically assumes that CM systems that are commercially available are installed, they cover only a subset of the WT subsystems and the balance of system (BOS) is still dependent on reactive or scheduled maintenance. It may be possible to conduct reactive maintenance for the onshore wind farms, but in case of OWFs, this could become extremely difficult due to weather windows, transportation costs and would involve higher downtimes due to their location. However, a study conducted by Nilsson *et al.* [14] in 2007 claims that scheduled maintenance is still the norm with OWFs and service is usually conducted once to twice a year.

In order to achieve better availability while reducing maintenance costs of OWFs, it is necessary to adapt a farm-level health management approach. While the CM or health monitoring is the knowledge of whether the system is healthy or faulty, the *health management is the capability to make intelligent, informed, appropriate decisions about maintenance and logistics actions based on diagnostics/prognostics information, available resources and operational demand* [15]. It is improving from current CM of specific zones of the WT to a solution that handles maintenance of the whole wind farm, intelligently. The CM systems should feed into maintenance planning rather than operate as end-products to assist in reactive maintenance tasks.

In order to achieve such farm-level health management, an architecture based on RCM and CBM methodologies is presented in this chapter. The architecture describes the necessary elements of a health management scheme that plans maintenance tasks and correspondingly the resources, schedules and logistics based on the asset health and failure predictions.

6.2 Maintenance methodologies

Maintenance in OWFs is usually undertaken by wind turbine manufacturers within the period of the first 1–5 years after the installation. After this period, the maintenance is performed by either the operator, third parties or maintenance contract with the OEMs are extended. The maintenance mainly comprises of two types: (1) scheduled maintenance, wherein maintenance is done either based on calendar (time) or number of operating hours and (2) reactive maintenance, wherein maintenance is done upon a failure.

Many of the wind farms are now equipped with CM systems along with the supervisory control and data acquisition (SCADA) systems for operation. As it was earlier estimated that gearbox and drive train together comprise up to 42% of the total downtime [16], these systems have received great attention in terms of CM research. Consequently, a number of methods targeting diagnostics for generator, gearbox and bearings have been developed primarily based on vibration sensing, temperature monitoring and oil debris analysis. CM systems for WT blades have been developed based on strain sensing, fiber optics and ultrasound. An elaborate review of the existing CM systems for WTs can be found in reviews by García Márquez *et al.* [17] and Takoutsing *et al.* [18], summarized in Figure 6.1. It may be noticed, the BOS components like pitch and yaw systems have not received much attention so far.

CM systems in wind farms often operate independently from the SCADA system. The maintenance personnel receive alarms based on asset condition and then maintenance task is manually determined. The advantage of existing CM systems can

Method	Blades	Rotor	Gearbox	Generator	Bearing	Tower
Vibration	●	●	●		●	●
Acoustic emission	●	●	●		●	
Ultrasonic techniques	●					
Oil analysis			●	●	●	
Strain	●	●				
Electrical effects	●			●		●
Shock pulse methods					●	
Process parameters	●			●		
Performance monitoring	●		●	●	●	
Radiographic inspections	●					
Thermography	●		●	●		

Figure 6.1 CM techniques for WTs: state-of-the-art (source: [19])

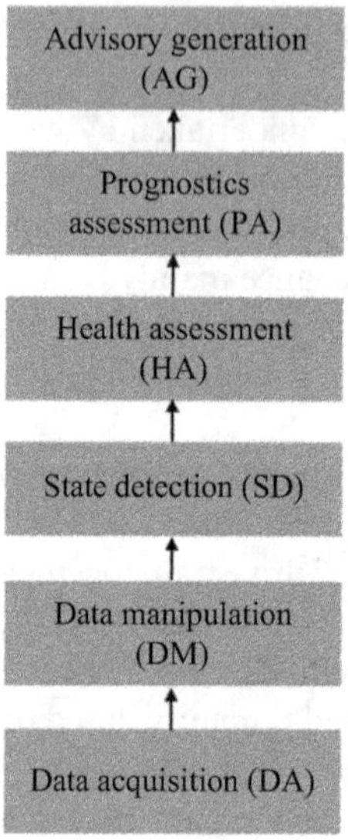

Figure 6.2 OSA-CBM architecture (source: [19])

be fully realized only when they contribute actively and systematically in maintenance planning. In order to achieve this, a comprehensive farm-level maintenance strategy is necessary.

6.2.1 Condition-based maintenance

CBM originated in the aerospace industry in order to reduce maintenance costs due to excessive scheduled maintenance. In essence it follows the idea of "if it's not broken, don't fix it." The condition of the equipment is monitored using CM systems that detect an *incipient* failure, and subsequently, a maintenance task is identified. CBM, by itself, does not give any guideline on which maintenance task to be performed and how to plan such maintenance in order to optimize the maintenance activities. The CBM concepts were standardized by MIMOSA as open system architecture for CBM (OSA-CBM), which covers the functional blocks [20], as shown in Figure 6.2. It is shown that the diagnostics and prognostics functions can support in *advisory generation* that could be used in either deferring a scheduled maintenance task or in performing a maintenance task to avoid an impending failure. A majority of the developed diagnostics for WTs can be utilized for CBM. However, prognostics for rotating machinery is still a nascent area due to the complexity of the problem and variables involved [21].

6.2.1.1 CBM candidate selection

CM is suitable for components that have sufficient degradation period, that is, from the time the CM systems detect an incipient failure or a *potential failure*, there should be enough time before the final *functional failure* occurs. Only then, a maintenance action can be planned and the failure averted. Also, it is important to assess whether CBM is a feasible solution for the component at hand, technically as well as monetarily. The following questions may be considered [22] for technical feasibility assessment:

- Is the failure mode observable through CM?
- Do state-of-the-art diagnostics and prognostics methods exist?
- Are the sensors and SCADA data already available for the diagnostics and prognostics?
- If additional sensors are necessary, is it possible to install them?
- Are the accuracy and reliability of the diagnostics and prognostics acceptable?
- Does CM reduce the risk of failure to an acceptable level?

After a successful analysis of technical feasibility is performed, cost–benefit analysis may be conducted. The cost of development of algorithms, sensors, installation and setup costs as well as the maintenance (if any required) of CM systems should be considered during this time. Only when these tests are successfully performed, does it justify to implement CBM for the system under consideration. Stringent technical feasibility tests are necessary as it has been shown that high false alarm rates would alter the cost–benefit analysis of CBM systems [23].

6.2.2 Reliability centered maintenance

Of the various maintenance methodologies, RCM is considered most suitable for farm-level health management. RCM focuses on ensuring overall availability of the system by choosing the most appropriate maintenance task for each fault condition. Originally developed in the aerospace sector [24], it was later successfully applied in a variety of industries like energy, oil & gas and chemical industries.

RCM strategy selects the most suitable maintenance method for each high priority failure modes of each component of the system. Based on this strategy, maintenance decisions are made at subsystem level, taking into account the failures of each component and then at system level. Hence, RCM gains the capability of zooming into component level and back to system level. SAE standards documents SAE JA-1011 [25], JA-1012 [22], NAVAIR 00-25-403 [26,27] are excellent resources for overall understanding of RCM process.

6.2.2.1 Prognostics and PF interval

All equipment eventually degrade with time. This degradation may be seen as a gradual progress in most of the mechanical and electrical systems and early signs of degradation may be detected during inspection. RCM defines the early signs of *potential failure* as "*an indication that a functional failure is about to occur or is in the process of occurring*" and *functional failure* as that state where "*a physical asset or a system is unable to perform a specific function to a desired level of performance*" [22]. The time between a potential failure and functional failure is defined as the *PF interval* as shown in Figure 6.3.

The duration of PF interval plays a significant role in maintenance planning. In this interval, a maintenance task has to be executed so that the availability of

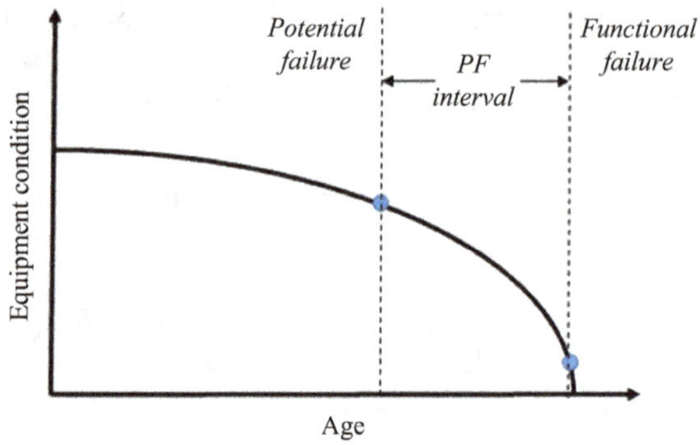

Figure 6.3 PF interval (source: [19])

the equipment is maintained. In case of CBM systems, the potential failure may be detected by the CM systems and associated diagnostics. The potential failure may be due to ageing or age-unrelated fault, which may result in accelerated degradation. After this stage, it is necessary to accurately quantify the fault and then predict the remaining useful life (RUL) or time to functional failure. Therefore, prognosis plays a very important role in maintenance planning and scheduling.

6.2.2.2 RCM for wind turbines

In choosing the right maintenance strategy for each component of the WT system, each subsystem of the WT should be analyzed at component level for functional failures. It should also be verified whether CBM is the right choice and if not then other maintenance methodologies may be considered, as shown in Figure 6.4. The idea of utilizing RCM for WT maintenance is put forward in a number of recently published literature. Besnard *et al.* [28] described the concept of "reliability centered asset management" that enhances the concepts of RCM with quantitative maintenance optimization in order to quantify a particular maintenance choice. Igba *et al.* [29] proposed a "systems approach" for implementing RCM for gearboxes. McMillan *et al.* [30] studied Danish concept of multi-MW onshore wind turbine toward RCM implementation and concluded that a "highly refined system of condition monitoring and maintenance management, similar to systems in aviation sector have to be rolled out." Baglee *et al.* [31] discussed application of RCM to WTs and explained usage of E-monitoring for this purpose. Dehghanian *et al.* [32–34] described a comprehensive assessment of electrical systems for RCM implementation including critical system identification and an approach for cost–benefit analysis.

Although there have been instances detailing implementation aspects of RCM for a selection of components in the literature, or a few select elements of RCM, to the authors' knowledge, little attention has been paid so far on realization of health management architecture for wind farms. Considering specific areas of the wind farm for RCM without a comprehensive view could lead to solutions that are incomplete or counterproductive as some of the functions or effects of faults may be overlooked [35].

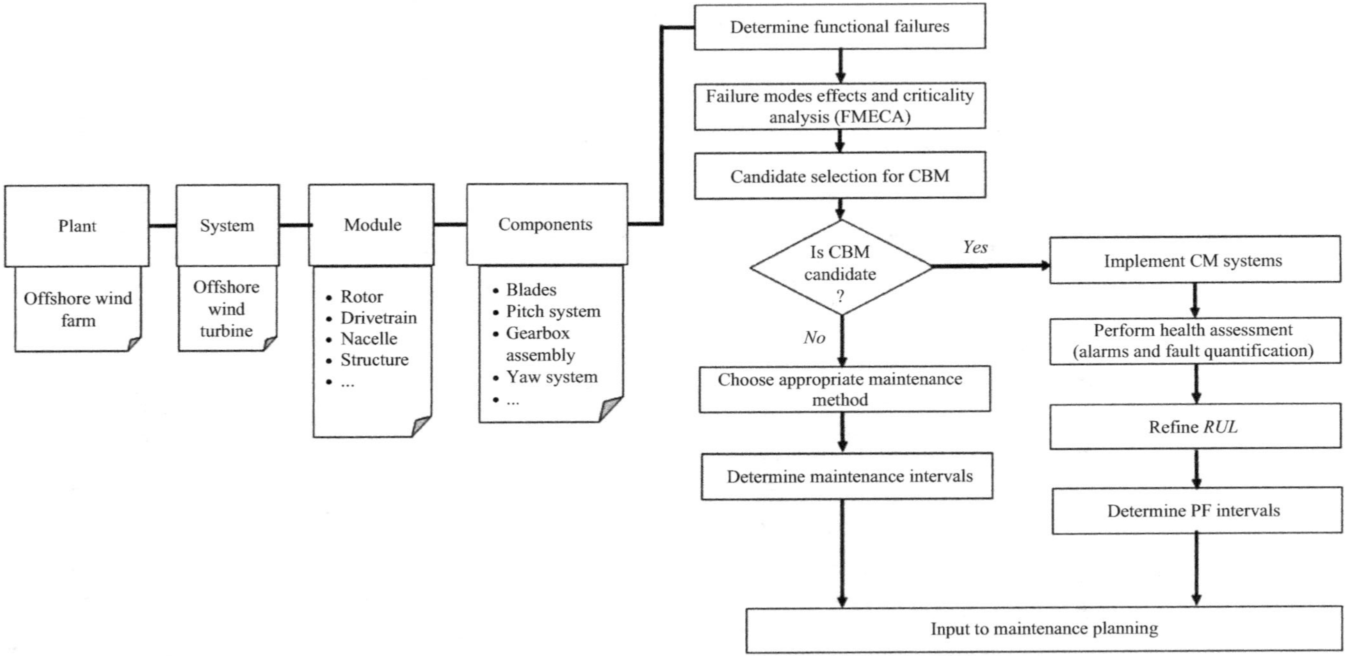

Figure 6.4 Selection of suitable maintenance procedure for each component (source: [19])

By defining the architecture, it would be possible to clearly identify the functions of each element, gaps to be filled, key technical questions to be answered and above all a system level understanding of maintenance scheme. This chapter presents a conceptual architecture for farm-level health management. The authors intend to disclaim that technical specification of each element in such architecture is not detailed as it is beyond the scope of this chapter and also yet to be realized. The intention is only to take a step closer to such a system.

6.3 Farm-level health management architecture

It is established that there exists a need for a farm-level health management especially for OWFs. In order to realize such a system, there are many elements that are to be accomplished. As it will be shown, research in wind energy community is heading in the right direction with majority of such areas being explored. However, the health management architecture as presented here involves maturation of each of the elements. Besides, it involves a substantial task of integrating all the elements. A conceptual architecture for fleet level health management is shown in Figure 6.5.

A few reasonable and practical assumptions are made during the process of development of the architecture as follows:

1. The wind farm has CM systems set up which may be independent from the SCADA systems.
2. The wind farm is connected to an onshore center that collects the SCADA operational data and CM data at their respective (time) resolutions.
3. The SCADA data and CM data from each wind turbine is transmitted reliably to the onshore operations center. Necessary protocols for secure error-free transmission and storage SCADA and CM data have been established.
4. The individual OEMs have supplied detailed design data, wiring diagrams, O&M manuals of the components to the maintenance personnel.
5. The conventional maintenance procedures are sufficiently documented and each existing maintenance task is clearly identified.
6. The CBM candidate selection activity has been performed, suitable maintenance tasks against functional failures are identified and RCM maintenance methodology is established.

Also, in the context of CBM, the diagnostics and prognostics information is classified into the following four stages as denoted in Figure 6.5:

- L1—Fault detection: The CM system is capable of differentiating between a healthy and faulty condition. At this stage, manual intervention is necessary to determine the fault.
- L2—Fault diagnosis: The exact fault is identified and a suitable maintenance action may be determined. However, at this stage, there is no information on *when* the maintenance action is necessary.

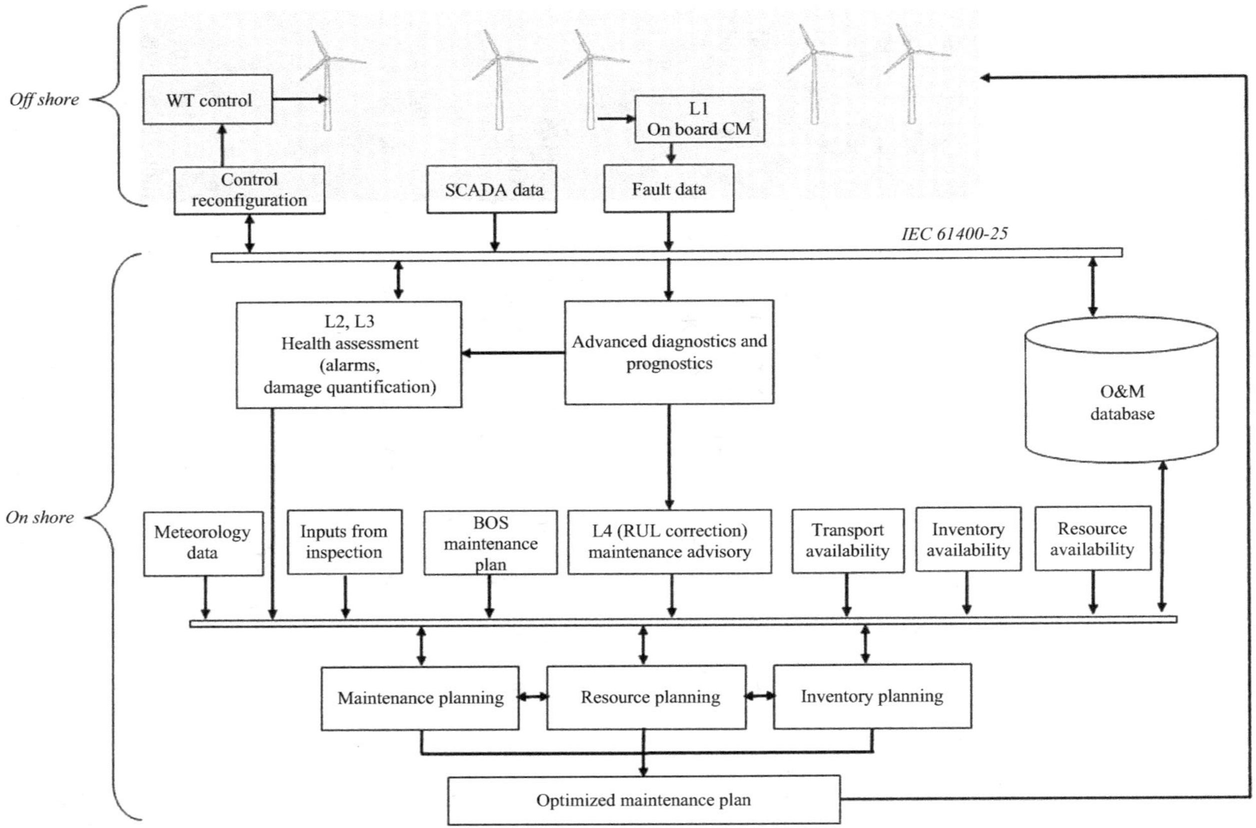

Figure 6.5 Farm-level health management architecture (source: [19])

- L3—Fault quantification: The exact fault is identified and its magnitude is determined. At this stage, it is possible to determine whether *immediate* action is necessary.
- L4—Fault prognosis: The exact fault is identified, its magnitude is determined and due course to critical stage is estimated with acceptable accuracy. The time to failure is identified and hence a maintenance window may be chosen.

The elements of the architecture are explained in the following subsections.

6.3.1 On-board CM systems

At every individual WT level, the *L1* information is available based on the CM systems. In diagnostic methods wherein high data rates from sensors are necessary such as acoustic emission techniques, it may be impractical to transfer data continuously, from each turbine to the onshore center. Hence, preliminary analysis can be performed at WT level, and only in case of anomalous behavior, a batch of data may be transmitted to the onshore center for advanced diagnostics and prognostics analysis. Candidate selection for CBM is a necessary task to be performed in order to evaluate technical feasibility and cost benefit of implementing CM. The rest of the functions will be covered under scheduled or run-to-fail maintenance, as per RCM.

6.3.2 SCADA and fault data

As all the WTs are connected to the onshore center, it is ensured that the SCADA data is continuously available for all the WTs in the OWF. The trip event data is intermittent and is supplied only upon incidents of faulty behavior. The capability of transferring, handling and storing data must be analyzed a priori. Communication protocols should be specified following standards such as IEC-61400-25 [36]. A prototype for generic WF data server as per IEC-61400-25 standards and data management architecture is described in [37].

6.3.3 Communication protocol

Communication protocols constitute the crucial link between the offshore and onshore and can have direct impact on the maintenance decision-making process. The communication protocols such as IEC-61400-25 provide a definite structure for data transmission from the turbines to the offshore center and hence avoid any ambiguity regarding the turbine as well as the type and location of the fault on the turbine. The IEC-61400-25 uses “logical device” for each WT and “logical nodes” for each zone of the turbine such as rotor, transmission and generator [36]. Thus, a fault within the yaw system of a particular WT, for example, is communicated with the specific turbine and location information. An extension of the communication standard to incorporate additional tags for OWFs is specified in [38].

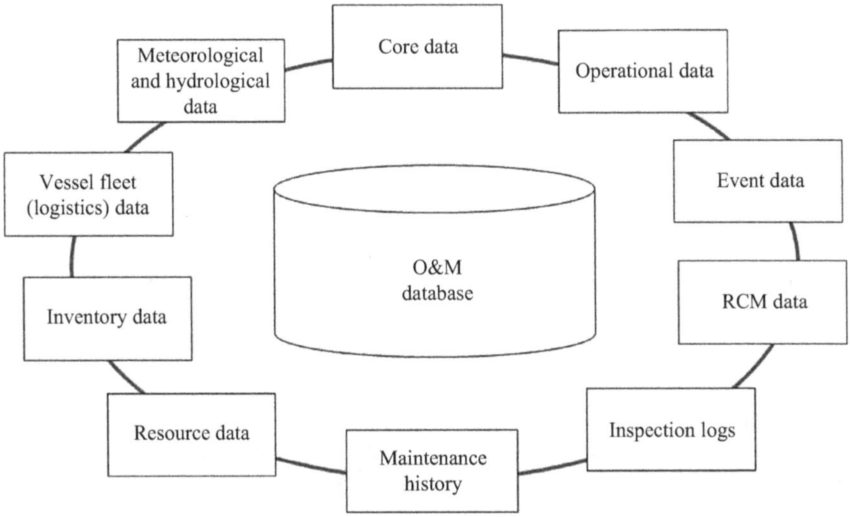

Figure 6.6 A database for farm-level health management (source: [19])

6.3.4 Advanced diagnostics and prognostics

The necessary fault diagnosis to determine exact fault post *L1* stage information is performed at the onshore center. Onshore location is beneficial for advanced diagnostics and prognostics as they are computationally intensive tasks that require high-performance computational resources and farm-level operational data. Detailed understanding of fault based on diagnostics and prognostics algorithms is performed here to derive *L2* and *L3* level information.

6.3.5 Operation and maintenance database

An O&M database should be designed to hold SCADA operational data, meteorological data, fault data and maintenance logs for the OWF. Detailed requirements of such database should be finalized and used in design. Faulstich *et al.* [39] described a concept of database for offshore WT analysis, which takes into account a number of factors beyond fault logs. This database stems from the German program Offshore Scientific Measurement and Evaluation Programme (OWMEP). A slightly modified version of the database, primarily focused for O&M use in a single OWF, is shown in Figure 6.6.

Here, the *core data* is defined similar to that in [39,40], as geographical and technical data of the turbines. The *operational data* is the continuous collection of SCADA data, of healthy turbine, and all fault events are collected in the *event data*. The *meteorological and hydrological data* consists of essential weather conditions like wind speed and wave height forecasts for the maintenance planning. The *RCM data* consists of the predefined maintenance tasks for all of the foreseen faults during RCM exercise and the associated procedures. The *inventory data* consists of detailed listing of the available spares and those that are ordered and to be delivered along with available time and quantity. Similarly, the *vessel fleet data* and *resource data* consists of the forecast of the transport vessels and resource availability for maintenance planning. All the information as shown in Figure 6.6 should be available online in an easily retrievable manner for intelligent maintenance plan generation.

6.3.6 Health assessment

The *L2* and *L3* level information derived using advanced diagnostics and prognostics is used to generate alerts for maintenance planning. Also, the fault quantification information may be utilized for control reconfiguration purposes [41].

6.3.7 Maintenance advisory

The RUL for systems that are chosen for CBM may be computed based on their current health status and fault quantification. This information overrides the design RUL information and hence refines the PF intervals. The modified subsystem level RUL information (*L4*) for all WT's in the OWF is provided to the maintenance planning.

6.3.8 Maintenance, resource and inventory planning

The maintenance planning for OWF should be a *living process*. Depending on the alarms from the CM systems and the modified RULs that are supplied, the maintenance plans should be reassessed timely. It has been shown that there are chances for significant savings if two different WT maintenance tasks are clubbed and conducted at the same time [14]. The maintenance, resource and inventory planning are interlinked and should be considered in that way for optimization. Adhikari *et al.* [42] describe the linkage between maintenance and resource management in aerospace scenario. Besides, for OWFs, there are a number of factors that are to be considered for efficient maintenance planning that are particularly unique to this area. Resource planning entails location of the maintenance crew, work shifts and number of maintenance teams as well as planning of transport vessels and helicopters and availability of spares [43–46]. Weather window forecasting is another important factor for determining maintenance schedules [47].

Although RCM exercise reveals the most suitable maintenance task for the given failure mode, it does not in itself determine the maintenance schedules or task prioritization. This has to be accomplished using methodologies like risk-based maintenance (RBM) [48,49] or opportunistic maintenance methods [50–52]. ECN's OMCE [53] and SINTEF's NOWICOB [54] are notable tools for assisting maintenance planning by estimating cost of maintenance scenario that have provision to include the constraints.

6.4 Issues, challenges and gaps

Wind sector has the advantage of being younger compared to the conventional energy sector, and this should be fully utilized by adapting the best practices and know-how from the energy and aerospace sector at the early stages of OWFs. Aspects like design for maintainability and reliability have to be given due consideration in design phase of the future WTs. Thorough and systematic analysis of the existing WT operational data and maintenance tasks will provide insights into aspects of the WTs that require design level changes toward improving reliability. Although some of the areas of such a systematic maintenance are receiving attention in terms of research and development, there are certain gaps to be filled.

6.4.1 Requirements, guidelines and standards

An OWF can comprise WTs from different manufacturers, several CM systems that run proprietary software lacking architectural commonalities. This could become a major problem for generalizing maintenance approach for an OWF. Although IEC 61400-25 [36] aims at standardizing the communications protocols, it is still dependent on the OWF and WT manufacturers, to decide the level at which they choose to apply the communication standard [37,55]. Besides, National Renewable Energy Laboratory (NREL) claims that the existing standards of IEC-61400 family are inadequate and to be modified for American offshore implementation [56]. NREL further claims that the European establishments complement the standards with guidelines from DNV [57] and GL [58] that are customized for local use. Besides, standards for health management are still in preparation even in aersopace sector [59], where research was being conducted for past two decades, it is worthwhile to consider development in wind energy sector.

In order to realize an effective farm-level maintenance strategy, it becomes necessary to follow a system engineering methodology. Various stakeholders have to be identified and the requirements of health management to be clearly defined. The elements of health management may be deployed in a modular fashion with suitable verification and validation against the identified system-level requirements. Finally, it must be noted that RCM itself is a "living process" and the developed system must be capable of accommodating future extensions for newer components, diagnostics and prognostics methods or changes in the maintenance method for components depending on their behavior through their lifetime.

6.4.2 Database creation and maintenance

The database mentioned in Section 6.3.5 involves significant efforts toward development and maintenance. Tasks such as standardization of data, efficient and error-free transfer of data to and from the database, standard maintenance and inspection logging formats based on the fault category, level and the equipment break-down structure are to be framed and strictly followed. This requires considerable resources, both in terms of personnel and infrastructure, and should be accounted for during the planning of the health management scheme.

6.4.3 Data analytics and big data

Large OWFs of the future entail significant historical data accumulated over time. Big data appears as a promising area for analysis of large data accumulated from several sources at variable rates [60,61]. Big data techniques such as multivariable analysis could provide insights into failure patterns [62]. Further, big data technology can be leveraged in gaining actionable knowledge for maintenance and prognostics by sifting through unstructured data such as notes made by maintenance personnel, inspection findings, opinions and sentiment analysis.

6.4.4 Advances in prognostics

Fault quantification and prognostics are currently at a nascent phase. There have been attempts in data-driven as well as model-based approaches to determine the damage accumulation, but until now, there is no single standard approach that has been reliably proven to perform consistently well, owing to the complexity of the problem as well as the time required to simulate such failures. Especially for OWFs, it has been noted that corrosion and associated fatigue wear play a significant role in equipment failure not only for the structure but also for the nacelle components [63]. These problems have to be addressed by prognostics in order to accurately predict the RULs for the CBM candidates and aid in effective maintenance planning. Also, capturing the effects of external factors such as a maintenance task performed or a lubrication task on the RUL predictions is still an open question. Besides, reliability of the developed prognostics methods is to be assessed before incorporating the results into maintenance decision making process. Metrics for qualifying prognostics methods such as described by Saxena *et al.* [64] are required for accuracy assurances.

6.4.5 Focus on balance-of-system

As it has been discussed earlier, the current CM systems are predominantly focused on drive train, main gearbox, generator and blades. There are also other modules of the WT that have been reported to have significant failure rates and down times such as yaw and pitch systems, electronics components, etc. These present additional challenges for diagnostics as they operate slowly and intermittently [65]. RCM strategy is a comprehensive approach to maintenance and during its application each module of the system should be carefully examined. The BOS candidates are to be evaluated following CBM candidate selection procedure as described in Section 6.2.1.1, and suitable maintenance strategy should be finalized in the RCM framework.

6.5 Conclusion

An outline of health management for OWFs has been detailed in this chapter with description of various important elements. The need for such farm level management is explained and benefits are discussed. Key gaps to be filled in order to realize such a system are identified. The proposed health management system is mainly based on the existing knowledge of fleet-level management in the aerospace sector. Health management is much broader than CM; there are a number of aspects beyond the prognostics capabilities that are to be designed in order to arrive at a comprehensive maintenance management scheme. A comprehensive maintenance program that is sensitive to the health of the assets and adapts maintenance schedule accordingly, depending upon resource availability, logistics and inventory, is key to cost optimization while ensuring reliability and availability. The advances in CM and diagnostics in wind energy are in the right direction, and many of them are building blocks for health management. Offshore wind faces a number of unique challenges that can be satisfactorily addressed by following a suitable systematic approach. RCM implementation appears to be the most suitable as it encompasses other maintenance strategies and is suitable for farm-level deployment.

Although development of a health management system is a major exercise involving multiple stakeholders and spanning across departments, it appears to be well worth the time and investment for the large OWFs in the future.

Acknowledgment

This work has been funded by Norwegian Centre for Offshore Wind Energy (NORCOWE) under grant no. 193821/S60 from Research Council of Norway (RCN). NORCOWE is a consortium with partners from industry and science, hosted by Christian Michelsen Research.

References

[1] J. Wilkes, I. Pineda, and G. Corbetta, "Wind energy scenarios for 2020," *A Report by European Wind Energy Association (EWEA)*, Jul. 2014.

[2] P. J. Tavner, H. Long, and Y. Feng, "Early experiences with UK round 1 offshore wind farms," *Proceedings of the ICE – Energy*, vol. 163, pp. 167–181, Nov. 2010.

[3] G. van Bussel and A. R. Henderson, "State of the art and technology trends for offshore wind energy: operation and maintenance issues," in *[Online] Delft University of Technology, European Commission*, 2001.

[4] L. Rademakers, H. Braam, and T. Verbruggen, "R&D needs for O&M of wind turbines," *Energy Center Netherlands (ECN) – Wind Energy, Report (RX-03-045)*, 2003.

[5] T. W. Verbruggen, "Wind turbine operation and maintenance based on condition monitoring," *ECN – Conmow Final Report*, 2003.

[6] J. B. Gayo, "Final publishable summary of results of project ReliaWind," in *Reliability-Focused Research on Optimizing Wind Energy System Design, Operation and Maintenance: Tools, Proof of Concepts, Guidelines and Methodologies for a New Generation*, FP7-Energy-2007-1-RTD, 2011.

[7] Supergen-Wind, UK. available at http://www.supergen-wind.org.uk/index.html."

[8] J. Ribrant and L. Bertling, "Survey of failures in wind power systems with focus on Swedish wind power plants during 1997–2005," in *Power Engineering Society General Meeting, 2007. IEEE*, pp. 1–8, 2007.

[9] B. Hahn, M. Durstewitz, and K. Rohrig, “Reliability of Wind Turbines,” in *Wind Energy* (J. Peinke, P. Schaumann, and S. Barth, eds.), ch. 62, pp. 329–332, Berlin, Heidelberg: Springer, 2007.

[10] P. Tavner, *Offshore Wind Turbines: Reliability, Availability and Maintenance. IET Power and Energy Series*. London: Institution of Engineering and Technology, 2012.

[11] B. Hahn, S. Faulstich, and V. Berkhout, “Well, How are They Running?,” in *Sea – Wind – Power*, pp. 159–167, Berlin, Heidelberg: Springer, 2017.

[12] D. Zappalá. C. J. Crabtree, and P. J. Tavner, “Survey of commercially available condition monitoring systems for wind turbines,” *Supergen Wind Report*, 2014.

[13] G. Van Bussel, “Offshore wind energy, the reliability dilemma,” in *Proceedings of the International Conference*, pp. 2–6, 2002.

[14] J. Nilsson and L. Bertling, “Maintenance management of wind power systems using condition monitoring systems—life cycle cost analysis for two case studies,” *IEEE Transactions on Energy Conversion*, vol. 22, pp. 223–229, 2007.

[15] “Integrated Product Support Element Guidebook,” *Defense Acquisition University, Department of Defense (US)*, pp. 265–301, Dec. 2011.

[16] L. Rademakers, H. Braam, T. Obdam, P. Frohbose, and N. Kruse, “Tools for estimating operation and maintenance costs of offshore wind farms: state of the art,” in *Proceedings of European Wind Energy Conference (EWEC)*, 2008.

[17] F. P. García Márquez, A. M. Tobias, J. M. Pinar Pérez, and M. Papaelias, “Condition monitoring of wind turbines: techniques and methods,” *Renewable Energy*, vol. 46, pp. 169–178, 2012.

[18] P. Takoutsing, R. Wamkeue, M. Ouhrouche, F. Slaoui-Hasnaoui, T. Tameghe, and G. Ekemb, “Wind turbine condition monitoring: state-of-the-art review, new trends, and future challenges,” *Energies*, vol. 7, no. 4, pp. 2595–2630, 2014.

[19] S. T. Kandukuri, K. G. Robbersmyr, and H. R. Karimi, “Towards farm-level health management of offshore wind farms for maintenance improvements,” *International Journal of Advanced Manufacturing Technology*, vol. 83, no. 9–12, 2016.

[20] M. Lebold and M. Thurston, “Open standards for condition-based maintenance and prognostic systems,” in *Maintenance and Reliability Conference (MARCON)*, pp. 6–9, May 2001.

[21] A. Heng, S. Zhang, A. C. C. Tan, and J. Mathew, “Rotating machinery prognostics: state of the art, challenges and opportunities,” *Mechanical Systems and Signal Processing*, vol. 23, no. 3, pp. 724–739, 2009.

[22] SAE, *JA-1012 A Guide to the Reliability-Centered Maintenance (RCM) Standard*. Warrendale, PA: SAE International, 2002 (Revised in 2011).

[23] A. May and D. McMillan, “Condition based maintenance for offshore wind turbines: the effects of false alarms from condition monitoring systems,” in *European Safety and Reliability Conference (ESREL)*, Sep. 2013.

[24] F. S. Nowlan and H. F. Heap, "Reliability-centered maintenance," *Tech. Rep., DTIC Document*, 1978.
[25] SAE, *JA1011 – Evaluation Criteria for Reliability-Centered Maintenance (RCM) Processes*. Warrendale, PA: SAE International, 1999 (Revised in 2009).
[26] NAVAIR 00-25-403, *Guide for the Naval Aviation Reliability-Centered Maintenance Process*, Naval Air Systems Command, U.S., 2005.
[27] A. M. Smith and G. R. Hinchcliffe, *RCM – Gateway to World Class Maintenance*. Butterworth – Heinemann, 1 ed., Nov. 2003.
[28] F. Besnard, K. Fischer, and L. Bertling, "Reliability-centred asset maintenance: a step towards enhanced reliability, availability, and profitability of wind power plants," in *Innovative Smart Grid Technologies Conference Europe (ISGT Europe), 2010 IEEE PES*, pp. 1–8, Oct. 2010.
[29] J. Igba, K. Alemzadeh, I. Anyanwu-Ebo, P. Gibbons, and J. Friis, "A systems approach towards reliability-centred maintenance (RCM) of wind turbines," *Procedia Computer Science*, vol. 16, no. 0, pp. 814–823, 2013. 2013 Conference on Systems Engineering Research.
[30] D. McMillan and G. W. Ault, *Towards Reliability Centered Maintenance of Wind Turbines*, pp. 183–194. Reliable and Sustainable Electric Power and Energy Systems Management, Springer India, 2014.
[31] D. Baglee and M. Knowles, *Developing RCM Strategy for Wind Turbines Utilizing Online Condition E-Monitoring*, pp. 11–20. Lecture Notes in Mechanical Engineering, London: Springer, 2014.
[32] P. Dehghanian, M. Fotuhi-Firuzabad, S. Bagheri-Shouraki, and A. A. Razi Kazemi, "Critical Component Identification in Reliability Centered Asset Management of Power Distribution Systems Via Fuzzy AHP," *IEEE Systems Journal*, vol. 6, pp. 593–602, Dec. 2012.
[33] P. Dehghanian, M. Fotuhi-Firuzabad, F. Aminifar, and R. Billinton, "A comprehensive scheme for reliability centered maintenance in power distribution systems – Part I: methodology," *IEEE Transactions on Power Delivery*, vol. 28, pp. 761–770, Apr. 2013.
[34] P. Dehghanian, M. Fotuhi-Firuzabad, F. Aminifar, and R. Billinton, "A comprehensive scheme for reliability-centered maintenance in power distribution systems – Part II: numerical analysis," *IEEE Transactions on Power Delivery*, vol. 28, pp. 771–778, Apr. 2013.
[35] J. Moubray, *The Case Against Streamlined RCM*, Aladon, UK, 2000.
[36] "IEC-61400-25: Wind turbine generator systems part 25: communications for monitoring and control of wind power plants," in *International Electrotechnical Commission (IEC)*. London: International Electrotechnical Commission (IEC), 2006.
[37] A. K. Olsen, B. Osdil, B. Poulsen, and K. O. H. Pedersen, "Prototype of generic server for wind power plants using IEC-61400-25 standard," *International Journal of Distributed Energy Resources*, vol. 3, no. 4, pp. 273–290, 2007.
[38] T. H. Nguyen, A. Prinz, T. Friisø, R. Nossum, and I. Tyapin, "A framework for data integration of offshore wind farms," *Renewable Energy*, vol. 60, pp. 150–161, Dec. 2013.

[39] S. Faulstich, P. Lyding, B. Hahn, and D. Brune, "A collaborative reliability database for maintenance optimization," in *European Wind Energy Conference, Warsaw*, 2010.

[40] L. Petersson, J.-O. Andersson, C. Orbert, and S. Skagermann, "RAMS – database for wind turbines," *Elforsk Report*, vol. 10, p. 67, 2010.

[41] S. Faulstich, P. Lyding, and B. Hahn, "Electrical subassemblies of wind turbines – a substantial risk for the availability," in *European Wind Energy Conference, Warsaw*, 2010.

[42] P. P. Adhikari, D. Makhecha, and M. Buderath, "A certifiable approach towards integrated solution for aircraft readiness management," in *European Conference of The Prognostics and Health Management Society*, 2014.

[43] F. Besnard, K. Fischer, and L. B. Tjernberg, "A model for the optimization of the maintenance support organization for offshore wind farms," *IEEE Transactions on Sustainable Energy*, vol. 4, pp. 443–450, Apr. 2013.

[44] E. E. Halvorsen-Weare, C. Gundegjerde, I. B. Halvorsen, L. M. Hvattum, and L. M. Nonå s, "Vessel fleet analysis for maintenance operations at offshore wind farms," *Energy Procedia*, vol. 35, pp. 167–176, 2013.

[45] A. Arvesen, C. Birkeland, and E. G. Hertwich, "The importance of ships and spare parts in LCAs of offshore wind power," *Environmental Science & Technology*, vol. 47, pp. 2948–2956, Mar. 2013.

[46] M. Shafiee, "Maintenance logistics organization for offshore wind energy: current progress and future perspectives," *Renewable Energy*, vol. 77, pp. 182–193, May 2015.

[47] M. O'Connor, T. Lewis, and G. Dalton, "Weather window analysis of Irish west coast wave data with relevance to operations & maintenance of marine renewables," *Renewable Energy*, vol. 52, pp. 57–66, Apr. 2013.

[48] J. D. Sorensen, "Framework for risk-based planning of operation and maintenance for offshore wind turbines," *Wind Energy*, vol. 12, pp. 493–506, Jul. 2009.

[49] J. J. Nielsen and J. D. Sorensen, "On risk-based operation and maintenance of offshore wind turbine components," *Reliability Engineering & System Safety*, vol. 96, pp. 218–229, Jan. 2011.

[50] F. Ding and Z. Tian, "Opportunistic maintenance for wind farms considering multi-level imperfect maintenance thresholds," *Renewable Energy*, vol. 45, pp. 175–182, Sep. 2012.

[51] P. P. Tambe, S. Mohite, and M. S. Kulkarni, "Optimisation of opportunistic maintenance of a multi-component system considering the effect of failures on quality and production schedule: a case study," *The International Journal of Advanced Manufacturing Technology*, vol. 69, pp. 1743–1756, Jul. 2013.

[52] T. Halim and L.-C. Tang, "Confidence interval for optimal preventive maintenance interval and its applications in maintenance planning," *The International Journal of Advanced Manufacturing Technology*, vol. 40, pp. 203–213, Dec. 2007.

[53] P. R. v.d. Pieterman, L. W. M. M. Rademakers, H. Braam, and T. S. Obdam, "Operation and maintenance cost estimator (OMCE) to estimate the future

O&M costs of offshore wind farms," in *European Wind Energy Conference*, Energy Research Center Netherlands, 2008.

[54] M. Hofmann and I. B. Sperstad, "NOWIcob – a tool for reducing the maintenance costs of offshore wind farms," *Energy Procedia*, vol. 35, pp. 177–186, 2013.

[55] G. Giebel, O. Gehrke, M. Mcgugan, and K. Borum, "Common access to wind turbine data for condition monitoring the IEC 61400-25 family of standards," in *Proceedings of the 27th Riso International Symposium on Materials Science*, 2006.

[56] S. Srinivas, W. Musial, B. Bailey, and M. Filippelli, "Assessment of offshore wind system design, safety and operation standards (nrel/tp-5000-60573)," *Tech. Rep., National Renewable Energy Laboratory (NREL)*, 2014.

[57] *DNV-OS-J103- Design of Floating Wind Turbine Structures*, Høvik, Norway: Det Norske Veritas AS, 2013.

[58] *GL – Guideline for Certification of Offshore Wind Turbines*, Hamburg, Germany: Germanischer Lloyd, 2012.

[59] *ARP-6290 – Guideline for the Development of Architectures for Integrated Vehicle Health Management Systems*, Warrendale, PA: Society of Automotive Engineers (SAE), 2013.

[60] W. Q. Meeker and Y. Hong, "Reliability meets big data: opportunities and challenges," *Quality Engineering*, vol. 26, pp. 102–116, Jan. 2014.

[61] J. Helsen, G. D. Sitter, and P. J. Jordaens, "Long-term monitoring of wind farms using big data approach," in *2016 IEEE Second International Conference on Big Data Computing Service and Applications (BigDataService)*, pp. 265–268, IEEE, Mar. 2016.

[62] A. P. Marugán, F. P. G. Márquez, and M. Papaelias, *Multivariable Analysis for Advanced Analytics of Wind Turbine Management*, pp. 319–328, Singapore: Springer, 2017.

[63] R. J. Wood, "Tribology and corrosion aspects of wind turbines," *Wind Energy – Challenges for Materials, Mechanics and Surface Science*, London: IOP, 2010.

[64] A. Saxena, J. Celaya, E. Balaban, *et al.*, "Metrics for evaluating performance of prognostic techniques," in *2008 International Conference on Prognostics and Health Management*, pp. 1–17, IEEE, Oct. 2008.

[65] S. T. Kandukuri, A. Klausen, H. R. Karimi, and K. G. Robbersmyr, "A review of diagnostics and prognostics of low-speed machinery towards wind turbine farm-level health management," *Renewable and Sustainable Energy Reviews*, vol. 53, pp. 697–708, Jan. 2016.

Chapter 7
Health monitoring of wind turbine: data-based approaches

Guang Wang[1], *Shen Yin*[2], *and Hamid Reza Karimi*[3]

7.1 Introduction

During the past decade, the use of wind energy as a clean and renewable resource has attracted global interest. A wind turbine is an aeroelastic mechanical system which converts kinetic wind energy to electrical power. On the other hand, wind turbines still suffer from potential problems, such as vibration, corrosion and temperature changes, e.g., in the motors, sensors, blades and gearbox, which could affect the production capability and may cause remarkable downtime of the entire system. The maintenance of faulty wind turbines is costly and even dangerous since many wind turbines are installed offshore. In order to detect abnormalities in the systems as early as possible to avoid potential irreversible damage, automatic detection of the faults is highly desirable from the application point of view, which motivates us to design a fault detection system for wind turbines.

In recent years, several fault detection systems have been successfully developed for wind turbines. Most of the proposed approaches rely on the physical model of wind turbines [1–10] and based on which, the well-established model-based fault diagnosis techniques [11–17] can be directly applied. However, modeling of a wind turbine is a difficult task in practice [18,19], which still limits the application of model-based approaches. Parallel to the research of model-based fault detection techniques, the data-driven methods are currently receiving considerable attention [20–24]. Different from model-based approaches which require the mathematical model of system known as a priori, data-driven methods only depend on the measured process data. Several basic data-driven methods, such as principle component analysis (PCA), dynamic PCA, independent component analysis, partial least squares and subspace aided approach (SAP), have been well developed [25–30]. SAP is mainly based on

[1]College of Engineering, Bohai University, China
[2]Research Institute of Intelligent Control and Systems, Harbin Institute of Technology, China
[3]Department of Mechanical Engineering, Politecnico di Milano, Italy

identifying the primary form of residual generators directly from the measured data. Residual generator is an important concept in the model-based fault detection framework. The basic idea of a residual generator is to generate a residual between the actual output and the estimated output. If noise is not considered, the generated residual should be nonzero (zero) when there is (no) fault. However, in real applications, the generated residual should be further evaluated, including threshold computation and decision making, to make a correct final decision. Several popular subspace-based fault detection approaches can be found in [31–34].

In order to promote the fault detection and other related technologies for wind turbines, Odgaard *et al.* developed a wind turbine simulator in [35] as a research competition for all the participants. In this framework, many effective designs have been proposed based on the benchmark physical model [36–38]. Recently, to improve the level of the wind turbine benchmark close to the actual statue, Odgaard *et al.* renewed the benchmark by added new challenges. Compared to the previous one, the latest benchmark is modified in many different ways [39–41]. First of all, the wind turbine model is more sophisticated and realistic, which may help to achieve better simulation abilities and make the problem more realistic. Another difference is that various wind inputs are introduced into this model. An IEC [42] von Karman turbulence model is used to generate three types of wind input files, of which the mean wind speeds at the 90-m hub height are 11, 14 and 17 m/s [43]. These turbulent wind input files can be directly applied in the latest benchmark. Moreover, extra relevant details are implanted into the fault scenarios. As a result, more sophisticated advanced fault detection techniques are required from both academic and practical aspects.

Since it is quite difficult to obtain a precise mathematical model of the wind turbine, the data-driven approaches seem more convenient from the application point of view. On the other hand, seen from a control theory perspective, a major problem in the wind turbine control system is that the wind turbine is driven by a disturbance, the wind. The wind speed is, however, measured with a large additive measurement noise, as well as a large risk of an offset [43]. As the wind speed can be treated as a disturbance input, robust residual generators should be constructed directly from the available process measurements, and most importantly, these residual generators shall be sensitive to output faults and insensitive to input disturbance.

For this purpose, a robust data-driven fault detection scheme is proposed for wind turbines. A schematic of the fault detection scheme is shown in Figure 7.1, which contains two steps, i.e., (a) residual generation and (b) residual evaluation including threshold computation and decision making. In the first step, a robust residual vector instead of a single residual signal is generated under a given performance index and an optimization criterion. In the second step, a proper evaluation approach as well as a suitable decision logic is given to make a correct final decision.

The rest of the chapter is organized as follows. Section 7.2 introduces the new wind turbine benchmark and the fault scenarios. Section 7.3 develops a robust fault detection scheme based on constructing robust residual generators directly from available process data. The proposed robust fault detection scheme is applied to the benchmark in Section 7.4. Finally, the chapter ends with the conclusion in Section 7.5.

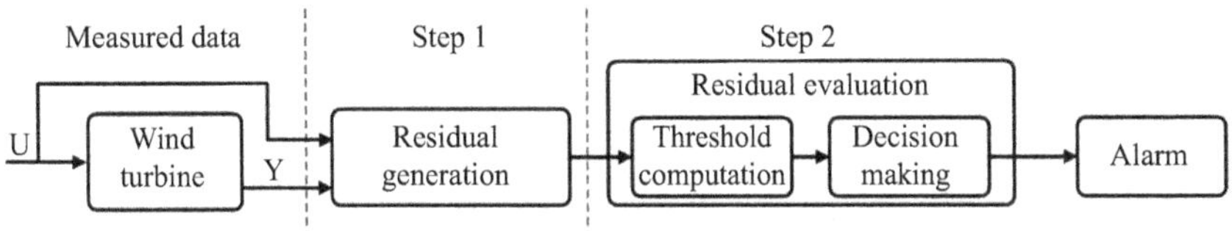

Figure 7.1 Block diagram of standard residual generation and decision logic structure

7.2 Benchmark system and faults description

7.2.1 Benchmark model

The wind turbine benchmark proposed by Odgaard *et al.* in [43] is described at the system level. It mainly consists of five subsystems: blade and pitch system, drive train system, generator and converter system, controller system and sensors. The benchmark simulates a three-bladed horizontal axis and variable speed wind turbine containing a pitch, torque controller and a yaw controller. It is disturbed by unknown wind disturbances and controlled in closed-loop with PI controllers. Combined with an IEC von Karman turbulence wind model and deliberately designed faults, the benchmark is complex and realistic. It is suitable for testing different detection and isolation schemes on the wind turbine. FAST [43] is an aeroelastic wind turbine simulator developed by NREL, and it is now widely used for studying the wind turbine control systems. There are several FAST models of varying types and sizes that can be available in the public sphere. In the benchmark, the NREL's 5 MW offshore baseline turbine is selected as the wind turbine model. The benchmark simulates a three-bladed, horizontal-axial, and variable-speed wind turbine containing pitch, torque controller and yaw controller. It is disturbed by unknown wind speed and controlled in closed-loop to follow the power reference. Figure 7.2 shows the block diagram of the benchmark.

Most components of the benchmark are implemented within the Simulink environment, in which 15 sensors are available for measuring the input and output variables, i.e., wind speed $v_{h,m}$, generator speed $w_{g,m}$, generator torque $\tau_{g,m}$, rotor speed $w_{r,m}$, generated electrical power $p_{g,m}$, pitch angle of ith blade $\beta_{i,m}$, blade root moment of ith blade $M_{i,m}$, azimuth angle at low speed side ϕ_m, tower top acceleration $\ddot{x}_m, \ddot{y}_m$ in the x and y directions and the yaw error Ξ_m. All of these sensors are modeled by adding a band-limited white noise. The detailed descriptions of sensors and noise power are summarized in Table 7.1.

7.2.2 Fault scenarios

In the benchmark model, ten sensor and actuator faults are considered along the time span of 630 s. Faults 1–6 are sensor faults, including measurements that are stuck, scaled or offset from the true values. Faults 7–10 are actuator faults, including Faults 7 and 8 in pitch actuators, Fault 9 in generator torque and Fault 10 in yaw actuator.

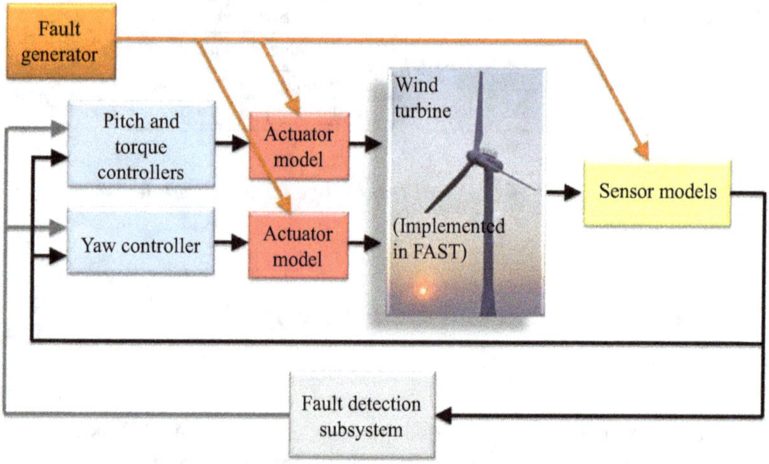

Figure 7.2 Block diagram of the wind turbines benchmark

Table 7.1 Available sensors and the added noise power [43]

Sensor type	Symbol	Unit	Noise power
Wind speed at hub height	$v_{h,m}$	m/s	0.0071
Rotor speed	$w_{r,m}$	rad/s	10^{-4}
Generator speed	$w_{g,m}$	rad/s	$2{\cdot}10^{-4}$
Generator torque	$\tau_{g,m}$	N m	0.9
Generated electrical power	$p_{g,m}$	W	10
Pith angle of *i*th blade	$\beta_{i,m}$	deg	$1.5{\cdot}10^{-3}$
Azimuth angle low speed side	ϕ_m	rad	10^{-4}
Blade root moment of *i*th blade	$M_{i,m}$	N m	10^3
Tower top acceleration in *x* direction	$\ddot{x}_m$	m/s^2	$5{\cdot}10^{-4}$
Tower top acceleration in *y* direction	$\ddot{y}_m$	m/s^2	$5{\cdot}10^{-4}$
Yaw error	Ξ_m	deg	$5{\cdot}10^{-2}$

Faults 7 and 8 are modeled by changing the parameters in the relevant pitch actuator model. Fault 9 is modeled by adding an offset on the generated generator torque and Fault 10 is modeled by setting the yaw angular velocity to zero rad/s. Detailed descriptions of these faults and their durations are summarized in Table 7.2.

7.3 Robust data-driven fault detection design

The robust fault detection scheme will be presented in this section. Based on the method proposed by Ding *et al.* [31], we first identify the parity space directly from the

Table 7.2 Fault scenarios [43]

No.	Fault	Type	Detail	Durations (s)
1	Blade root bending moment sensor	Scaling	$M_{2,m}$ scaled by 0.95	20–45
2	Accelerometer	Offset	$-0.5\,\mathrm{m/s^2}$ offset on $\ddot{x}_m$ and $\ddot{y}_m$	75–100
3	Generator speed sensor	Scaling	$w_{g,m}$ scaled by 0.95	130–155
4	Pitch angle sensor	Stuck	$\beta_{1,m}$ holds to 1 deg	185–210
5	Generator power sensor	Scaling	$P_{g,m}$ scaled by 1.1	240–265
6	Low speed shaft position encoder	Bit error	random offset on ϕ_m	295–320
7	Pitch actuator	Abrupt change in dynamics	$\omega_{n,2} = 5.73$, $\zeta_{n,2} = 0.45$	350–410
8	Pitch actuator	Slow change in dynamics	$\omega_{n,2} = 3.42$, $\zeta_{n,2} = 0.9$	440–465
9	Torque offset	Offset	1,000 N m offset on $\tau_{g,m}$	495–520
10	Yaw drive	Stuck drive	yaw angular velocity set to 0 rad/s	550–575

measured data. Then, we select the optimal parity vectors under a given performance index as well as an optimization criterion. It follows the construction of robust residual generators using the observer-based residual generation technique. At last, the robust fault detection scheme is summarized into an algorithm.

7.3.1 Identify parity space directly from measured data

A linear time-invariant (LTI) system is usually modeled in the following discrete-time state-space form:

$$x(k+1) = Ax(k) + Bu(k) + w(k) \tag{7.1}$$

$$y(k) = Cx(k) + Du(k) + v(k) \tag{7.2}$$

where $x \in \mathbf{R}^n$ is the vector of the state variables, $u \in \mathbf{R}^l$ is the input signal and $y \in \mathbf{R}^m$ is the measured output signal. $w \in \mathbf{R}^n$ and $v \in \mathbf{R}^m$ are process noise and measurement noise, respectively. As standard assumptions, w and v are zero-mean and normal distributed white noise, and they are independent of the input vector u and the initial state vector $x(0)$. We define the following block Hankel matrices for outputs:

$$Y_p = \begin{bmatrix} y(k-s) & y(k-s+1) & \cdots & y(k-s+N-1) \\ y(k-s+1) & y(k-s+2) & \cdots & y(k-s+N) \\ \vdots & \vdots & \vdots & \vdots \\ y(k) & y(k+1) & \cdots & y(k+N-1) \end{bmatrix}$$

$$Y_f = \begin{bmatrix} y(k+1) & y(k+2) & \cdots & y(k+N) \\ y(k+2) & y(k+3) & \cdots & y(k+N+1) \\ \vdots & \vdots & \vdots & \vdots \\ y(k+s+1) & y(k+s+2) & \cdots & y(k+s+N) \end{bmatrix}$$

where s and N are parameters that defined by users. Usually, s is selected with $s \geq n$ and N is chosen large enough. The matrices U_p, U_f, W_f and V_f have similar Hankel structures with Y_p and Y_f.

By iterating (7.1) and (7.2), we get an extended state-space model as follows:

$$Y_f = \Gamma_s X + H_s^u U_f + H_s^w W_f + V_f \tag{7.3}$$

where

$$\Gamma_s = \begin{bmatrix} C \\ CA \\ \vdots \\ CA^s \end{bmatrix}$$

$$H_s^u = \begin{bmatrix} D & 0 & \cdots & 0 \\ CB & D & \cdots & 0 \\ \vdots & \ddots & \ddots & \vdots \\ CA^{s-1}B & \cdots & CB & D \end{bmatrix}$$

$$H_s^w = \begin{bmatrix} 0 & 0 & \cdots & 0 \\ C & 0 & \cdots & 0 \\ \vdots & \ddots & \ddots & \vdots \\ CA^{s-1} & \cdots & C & 0 \end{bmatrix}$$

$$X = \begin{bmatrix} x(k+1) & x(k+2) & \cdots & x(k+N) \end{bmatrix}$$

A parity-space-based residual generator is expressed as follows:

$$r(k) = \alpha_s(y_s(k) - H_s^u u_s(k)), \ \alpha_s \in P_s \tag{7.4}$$

where

$$P_s = \{\alpha_s | \alpha_s \Gamma_s = 0\} \tag{7.5}$$

$$y_s(k) = \begin{bmatrix} y(k-s) \\ y(k-s+1) \\ \vdots \\ y(k) \end{bmatrix} \tag{7.6}$$

$$u_s(k) = \begin{bmatrix} u(k-s) \\ u(k-s+1) \\ \vdots \\ u(k) \end{bmatrix} \tag{7.7}$$

P_s and α_s are called the parity space and the parity vector, respectively. $r(k)$ is a residual signal which equals to zero when there is no fault.

If we denote $\Gamma_s^\perp$ as the left null space of Γ_s, it is clear that $\alpha_s \in \Gamma_s^\perp$. Thus, the construction of a parity-space-based residual generator can be achieved by identification of $\Gamma_s^\perp$ and $\Gamma_s^\perp H_s^u$ from measured data. For this purpose, we have the following identification process:

Let

$$Z_p = \begin{bmatrix} Y_p \\ U_p \end{bmatrix}, \quad Z_f = \begin{bmatrix} Y_f \\ U_f \end{bmatrix}$$

From (7.3), we have

$$\Gamma_s^\perp Y_f = \Gamma_s^\perp H_s^u U_f + \Gamma_s^\perp (H_s^w W_f + V_f) \tag{7.8}$$

By moving the input term to left-hand side and replacing the noise term by E_f, (7.8) becomes

$$\Gamma_s^\perp \left[I, \ -H_s^u \right] Z_f = \Gamma_s^\perp E_f \tag{7.9}$$

With Z_p^T as the instrumental variables [44], the following equation holds

$$\lim_{N\to\infty} \frac{1}{N} \Gamma_s^\perp \left[I, \ -H_s^u \right] Z_f Z_p^T = 0 \tag{7.10}$$

Thus, $\Gamma_s^\perp [I, -H_s^u]$ belongs to the left null space of $(1/N) Z_f Z_p^T |_{N\to\infty}$. Performing singular value decomposition on $(1/N) Z_f Z_p^T$ gives

$$\frac{1}{N} Z_f Z_p^T = U_z \begin{bmatrix} \Sigma_{z,1} & 0 \\ 0 & \Sigma_{z,2} \end{bmatrix} V_z, \quad \Sigma_{z,2} \simeq 0 \tag{7.11}$$

where

$$U_z = \begin{bmatrix} U_{z,11} & U_{z,12} \\ U_{z,21} & U_{z,22} \end{bmatrix} \in \mathbf{R}^{(l+m)s\times(l+m)s}$$

and

$$U_{z,12} \in \mathbf{R}^{s\times(sm-n)}, \quad U_{z,22} \in \mathbf{R}^{ls\times(sm-n)}$$

It is well known that $[U_{z,12}^T, U_{z,22}^T]$ also belongs to the left null space of $(1/N) Z_f Z_p^T$. In [31], $\Gamma_s^\perp [I, -H_s^u]$ is equal to $[U_{z,12}^T, U_{z,22}^T]$ that

$$\Gamma_s^\perp = U_{z,12}^T, \quad \Gamma_s^\perp H_s^u = -U_{z,22}^T \tag{7.12}$$

To make sure $\Gamma_s^\perp$ and $\Gamma_s^\perp H_s^u$ are available, the following input excitation condition should be satisfied [45]:

$$rank\left(\frac{1}{N} \begin{bmatrix} X \\ U_f \end{bmatrix} Z_p^T \right) = n + sl \tag{7.13}$$

The observer-based residual generator (also called diagnostic observer) is another well-established technique. It is in the form of the following equations:

$$z(k+1) = A_z z(k) + B_z u(k) + Ly(k) \tag{7.14}$$

$$r(k) = gy(k) - c_z z(k) - d_z u(k) \tag{7.15}$$

where A_z, B_z, c_z, g_z and L together with the transformation matrix T can be calculated from the following Luenberger equations:

$$TA - A_z = LC, \quad c_z T = gC, \quad B_z = TB - LD, \quad dz = gD \tag{7.16}$$

However, the parity-space-based generator and the observer-based residual generator are equivalent [13]. Consider a parity vector $\alpha_s = [\alpha_{s,0}, \alpha_{s,1}, \ldots, \alpha_{s,s}]^T \in \mathbf{R}^{(s+1)m}$, with $\alpha_{s,i} \in \mathbf{R}^m$ $(i = 1, 2, \ldots, s)$, then the following equations hold:

$$A_z = \begin{bmatrix} 0 & 0 & \cdots & 0 \\ 1 & 0 & \cdots & 0 \\ \vdots & \ddots & \ddots & \vdots \\ 0 & \cdots & 1 & 0 \end{bmatrix}, \quad L = -\begin{bmatrix} \alpha_{s,0} \\ \alpha_{s,1} \\ \vdots \\ \alpha_{s,s} \end{bmatrix} \tag{7.17}$$

$$c_z = \begin{bmatrix} 0 & \cdots & 0 & 1 \end{bmatrix}, \quad g = \alpha_{s,s} \tag{7.18}$$

and B_z, d_z in terms of $\alpha_s H_s^u$:

$$B_z = \begin{bmatrix} \alpha_s H_{s,0}^u \\ \alpha_s H_{s,1}^u \\ \vdots \\ \alpha_s H_{s,s-1}^u \end{bmatrix}, \quad d_z = \alpha_s H_{s,s}^u \tag{7.19}$$

where

$$H_s^u = \begin{bmatrix} H_{s,0}^u & H_{s,1}^u & \cdots & H_{s,s}^u \end{bmatrix}$$

$$H_{s,i}^u = \begin{bmatrix} 0 \\ \vdots \\ 0 \\ D \\ CB \\ \vdots \\ CA^{s-i-1}B \end{bmatrix}, \quad H_{s,s}^u = \begin{bmatrix} 0 \\ \vdots \\ 0 \\ D \end{bmatrix}$$

for $i = 0, 1, \ldots, s-1$.

With the identification of $\Gamma_s^\perp$ and $\Gamma_s^\perp H_s^u$ directly from the measured data, a parity vector $\alpha_s \in \Gamma_s^\perp$ and the corresponding vector $\alpha_s H_s^u \in \Gamma_s^\perp H_s^u$ can be selected to construct a residual generator either in the form of (7.4) or in the forms of (7.14) and (7.15). In the next section, we will discuss the optimal selections of α_s and $\alpha_s H_s^u$ under a given performance index, so that the construed residual generator could be nonsensitive to unknown disturbances and sensitive to sensors faults simultaneously.

7.3.2 Select optimal parity vector from parity space

In order to discuss the robustness against disturbances and sensitivity to faults, the following discrete-time LTI system is considered:

$$x(k+1) = Ax(k) + Bu(k) + E_d d(k) + E_f f(k) \tag{7.20}$$

$$y(k) = Cx(k) + Du(k) + F_d d(k) + F_f f(k) \tag{7.21}$$

where $d(k)$ and $f(k)$ represent disturbances and faults, respectively. By a similar iterative process to (7.1) and (7.2), we can obtain the following equation:

$$y_s(k) = \Gamma_s x(k-s) + H_s^u u_s(k) + H_s^f f_s(k) + H_s^d d_s(k) \tag{7.22}$$

where H_s^f and H_s^d have similar structures with H_s^w and H_s^u, respectively. $f_s(k)$ and $d_s(k)$ have similar structures with $y_s(k)$ and $u_s(k)$, respectively. From (7.4), (7.5) and (7.22), a residual generator can be recalculated as

$$r(k) = \alpha_s(H_s^d d_s(k) + H_s^f f_s(k)), \quad \alpha_s \in P_s \tag{7.23}$$

The main task of this section is to select an optimal α_s from the parity space P_s under a given performance index to make the residual generator to be nonsensitive to unknown disturbances while sensitive to sensors faults. For this purpose, the ratio of sensitivity to robustness proposed by Frank in [46] is used as the performance index.

$$J_{S/R} = \max_{\alpha_s} \frac{\alpha_s H_s^f (\alpha_s H_s^f)^T}{\alpha_s H_s^d (\alpha_s H_s^d)^T} \tag{7.24}$$

The solution of the maximization problem in (7.24) is as follows [13].

First of all, we perform a singular value decomposition on H_s^d:

$$H_s^d = U\Sigma V^T \tag{7.25}$$

with

$$UU^T = I_{\alpha\times\alpha}, \; VV^T = I_{\beta\times\beta} \tag{7.26}$$

$$\Sigma = \left[\, diag(\sigma_1, \ldots, \sigma_\alpha) \;\, 0_{\alpha\times(\beta-\alpha)} \,\right] \tag{7.27}$$

$$S = diag(\sigma_1, \ldots, \sigma_\alpha) \tag{7.28}$$

where σ_i $(i = 1, \ldots, \alpha)$ are nonzero singular values.

Suppose that $\alpha_{s,opt}$ is the optimal solution of (7.24), then it holds

$$\alpha_{s,opt}(J_{S/R} H_s^d (H_s^d)^T - H_s^f (H_s^f)^T)(\alpha_{s,opt})^T = 0.$$

Let

$$\alpha_{s,opt} = \bar{\alpha}_s S^{-1} U^T$$

we have

$$J_{S/R}\bar{\alpha}_s(\bar{\alpha}_s)^T - \bar{\alpha}_s S^{-1} U^T H_s^f (\bar{\alpha}_s S^{-1} U^T H_s^f)^T = 0.$$

Therefore, choosing the nominal eigenvector corresponding to the maximal eigenvalue of $S^{-1}U^T H_s^f (S^{-1}U^T H_s^f)^T$ as $\bar{\alpha}_s$, i.e.,

$$\bar{\alpha}_s(\lambda_{max} - S^{-1}U^T H_s^f (H_s^f)^T US^{-1}) = 0, \bar{\alpha}_s(\bar{\alpha}_s)^T = 1 \tag{7.29}$$

we can get the results

$$J_{S/R} = \max_{\alpha_s} \frac{\alpha_s H_s^f (\alpha_s H_s^f)^T}{\alpha_s H_s^d (\alpha_s H_s^d)^T} = \lambda_{max} \tag{7.30}$$

In [13], it has been proven that (7.29) can be brought into the following form:

$$v^*(\lambda^* H_s^d (H_s^d)^T - H_s^f (H_s^f)^T) = 0. \tag{7.31}$$

which indicates that the maximization problem in (7.24) comes down to a generalized eigenvalue–eigenvector problem given by (7.31). The value of the performance index is the maximal eigenvalue, and the corresponding eigenvector is the direction that used to select an optimal parity vector from the parity space.

In this chapter, we set $H_s^f = \Gamma_s^\perp$ and $H_s^d = \Gamma_s^\perp H_s^u$. Thus, the following performance index can be used:

$$J = \max_{\alpha_s} \frac{\alpha_s \Gamma_s^\perp (\alpha_s \Gamma_s^\perp)^T}{\alpha_s \Gamma_s^\perp H_s^u (\alpha_s \Gamma_s^\perp H_s^u)^T} \tag{7.32}$$

which is equivalent to solve the eigenvalue–eigenvector problem in the following equation:

$$v^*(\lambda^* \Gamma_s^\perp H_s^u (\Gamma_s^\perp H_s^u)^T - \Gamma_s^\perp (\Gamma_s^\perp)^T) = 0 \tag{7.33}$$

Assume that v_{max}^* is the eigenvector corresponding to the maximal eigenvalue λ_{max}^* of (7.33), the optimal selections for α_s and $\alpha_s H_s^u$ are

$$\alpha_s = v_{max}^* \Gamma_s^\perp, \quad \alpha_s H_s^u = v_{max}^* \Gamma_s^\perp H_s^u. \tag{7.34}$$

It follows then a robust residual generator is constructed using the above optimal α_s and $\alpha_s H_s^u$. However, a single residual signal usually cannot reveal the whole structure of the detection system which means there is a risk of missing some fault information. To overcome this potential problem, the generating of a residual vector, instead of a single residual signal, is the research hotspot in recent years [47].

7.3.3 Construct robust residual generators

In this section, the observer-based residual generator (diagnostic observer) is used to discuss the above problem. The main objective is to generate m linearly independent residual signals using m diagnostic observers under the constraint that the m diagnostic observers should span the whole n-dimensional state-space. Suppose that we have selected m parity vectors $\alpha_s^i = \begin{bmatrix} \alpha_{s,0}^i & \alpha_{s,1}^i & \cdots & \alpha_{s,s}^i \end{bmatrix}^T \in \mathbf{R}^{(s+1)m}, \alpha_s^i \in \Gamma_s^\perp$, $i = 1, 2, \dots, m$. Associated with α_s^i, m corresponding vectors $\alpha_s^i H_s^u$ are also selected from $\Gamma_s^\perp H_s^u$. Afterwards, m diagnostic observers are constructed as follows:

$$z^i(k+1) = A_z^i z^i(k) + B_z^i u(k) + L_z^i y(k) \tag{7.35}$$

$$r^i(k) = g^i y(k) - c_z^i z^i(k) - d_z^i u(k) \tag{7.36}$$

for $i = 1, 2, \dots, m$.

If the selected α_s^i $(i = 1, 2, \ldots, m)$ satisfy

$$rank = \begin{bmatrix} \alpha_{s,0}^1 \\ \vdots \\ \alpha_{s,s}^1 \\ \vdots \\ \alpha_{s,0}^m \\ \vdots \\ \alpha_{s,s}^m \end{bmatrix} = m \tag{7.37}$$

then the new output variables $z^i(k)$ $(i = 1, 2, \ldots, m)$ are linearly independent and there is no loss in system observability [47]. Moreover, we have

$$z(k) = \begin{bmatrix} z^1(k) \\ \vdots \\ z^m(k) \end{bmatrix} = T_s x(k)$$

$$T_s = \begin{bmatrix} T^1 \\ T^2 \\ \vdots \\ T^m \end{bmatrix}$$

$$rank(T_s) = n$$

where $T^1, T^2, \ldots, T^m$ are the transformation matrices [see (7.16)] respect to each diagnostic observer. Thereby, the m diagnostic observers span the whole n-dimensional state-space.

To generate a robust residual vector

$$r(k) = \begin{bmatrix} r^1(k) & r^2(k) & \cdots & r^m(k) \end{bmatrix} \tag{7.38}$$

we propose the following ***optimization criterion*** for selecting α_s^i $(i = 1, 2, \ldots, m)$:

α_s^i $(i = 1, 2, \ldots, m)$ correspond to the previous largest eigenvalues of (7.33) and satisfy the rank condition (7.37), simultaneously.

With the above criterion, m robust residual generators are constructed and a robust residual vector can be generated.

7.3.4 A designed robust fault detection scheme

A successful fault detection system consists of two steps as shown in Figure 7.1. The first step (i.e., residual generation) has been completed and a robust residual vector is generated. In the second step, the residual signals contained in the residual vector will be further evaluated before a final decision is made. It is worth noting that the evaluation method should be carefully chosen as it is very critical for a correct decision making. The standard generalized likelihood ratio (GLR) [11,48] is a proper evaluation method for most cases.

To the end, we summarize the main steps of the designed robust fault detection scheme into Algorithm 1.

Algorithm 1: Robust fault detection scheme

Step 1: Collect measured data and form $(1/N)Z_f Z_p$;
Step 2: Perform the singular value decomposition on $(1/N)Z_f Z_p$;
Step 3: Set $\Gamma_s^{\perp}$ and $\Gamma_s^{\perp} H_s^u$ as (7.12);
Step 4: Set the performance index as (7.32);
Step 5: Calculate the eigenvalues and the corresponding eigenvectors of (7.33);
Step 6: Select α_s^i and $\alpha_s^i H_s^u$ $(i = 1, 2, \ldots, m)$ following the optimization criterion described in Section 7.3.3;
Step 7: Generate a robust residual vector with (7.35), (7.36) and (7.38);
Step 8: Evaluate the residual signals using a proper evaluation method (e.g., GLR) and compare with the threshold;
Step 9: Make a final decision following an appropriate decision logic.

7.4 Benchmark simulation

In this section, the proposed fault detection scheme will be applied to the wind turbine benchmark. The benchmark is provided by a Simulink™ Model, in which 15 sensors are available for measuring the input and output data. As it is required, no other measurements should be assumed available and any modifications to the model are not allowed. As previously mentioned, ten sensor and actuator faults are designed along the time span of 630 s, all of which are expected to be detected. However, the benchmark is modeled in closed-loop with PI controllers which can be regarded as the fault tolerant control [49–53] to a certain extent. As a result, some of the faults may be quite difficult to be detected due to the measured data with no obvious abnormalities. This chapter only makes efforts on the detections of Faults 3–10, except Faults 1 and 2. Since we only try to detect the last eight of the ten faults, not all the available sensors are necessary for our detection system. Next, we will choose and determine the input and output variables of the system. In the Simulink Model, variables $w_{g,m}$ and Ξ_m are fed back to the input terminals, and they play a role on the controllers; therefore, these two variables are regarded as input variables. In addition, the wind speed $v_{h,m}$ is also treated as an input variable. The rest of the variables, like $w_{r,m}$, $w_{g,m}$, $\tau_{g,m}$, $\beta_{1,m}$, $\beta_{2,m}$, $\beta_{3,m}$, $\ddot{x}_m$, $\ddot{y}_m$, and $P_{g,m}$, are chosen as output variables. Then, the input $u(k)$ and output $y(k)$ are as follows:

$$u(k) = \begin{bmatrix} v_{h,m} \\ w_{g,m} \\ \Xi_m \end{bmatrix}, \quad y(k) = \begin{bmatrix} w_{r,m} \\ w_{g,m} \\ \tau_{g,m} \\ P_{g,m} \\ \beta_{1,m} \\ \beta_{2,m} \\ \beta_{3,m} \end{bmatrix}$$

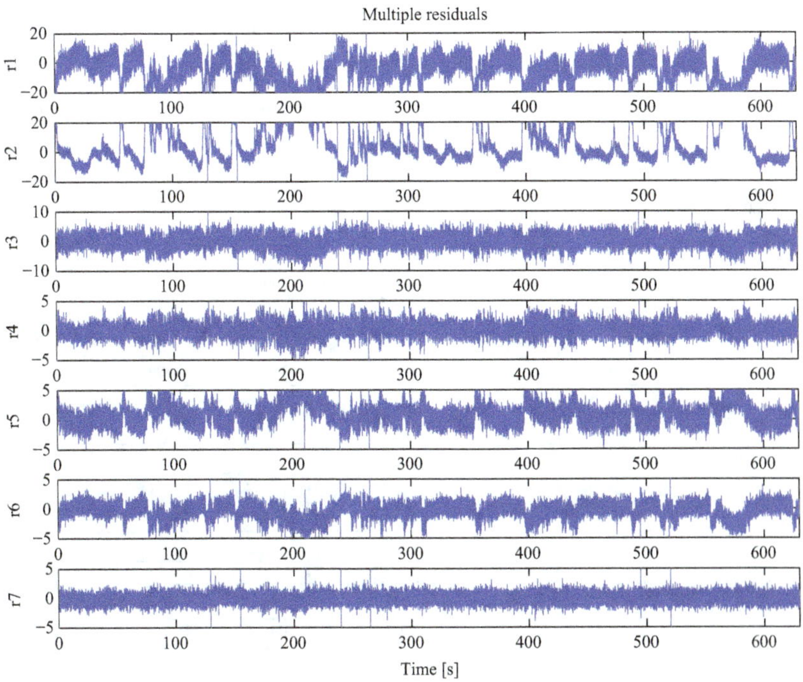

Figure 7.3 Residuals generation at wnd = 11 m/s: general approach

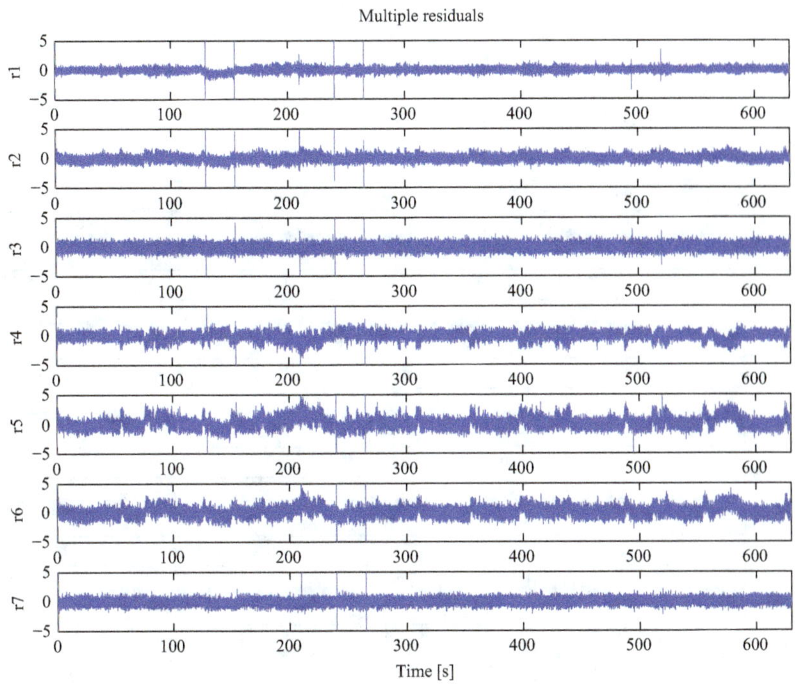

Figure 7.4 Residuals generation at wnd = 11 m/s: robust approach

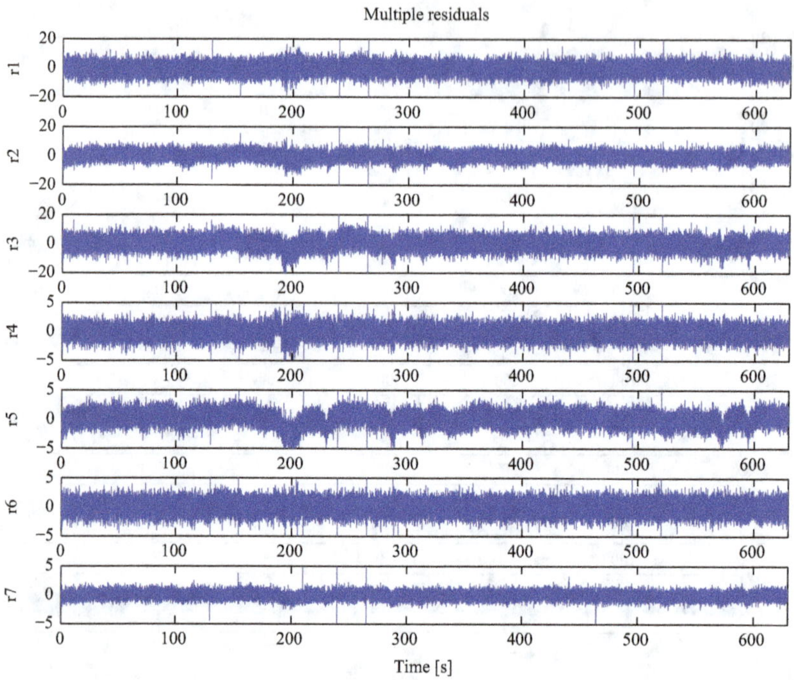

Figure 7.5 Residuals generation at wnd = 14 m/s: general approach

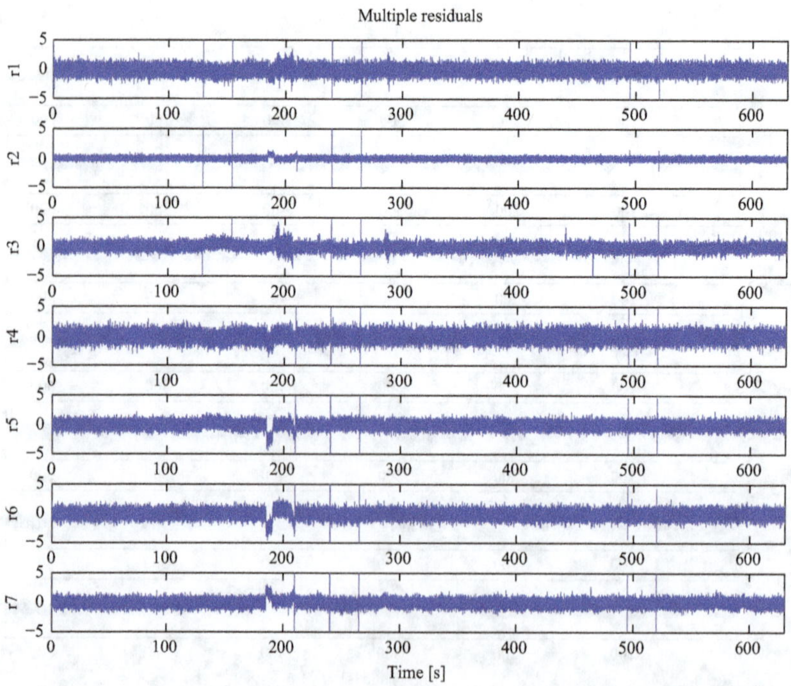

Figure 7.6 Residuals generation at wnd = 14 m/s: robust approach

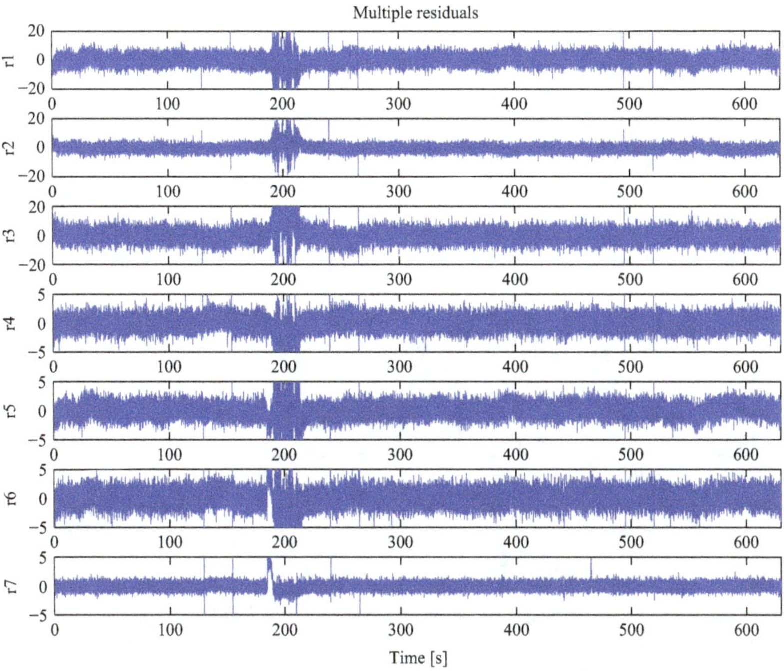

Figure 7.7 Residuals generation at wnd = 17 m/s: general approach

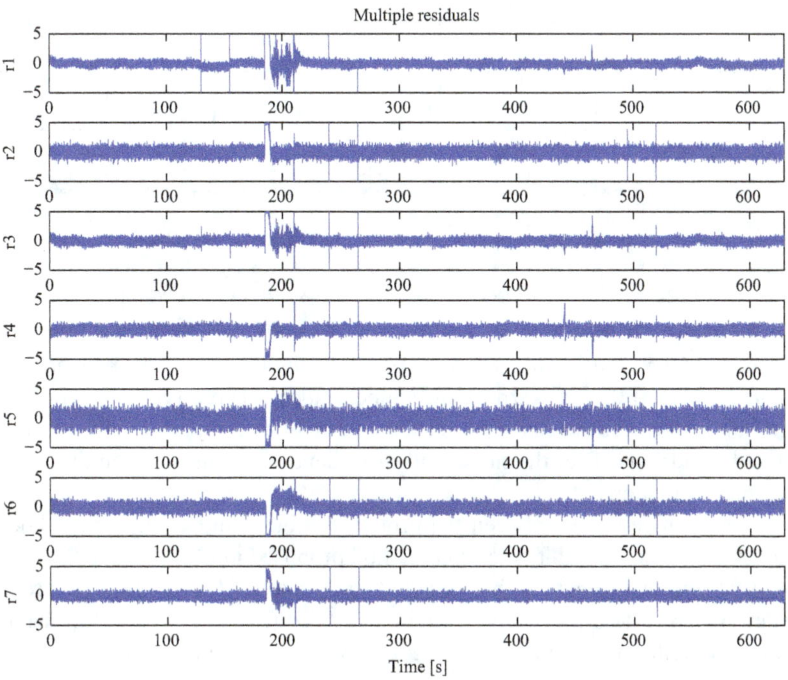

Figure 7.8 Residuals generation at wnd = 17 m/s: robust approach

Table 7.3 Four different fault detection schemes

Scheme	**Generator**	**Evaluate**	**Decision**
General-single	Single residual	GLR	Logic I
General-vector	Residual vector	GLR	Logic II
Robust-single	Single residual	GLR	Logic III
Robust-vector	Residual vector	GLR	Logic IV

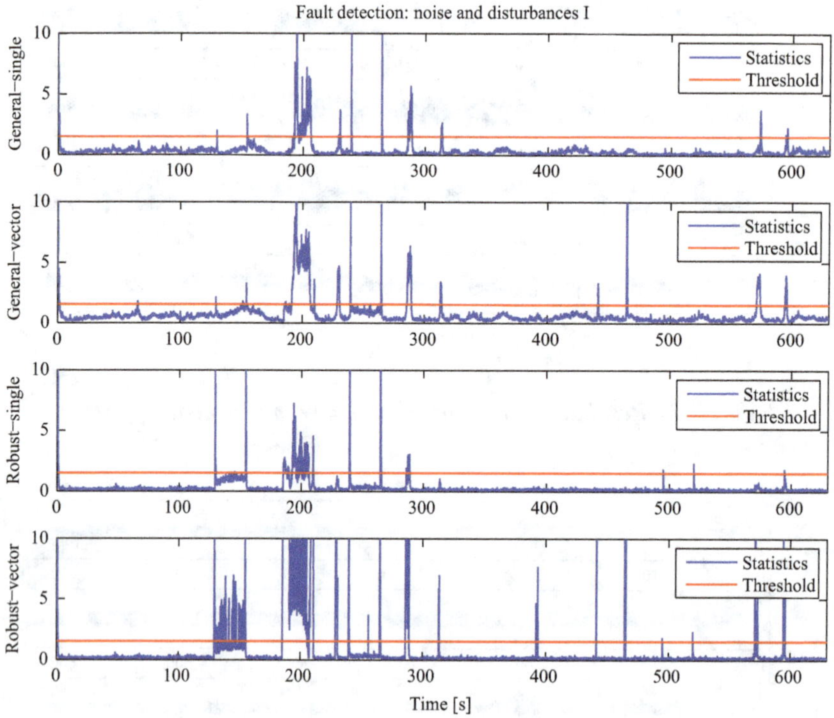

Figure 7.9 Fault detection under noise and disturbances I

As mentioned previously, the wind input files supplied with the benchmark are generated by an IEC von Karman turbulence model. Three types of wind inputs are available, the mean speeds of which at the 90-m hub height are 11, 14, and 17 m/s. For fixed wind input file, the noise and disturbances are also constantly changing during the simulation.

A comparative study between the proposed robust fault detection method and the general data-driven fault detection method proposed in [31] is made. In fact, the major difference between the robust method and the general method is in the residual generation step, where the robust method selects α_s^i $(i = 1, 2, \ldots, m)$ following the optimization criterion described in Section 7.3.3, while the general method selects

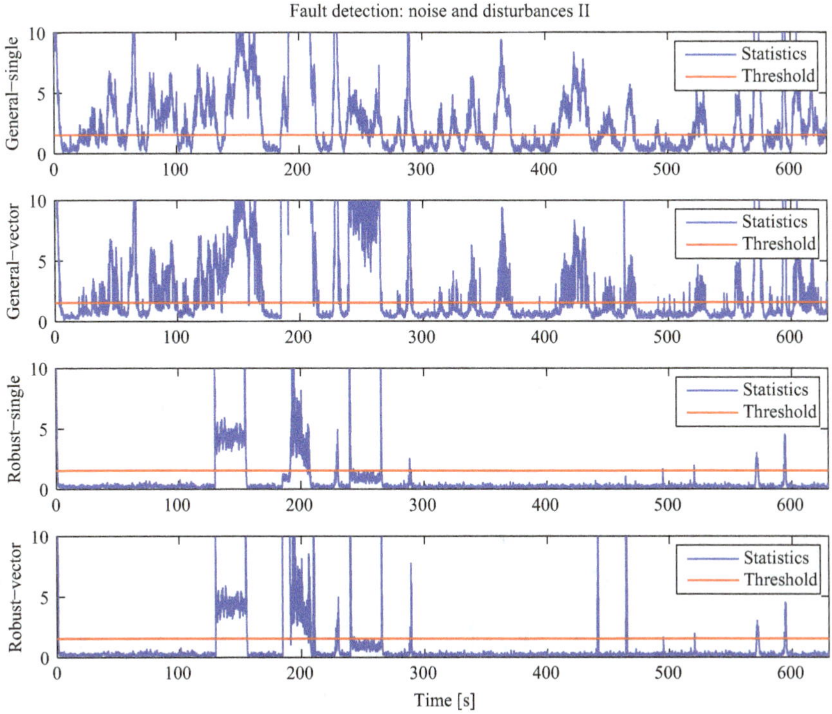

Figure 7.10 Fault detection under noise and disturbances II

α_s arbitrarily. However, taking into account the generating of a single residual or a residual vector, we are able to derive four fault detection methods altogether, which are the general method generating a single residual, the general method generating a residual vector, the robust method generating a single residual (corresponding to the maximal eigenvalue) and the robust method generating a residual vector (see Algorithm 1). Choosing a proper evaluation approach as well as a proper decision logic for each method in the second step, we can finally get four fault detection schemes, and we name them general-single, general-vector, robust-single and robust-vector, respectively. On equity grounds, the input–output variables and identification parameters are the same in all the methods. In addition, the collected data satisfies excitation condition equation (7.13) and the selected parity vectors α_s^i $(i = 1, 2, \ldots, m)$ satisfy the rank condition equation (7.37).

First of all, we compare the residual generation of the robust method and the general method under three different types of wind inputs. Both methods will generate a residual vector containing m residual signals for each type of wind input. Figures 7.3–7.8 show the results, in which Figures 7.3, 7.5, and 7.7 show the m residual signals generated by the general method under wind speeds of 11, 14, and 17 m/s, respectively; Figures 7.4, 7.6, and 7.8 show the m residual signals generated by the

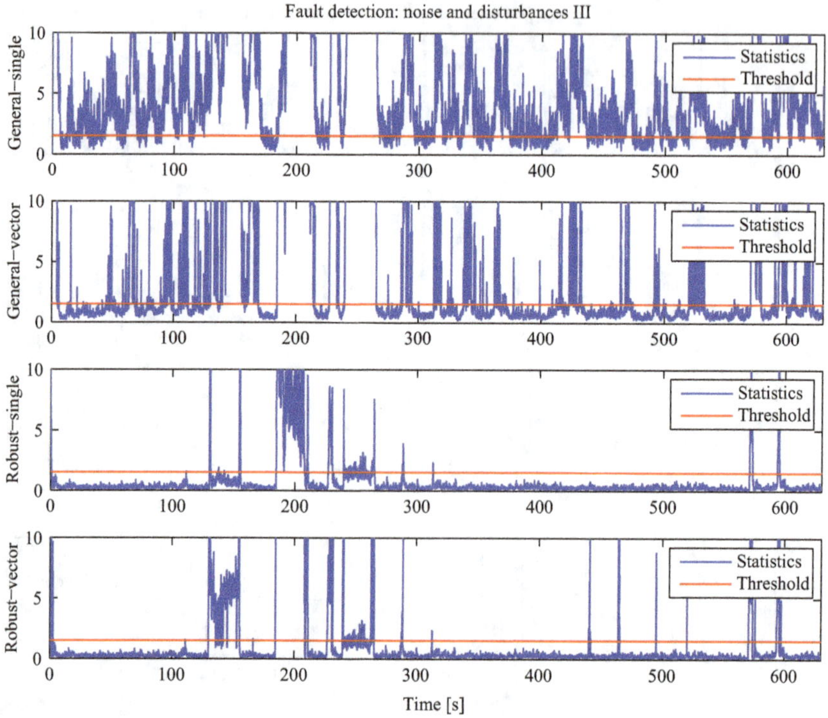

Figure 7.11 Fault detection under noise and disturbances III

robust method under wind speeds of 11, 14, and 17 m/s, respectively. As we can see from Figures 7.3, 7.5, and 7.7, the power of residual signals in the fault-free cases is quite large, ranging from $[-5, 5]$ to $[-20, 20]$, which may bring trouble to the residual evaluation and cause false detection or missed detection in the final decision. Especially in Figure 7.3, the residual signals $r1$ and $r2$ are totally disturbed by the disturbances and the noise. However, in contrast, seen from Figure 7.4, 7.6, and 7.8, the power of all the residual signals generated by the robust method is restrained to around zero in fault-free cases. In addition, the disturbances in these residual signals are quite small. These advantages reveal that the proposed robust method can effectively restrain the input disturbances.

Next, we compare the fault detection of the robust method and the general method under a fixed wind input. As mentioned before, the noise and the disturbances will be constantly changing during the simulation. For a comprehensive comparison, four fault detection schemes mentioned previously are considered. All these schemes take GLR as the evaluation approach. Besides, we give a decision logic to each scheme, just as follows:

- I: *if the evaluated residual signal exceeds the threshold then alarm*
- II: *if half of the m evaluated residual signals exceed the threshold then alarm*
- III: *if the evaluated residual signal exceeds the threshold then alarm*

Table 7.4 Fault detection results of four schemes, in which "✓" means accurate detection, "○" means delay detection, "×" means missed detection, "∗" means unreliable or false detection

A	Fault							
	3	4	5	6	7	8	9	10
General-single	✓	○	✓	✓	×	×	×	✓
General-vector	✓	○	✓	✓	×	✓	×	✓
Robust-single	✓	✓	✓	✓	×	×	✓	×
Robust-vector	✓	✓	✓	✓	○	✓	✓	○
B	***Fault***							
	3	4	5	6	7	8	9	10
General-single	∗	∗	∗	∗	∗	∗	∗	∗
General-vector	∗	∗	∗	∗	∗	∗	∗	∗
Robust-single	✓	○	✓	✓	×	×	✓	○
Robust-vector	✓	✓	✓	✓	×	✓	✓	○
C	***Fault***							
	3	4	5	6	7	8	9	10
General-single	∗	∗	∗	∗	∗	∗	∗	∗
General-vector	∗	∗	∗	∗	∗	∗	∗	∗
Robust-single	✓	✓	✓	✓	×	×	×	○
Robust-vector	✓	✓	✓	✓	×	✓	✓	○

- IV: *if the evaluated residual signal corresponding to the maximal eigenvalue exceeds the threshold or half of the rest $m-1$ evaluated residual signals exceed the threshold then alarm*

All the detection schemes are summarized in Table 7.3. In order to reduce false alarm rates, we apply the following filter algorithm to all the detection schemes to smooth the evaluated residual signals:

$$\hat{r}_k = \gamma \hat{r}_{k-1} + (1-\gamma) r_k \tag{7.39}$$

where r_k is the evaluated residual signal before smoothing, $\hat{r}_k$ is the evaluated residual signal after smoothing, $\gamma = 0.95$. We take the wind input with the mean speed of 14 m/s at the 90-m hub height as an example. Three different disturbances and noise are involved during the simulations. Figures 7.9–7.11 show the detection results. We summarize these detection results into Table 7.4, in which Table 7.4A contains the results of Figure 7.9, Table 7.4B contains the results of Figure 7.10 and Table 7.4C contains the results of Figure 7.11.

From Table 7.4A, we can see that the robust-vector scheme detected all the expected eight faults, while the other three schemes missed two or three of them.

In Figures 7.10 and 7.11, more serious disturbances and noise were contained in the system; as a result, the general-single and general-vector schemes totally failed and all the detection results were unreliable or false. On the contrary, the two robust schemes still kept stable. In Table 7.4B, the robust-single scheme failed to detect Faults 7, 8; in Table 7.4C, it failed to detect Faults 7–9. The robust-vector scheme, however, just missed Fault 7 in both Table 7.4B and C. Based on these results and analyses, we conclude that the proposed fault detection scheme (i.e., robust-vector scheme) has the best performance and is the most robust scheme among all the considered schemes.

As there is no requirement for the mathematical model of the wind turbine as well as the effective inhibition on disturbances in wind input, the proposed scheme is suitable to some extent for the fault detection of wind turbines. However, this scheme needs further improvements to detect Faults 1 and 2 which are not considered currently, and further reductions of both the false detection rate and the missed detection rate.

7.5 Conclusions

This chapter presented a robust data-driven fault detection scheme with the application to a wind turbine benchmark. The proposed scheme is based on robust residual generators constructed directly from available process measurements. For this purpose, a parity space is first identified from the measured data, and optimal parity vectors are selected from the parity space according to a given performance index and an optimization criterion to generate a robust residual vector. A proper evaluation approach as well as a suitable decision logic is further given to make a correct final decision. The effectiveness of the proposed scheme is finally demonstrated by the results obtained from the simulation of a wind turbine benchmark model.

References

[1] Lu B, Li Y, Wu X, *et al.* A review of recent advances in wind turbine condition monitoring and fault diagnosis. In: Power Electronics and Machines in Wind Applications, 2009. PEMWA 2009. IEEE. IEEE; 2009. p. 1–7.

[2] Odgaard PF, Stoustrup J. Unknown input observer based scheme for detecting faults in a wind turbine converter. In: Proceedings of the 7th IFAC Symposium on Fault Detection, Supervision and Safety of Technical processes. vol. 1; 2009. p. 161–166.

[3] Polinder H, Lendenmann H, Chin R, *et al.* Fault tolerant generator systems for wind turbines. In: Electric Machines and Drives Conference, 2009. IEMDC'09. IEEE International. IEEE; 2009. p. 675–681.

[4] Johnson KE, Fleming PA. Development, implementation, and testing of fault detection strategies on the National Wind Technology Centers controls advanced research turbines. Mechatronics. 2011;21(4):728–736.

[5] Hameed Z, Hong Y, Cho Y, *et al.* Condition monitoring and fault detection of wind turbines and related algorithms: a review. Renewable and Sustainable energy Reviews. 2009;13(1):1–39.

[6] Hatch C. Improved wind turbine condition monitoring using acceleration enveloping. Orbit. 2004;61:58–61.
[7] Yang W, Tavner P, Wilkinson M. Condition monitoring and fault diagnosis of a wind turbine synchronous generator drive train. IET Renewable Power Generation. 2009;3(1):1–11.
[8] Yang W, Tavner PJ, Crabtree CJ, *et al.* Cost-effective condition monitoring for wind turbines. IEEE Transactions on Industrial Electronics. 2010;57(1): 263–271.
[9] Caselitz P, Giebhardt J, Mevenkamp M, *et al.* Application of condition monitoring systems in wind energy converters. In: EWEC-Conference-. Bookshop for Scientific Publications; 1997. p. 579–582.
[10] Schlechtingen M, Santos IF. Comparative analysis of neural network and regression based condition monitoring approaches for wind turbine fault detection. Mechanical Systems and Signal Processing. 2011;25(5):1849–1875.
[11] Basseville M, Nikiforov I V. Detection of abrupt changes: theory and application. vol. 104. Prentice Hall, Englewood Cliffs, NJ; 1993.
[12] Gustafsson F, Gustafsson F. Adaptive filtering and change detection. vol. 1. Wiley, Londres; 2000.
[13] Ding SX. Model-based fault diagnosis techniques: design schemes, algorithms, and tools. Springer, Heidelberg; 2008.
[14] Paviglianiti G, Pierri F, Caccavale F, *et al.* Robust fault detection and isolation for proprioceptive sensors of robot manipulators. Mechatronics. 2010;20(1):162–170.
[15] Dong H, Wang Z, Gao H. Fault detection for Markovian jump systems with sensor saturations and randomly varying nonlinearities. IEEE Transactions on Circuits and Systems I: Regular Papers. 2012;59(10):2354–2362.
[16] Gao H, Chen T, Wang L. Robust fault detection with missing measurements. International Journal of Control. 2008;81(5):804–819.
[17] Zhao Y, Lam J, Gao H. Fault detection for fuzzy systems with intermittent measurements. IEEE Transactions on Fuzzy Systems. 2009;17(2): 398–410.
[18] Van Engelen T. Control design based on aero–hydro–servo–elastic linear models from TURBU (ECN). In: Proceedings of European Wind Energy Conference, Milan, Italy, May; 2007. p. 7–10.
[19] Van Der Veen G, Van Wingerden J, Fleming P, *et al.* Global data-driven modeling of wind turbines in the presence of turbulence. Control Engineering Practice. 2013;21(4):441–454.
[20] Dong J, Verhaegen M. Data driven fault detection and isolation of a wind turbine benchmark. IFAC Proceedings Volumes. 2011;44(1):7086–7091.
[21] Dai X, Gao Z. From model, signal to knowledge: a data-driven perspective of fault detection and diagnosis. IEEE Transactions on Industrial Informatics. 2013;9(4):2226–2238.
[22] Yin S, Wang G, Karimi HR. Data-driven design of robust fault detection system for wind turbines. Mechatronics. 2014;24(4):298–306.

[23] Yin S, Ding SX, Xie X, *et al.* A review on basic data-driven approaches for industrial process monitoring. IEEE Transactions on Industrial Electronics. 2014;61(11):6418–6428.
[24] Yin S, Yang X, Karimi HR. Data-driven adaptive observer for fault diagnosis. Mathematical Problems in Engineering. 2012;1–21:2012.
[25] Yin S, Ding SX, Haghani A, *et al.* A comparison study of basic data-driven fault diagnosis and process monitoring methods on the benchmark Tennessee Eastman process. Journal of Process Control. 2012;22(9):1567–1581.
[26] Li W, Qin SJ. Consistent dynamic PCA based on errors-in-variables subspace identification. Journal of Process Control. 2001;11(6):661–678.
[27] Yin S, Wang G, Yang X. Robust PLS approach for KPI-related prediction and diagnosis against outliers and missing data. International Journal of Systems Science. 2014;45(7):1375–1382.
[28] Yin S, Wang G, Gao H. Data-driven process monitoring based on modified orthogonal projections to latent structures. IEEE Transactions on Control Systems Technology. 2016;24(4):1480–1487.
[29] Jiao J, Yu H, Wang G. A quality-related fault detection approach based on dynamic least squares for process monitoring. IEEE Transactions on Industrial Electronics. 2016;63(4):2625–2632.
[30] Wang G, Yin S. Quality-related fault detection approach based on orthogonal signal correction and modified PLS. IEEE Transactions on Industrial Informatics. 2015;11(2):398–405.
[31] Ding S, Zhang P, Naik A, *et al.* Subspace method aided data-driven design of fault detection and isolation systems. Journal of process control. 2009;19(9):1496–1510.
[32] Dong J, Verhaegen M, Gustafsson F. Robust fault detection with statistical uncertainty in identified parameters. IEEE Transactions on Signal Processing. 2012;60(10):5064–5076.
[33] Döhler M, Mevel L. Subspace-based fault detection robust to changes in the noise covariances. Automatica. 2013;49(9):2734–2743.
[34] Chen Z, Fang H, Chang Y. Weighted data-driven fault detection and isolation: a subspace-based approach and algorithms. IEEE Transactions on Industrial Electronics. 2016;63(5):3290–3298.
[35] Odgaard PF, Stoustrup J, Kinnaert M. Fault tolerant control of wind turbines–a benchmark model. IFAC Proceedings Volumes. 2009;42(8):155–160.
[36] Chen W, Ding SX, Sari A, *et al.* Observer-based FDI schemes for wind turbine benchmark. In: Proceedings of IFAC World Congress; 2011. p. 7073–7078.
[37] Laouti N, Sheibat-Othman N, Othman S. Support vector machines for fault detection in wind turbines. In: Proceedings of IFAC World Congress; 2011. p. 7067–7072.
[38] Ozdemir AA, Seiler P, Balas GJ. Wind turbine fault detection using counter-based residual thresholding. In: Proceedings of IFAC World Congress; 2011. p. 8289–8294.

[39] Odgaard PF, Johnson KE. Wind turbine fault detection and fault tolerant control-an enhanced benchmark challenge. In: American Control Conference (ACC), 2013. IEEE; 2013. p. 4447–4452.

[40] Odgaard PF, Stoustrup J, Kinnaert M. Fault-tolerant control of wind turbines: a benchmark model. IEEE Transactions on Control Systems Technology. 2013;21(4):1168–1182.

[41] Odgaard PF, Stoustrup J. A benchmark evaluation of fault tolerant wind turbine control concepts. IEEE Transactions on Control Systems Technology. 2015;23(3):1221–1228.

[42] Büthe T. Engineering uncontestedness? The origins and institutional development of the International Electrotechnical Commission (IEC). Business and Politics. 2010;12(3):1–62.

[43] Odgaard PF, Johnson KE. Wind turbine fault detection and fault tolerant control – a second challenge. In: 8th IFAC Symposium on Fault Detection, Supervision and Safety of Technical Processes; 2012. p. 127.

[44] Wang J, Qin SJ. Principal component analysis for errors-in-variables subspace identification. In: Decision and Control, 2001. Proceedings of the 40th IEEE Conference on. vol. 4. IEEE; 2001. p. 3936–3941.

[45] Wang J, Qin SJ. A new subspace identification approach based on principal component analysis. Journal of Process Control. 2002;12(8):841–855.

[46] Frank PM. Enhancement of robustness in observer-based fault detection. International Journal of Control. 1994;59(4):955–981.

[47] Ding SX, Yin S, Wang Y, *et al.* Data-driven design of observers and its applications. In: Proceedings of the 18th IFAC World Congress, Milano, Italy; 2011.

[48] Qin SJ, Li W. Detection and identification of faulty sensors in dynamic processes. AIChE Journal. 2001;47(7):1581–1593.

[49] Blanke M. Diagnosis and fault-tolerant control. Springer, Verlag; 2003.

[50] Li H, Wu L, Si Y, *et al.* Multi-objective fault-tolerant output tracking control of a flexible air-breathing hypersonic vehicle. Proceedings of the Institution of Mechanical Engineers, Part I: Journal of Systems and Control Engineering. 2010;224(6):647–667.

[51] Li H, Gao H, Liu H, *et al.* Fault-tolerant H¡Þ control for active suspension vehicle systems with actuator faults. Proceedings of the Institution of Mechanical Engineers, Part I: Journal of Systems and Control Engineering. 2012;226(3):348–363.

[52] Li H, Liu H, Gao H, *et al.* Reliable fuzzy control for active suspension systems with actuator delay and fault. IEEE Transactions on Fuzzy Systems. 2012;20(2):342–357.

[53] Liu M, Shi P, Zhang L, *et al.* Fault-tolerant control for nonlinear Markovian jump systems via proportional and derivative sliding mode observer technique. IEEE Transactions on Circuits and Systems I: Regular Papers. 2011;58(11):2755–2764.

Chapter 8

Fault diagnostics for electrically operated pitch systems in offshore wind turbines

Surya Teja Kandukuri[1], Van Khang Huynh[1], Hamid Reza Karimi[2], and Kjell G. Robbersmyr[1]

8.1 Introduction

Offshore wind is among promising renewable energy industries growing at a rapid pace. Although offshore wind farms are advantageous, it has been witnessed that their maintenance costs are high and reliability is low compared to their onshore counterparts. This is due to the harsh weather conditions and limited weather windows for performing maintenance activities. As the offshore wind farms grow larger and further offshore, it becomes imperative to adopt a maintenance strategy in order to operate them reliably and profitably [1]. Condition monitoring and condition-based maintenance are necessary tasks in realising such maintenance strategy. In so far, the condition monitoring systems, that are commercially available, are focused primarily on the drivetrain, generator, main bearing and blades [2]. Upgrading the maintenance method to condition-based maintenance at component level can enhance such a maintenance strategy. Towards this end, the electrically operated pitch systems in offshore wind turbines are assessed for condition-based maintenance [3].

The pitch system in a wind turbine operates intermittently, depending on the wind conditions in order to reduce structural loads and improve the turbine efficiency. These are either hydraulically or electrically actuated. It is estimated that the market is equally split between the two types of actuation [4]. In this chapter, the electrically operated pitch systems are studied. An electrically operated pitch system consists of a multi-stage planetary gearbox with an electric motor as prime mover. The gearbox drives the blade bearing (a large slew bearing) through a pinion. The bearing has a gear on the inner perimeter of the inner ring, and the blade is bolted to the inner ring. The outer ring is rigidly bolted to the hub. Due to this arrangement, the blade root loads are transferred to the blade bearing. The pitch system therefore experiences the periodic gravitational loads and aperiodic wind disturbance and gyroscopic loads

[1]Department of Engineering Sciences, University of Agder, Norway
[2]Department of Mechanical Engineering, Politecnico di Milano, Italy

that result in wide variations in operating profiles, which could lead to failures in the motor, gearbox and blade bearing.

The pitch systems have been reported among the wind turbine components with frequent failures [5]. More recently, Lin *et al.* [6] provide insights into the common failure modes of electric pitch systems experienced in China. It has been reported that faults due to over-heating and poor ventilation, such as the stator windings failure and broken rotor bars (BRBs), are found to be common among pitch motors besides bearing faults.

There has been great attention in the literature on the control problem of the pitching mechanism [7], pitch actuator fault tolerance [8] and fault detection [9]. However, to the authors' knowledge, there are little published efforts that are focused on predictive maintenance approach for pitch systems. A predictive maintenance approach should include incipient fault diagnosis, fault severity assessment and remaining useful life estimation so that a maintenance action may be planned. For instance, detecting a bearing fault before it jams the pitch actuator can help in maintenance planning. It can be understood that the health monitoring of pitch drive mechanisms has been a low priority because of the fact that they are easily replaceable in onshore turbines. However, the need for such predictive maintenance of these systems is becoming stronger in the case of offshore as such reactive maintenance measures are expensive, owing to the cost of transportation and weather conditions.

In this chapter, the main objective is to determine the feasibility and applicability of current signature analysis for pitch motors in typical operating profiles. In order to determine pitch system operation profiles, the 5 MW reference wind turbine is simulated in FAST analysis tools developed by National Renewable Energy Laboratory [10]. The pitch systems however pose significant challenge in terms of intermittent, start–stop operating profiles and low speed operations. The main contribution of this chapter is therefore twofold: (1) to develop a detailed physical modelling of various motor faults and study their effect on motor currents in pitch system operating profiles and (2) to determine the feasibility of current signature analysis in such operating profiles.

The rest of this chapter is organised as follows: In Section 8.2, determination of the typical pitch profiles from FAST analysis tool is described. Further, a detailed modelling of induction motor with implementation of various fault conditions is described in Section 8.3. In Section 8.4, the motor current signature analysis (MCSA) is tested for pitch motor diagnostics in various wind turbine operating profiles. Finally, accuracy of the fault detection algorithms and steps towards implementation in wind farms are discussed.

8.2 FAST analysis

The wind turbine blade root loads are studied based on FAST analysis tools. Various full-field wind conditions and wave conditions are generated using TURBSIM software for simulation purposes [11]. The 5 MW reference wind turbine is then simulated to determine the command pitch angles based on GH-Bladed dll style controller [10].

In order to gain insight into the pitch system operation, the specifications of the pitch motor, gearbox and blade bearing are required for simulation study. However, this information is not generic and varies with turbine model. The selected reference wind turbine uses blade structure definition based on LM Glasfiber's LM64-5 blade, with a reduced length of 61.2 m. The bolt circle diameter for a 61.2-m long blade LM-61.2P is obtained from the LM Wind Powers website as 2.3 m [12]. A 4-point contact bearing KD-320 series, with inner ring gear and inner bolt circle diameter of 2.43 m is chosen from Rothe Erde ThyssenKrupp slew bearings catalogue [13], which is designed to withstand a simultaneous axial loads of up to 3,600 kN and moment load of 2,600 kN m. The inner race gear has 144 gear teeth. The pitch drive is required to generate torque equal to the rotating torque of the slew bearing. The starting torque of the slew bearing is given by semi-empirical relationship [13] as

$$T_{start} = \frac{\mu}{2}(4.4M_k + F_aD_l + 2.2(F_rD_l)) \tag{8.1}$$

where F_a is the axial load (kN), F_r is the radial load (kN), M_k is the resultant tilting moment (kN m), D_l is the bearing race diameter (m) and $\mu = 0.004$ is the coefficient of friction of the bearing. These axial and radial loads are the blade-root forces and moments obtained from the FAST simulation. Although, in reality, the rolling torque of the bearing is much lesser than T_{start}, it is complex to be calculated as it depends on factors such as elasto-hydrodynamics of lubrication, bearing-seal friction and contact mechanics [14]. Therefore, T_{start} is assumed as the resultant bearing torque that is to be counteracted by the pitch drive. This is a conservative but acceptable assumption as the objective here is not to design a pitch bearing but to evaluate feasibility of diagnostics.

Further, a multistage planetary gearbox RPR250DC from Brevini [15] is chosen with a gear ratio of 1:1,800. Finally, a two-pole induction motor is chosen as the prime-mover. The FAST simulation, presently, does not include pitch actuator dynamics. These are typically approximated by a second-order system [16] as

$$\begin{aligned} \dot{x}_p &= \begin{bmatrix} -2\omega_n\eta & -\omega_n^2 \\ 1 & 0 \end{bmatrix} x_p + \begin{bmatrix} 1 \\ 0 \end{bmatrix} u_p \\ y_p &= \begin{bmatrix} 1 & 0 \\ 0 & \omega_n^2 \end{bmatrix} x_p \end{aligned} \tag{8.2}$$

where the natural frequency is $\omega_n = 0.88$ rad/s and the damping ratio is $\eta = 0.9$. The state vector x_p has the velocity and position of the pitch actuation and u_p has the pitch actuator command obtained from FAST. Under normal operations (apart from emergency shutdown), the pitch rate is limited to $\pm 5°$/s and the acceleration to $\pm 30°/s^2$.

The 5 MW reference wind turbine is simulated in offshore monopile configuration with a turbulent full-field wind profile generated by TURBSIM. The Kaimal spectrum is simulated with a mean wind speed of 18 m/s and turbulence class A, as per IEC standards [17]. The wave profile is selected with a significant wave height $H_s = 6$ m and the time period of $T_p = 10$ s. The resultant pitch angle, speed profile,

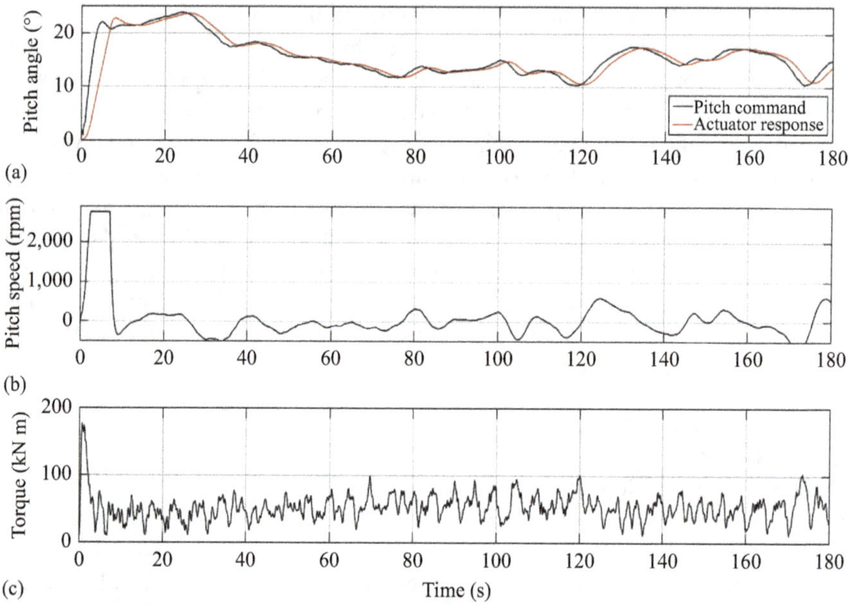

Figure 8.1 Time response of pitch angle (a), pitch motor speed (b) and bearing torque (c) (source: [3])

and maximum starting torque at the bearing are described in Figure 8.1. It can be seen that for most part of the operating period, the motor runs at very low speeds, lesser than 1,000 rpm, while the loads are continuously varying. The objective is, therefore, to verify the detectability of current signatures in such operating conditions.

8.3 Induction motor behaviour in faulty conditions

The induction motors are common prime movers in many industrial scenarios as they are inexpensive and robustly built to withstand a wide range of operating and environmental conditions. A three-phase induction motor typically consists of sinusoidally distributed stator windings and a squirrel cage rotor with a number of rotor bars that are shorted at both ends with end rings. It was observed that bearing faults contribute to 41% of the total failures, while stator turn faults (STFs) contribute to 37%, rotor related failures 10% [18]. In this chapter, the behaviour of induction motor is studied under all of the aforementioned fault conditions in closed-loop under pitch system operating conditions.

The concept of MCSA is well-established in literature, see for instance [19,20]. The principal idea in MCSA is that the mechanical faults inside an electrical motor tend to disturb the air gap, leading to a periodic variation in the magnetic field as the rotor rotates. Similarly, the electrical faults such as the stator winding or rotor-bar faults result in changes in electrical circuit. These variations in turn, affect the stator current and can be detected in the Fourier spectrum of single line current at specific frequencies.

However, it has been observed that the single line current signature may not be accurate in various load and speed-varying conditions. The analysis of the quadrature axes currents (i_d, i_q) and the extended Park vector analysis (EPVA) was found to be more reliable [21,22].

The motor is typically modelled using *dq*-modelling approach [23] with rotor represented by a fictitious lumped winding for performance analysis and control design. In order to study the behaviour of induction motor in faulty conditions, a detailed model based on modified winding function theory (MWFTh) has been utilised. This modelling approach provides the ability to incorporate the electrical and mechanical faults into the model and analyse the faulty behaviour. This gives insight into the effect of various factors on the capability of MCSA such as the severity of fault on the machine behaviour, the behaviour of faulty machines in closed-loop, varying speeds and varying load conditions. All of these are necessary considerations for diagnostics of wind turbine pitch systems.

The electrical circuit of the induction motor is rewritten by considering all magnetic interactions between each stator winding and rotor bar through the machine air gap. Based on the winding function theory [24], the mutual inductance between two stator windings corresponding to phases A and B can be expressed as

$$L_{AB} = \frac{\mu_0 rl}{g} \int_0^{2\pi} N_A(\phi, \theta) N_B(\phi, \theta) d\phi \tag{8.3}$$

where N_A represents the winding function of the phase A winding and N_B represents that of phase B. The ϕ represents the spatial angle inside the machine and θ represents the rotor position. μ_0 represents the permissivity of air, r is the radius of the rotor, l is the effective length of rotor and g is the length of air gap.

In case of the sinusoidally distributed stator windings, the winding function for phase A of the rotor is given by

$$N_A(\phi, \theta) = \frac{2N_s}{p} cos\phi \tag{8.4}$$

where N_s is the number of stator winding turns per phase and p is the number of poles. The winding functions for the phases B and C lag phase A by 120° and 240°, respectively. The rotor bar winding function is given by

$$N_i(\phi, \theta) = \begin{cases} -\frac{\alpha_r}{2\pi} & 0 \leq \phi < \theta_i \\ 1 - \frac{\alpha_r}{2\pi} & \theta_i \leq \phi < \theta_{i+1} \\ -\frac{\alpha_r}{2\pi} & \theta_{i+1} \leq \phi < 2\pi \end{cases} \tag{8.5}$$

where $i \in N_r$, the number of rotor bars and $\theta_i = \theta + (i-1)\alpha_r$ and $\alpha_r = 2\pi/N_r$. The induction motor dynamics can be represented by the equations

$$\begin{aligned} \{V_s\}_{(3\times1)} &= [R_s]_{(3\times3)}\{i_s\}_{(3\times1)} \\ &\quad + \tfrac{d}{dt}\left([L_s]_{(3\times3)}\{i_s\}_{(3\times1)} + [L_{sr}]_{3\times N_r}\{i_r\}_{N_r\times1}\right) \\ \{0\}_{(N_r\times1)} &= [R_r]_{(N_r\times N_r)}\{i_r\}_{(N_r\times1)} \\ &\quad + \tfrac{d}{dt}\left([L_r]_{(N_r\times N_r)}\{i_r\}_{(N_r\times1)} + [L_{rs}]_{N_r\times3}\{i_s\}_{3\times1}\right) \end{aligned} \tag{8.6}$$

where V_s represents the three-phase power supply, i_s is the stator current, i_r represents the rotor bar currents in N_r bars, R_s is the stator resistance of the three phase windings and R_r represents the individual rotor bar resistance. The self and mutual inductances of stator windings are captured in L_s and L_r, respectively, while the mutual inductance between each stator winding and rotor bar are captured in L_{sr} and $L_{rs} = L'_{sr}$.

Further, the mutual inductance L_{sr} is variable with respect to the rotor angular displacement θ_r; therefore,

$$\frac{d}{dt}(L_{sr}i_r) = \frac{p}{2}\omega_r\frac{\partial L_{sr}}{\partial\theta_r}i_r + L_{sr}\frac{di_r}{dt} \tag{8.7}$$

The rotor electromechanical torque is given by

$$T_{em} = \frac{p}{2}i'_s\frac{\partial L_{sr}}{\partial\theta_r}i_r \tag{8.8}$$

and the angular speed and angular displacement of the rotor are given by

$$\begin{aligned} \frac{d\omega_r}{dt} &= \frac{p}{2J}(T_{em} - T_l) \\ \frac{d\theta_r}{dt} &= \omega_r \end{aligned} \tag{8.9}$$

T_l is the external mechanical load on the rotor.

Now the state-space of the electrical motor may be written as

$$A_1\dot{x}(t) = A_2x(t) + u(t) \tag{8.10}$$

where

$$x(t) = \{\, i_s \quad i_r \quad \omega_r \quad \theta_r \,\}' \tag{8.11}$$

and the input vector $u(t)$ is given by

$$u(t) = \{\, v_s \quad 0_{N_r} \quad \tfrac{p}{2J}(T_{em} - T_l) \quad 0 \,\}' \tag{8.12}$$

with the matrices

$$A_1 = \begin{bmatrix} L_s & L_{sr} & 0 & 0 \\ L_{rs} & L_r & 0 & 0 \\ 0 & 0 & 1 & 0 \\ 0 & 0 & 0 & 1 \end{bmatrix}$$
$$A_2 = \begin{bmatrix} -R_s & -\omega_r \frac{\partial L_{sr}}{\partial \theta_r} & 0 & 0 \\ -\omega_r \frac{\partial L_{rs}}{\partial \theta_r} & -R_r & 0 & 0 \\ 0 & 0 & 0 & 0 \\ 0 & 0 & 1 & 0 \end{bmatrix} \tag{8.13}$$

The dynamic model of induction motor based on winding function theory is thereby developed in the manner as described in [25]. In the case of mechanical faults, such as AGE or a bearing fault, the air gap g is no longer constant. The mutual inductance is now based on *modified* winding function [26] given by

$$L_{AB} = \mu_0 rl \left[\int_0^{2\pi} n_A(\phi,\theta) n_B(\phi,\theta) g^{-1}(\phi,\theta) d\phi \right. \\ \left. -2\pi \langle M_A(\phi,\theta)\rangle \langle M_B(\phi,\theta)\rangle \langle g^{-1}(\phi,\theta)\rangle \right] \tag{8.14}$$

where $M_i(\phi,\theta)$ represents the corresponding modified winding function of the winding 'i' and $\langle M_i(\phi,\theta)\rangle$ represents its mean value described as follows:

$$\langle M_i(\phi,\theta)\rangle = \frac{1}{2\pi \langle g^{-1}(\phi,\theta)\rangle} \int_0^{2\pi} n_i(\phi,\theta) g^{-1}(\phi,\theta) d\phi \tag{8.15}$$

In the case of a dynamic air gap eccentricity (AGE) fault, such as with unbalanced load or bent shaft, the air gap varies spatially, depending on the speed of rotation. Thus, the new air gap is defined as [27]:

$$g^{-1}(\phi,\theta) = \frac{1}{g(1 - e\,cos(\phi - \theta))} \tag{8.16}$$

where e is a tunable parameter to approximate the degree of severity of the fault. In the case of a bearing fault [28], the air gap function varies periodically, whenever the rolling element passes the crack on the bearing outer ring. This is approximated by a series of Dirac delta functions as

$$g^{-1}(\phi,\theta) = \left[g \left(1 - e\,cos(\phi - \theta) \sum_{k=-\infty}^{\infty} \delta \left(t - \frac{k}{f_c} \right) \right) \right]^{-1} \tag{8.17}$$

where f_c is the corresponding fault frequency for inner race, outer race and rolling element faults.

In the case of electrical faults, such as the BRB fault, the bars crack and their resistance increases largely compared to the healthy bars. As the bars are not insulated from the rotor in low power induction motors, the circuit does not break completely [26]. Therefore, the BRB fault can be simulated by merely increasing the resistance of rotor bars. In the case of a STF, the shorted windings form an additional stator winding loop. For instance, if the stator winding of phase A has shorted coils, these coils form a separate circuit with fault current flowing through it. If the winding A has a percentage of windings that are shorted, the inductance and resistance of coil A are proportionally decreased and its mutual inductances with other stator windings as well as rotor-bars are recalculated accordingly. The stator now has *four* windings instead of *three*, and the system of (8.6) can be represented as

$$\begin{aligned}\{V_s\}_{(4\times 1)} &= [R_s]_{(4\times 4)}\{i_s\}_{(4\times 1)} \\ &\quad + \tfrac{d}{dt}\left([L_s]_{(4\times 4)}\{i_s\}_{(4\times 1)} + [L_{sr}]_{4\times N_r}\{i_r\}_{N_r\times 1}\right) \\ \{0\}_{(N_r\times 1)} &= [R_r]_{(N_r\times N_r)}\{i_r\}_{(N_r\times 1)} \\ &\quad + \tfrac{d}{dt}\left([L_r]_{(N_r\times N_r)}\{i_r\}_{(N_r\times 1)} + [L_{rs}]_{N_r\times 4}\{i_s\}_{4\times 1}\right)\end{aligned} \tag{8.18}$$

For the cases of AGE and BRB, the self and mutual inductance matrices L_s, L_r and L_{sr} are recalculated as described in (8.14). Notice that in the case of AGE and BRB faults, the inductances are calculated at every time step as they vary based on the rotor angular position. The induction motor is thus simulated in various faulty conditions in a similar fashion as described in [25,29].

A 4 kW three-phase two-pole, 220 V star connected induction motor is simulated in all the different fault cases described, the parameters of the motor are specified in Table A1. The motor is started at no-load, and then a load of 8 N m is added after 1.2 s. The stator phase A current is then analysed through Fast Fourier transform (FFT).

In the case of healthy motor, directly fed from the mains with a steady, balanced 220 V supply, the Fourier spectrum of the current shows only one peak at the supply frequency 50 Hz. As this is a simulation, the motor is assumed to be perfectly healthy, without any manufacturing defects. In case of the AGE fault, it has been noticed that the presence of spectral components corresponding to the fault increase in magnitude with increase in degree of eccentricity [30]. Two different fault harmonics are seen in Fourier spectrum, one near the supply frequency given by

$$f_{AGE_1} = |f_s \pm kf_{rot}|, \quad k = 1, 2, 3\ldots \tag{8.19}$$

and another at the principal slot harmonic given by

$$f_{AGE_2} = \left[\frac{(kN_r \pm n_d)(1-s)}{p} \pm 1\right], \quad n_d = 1, 2, 3\ldots \tag{8.20}$$

where f_s is the supply frequency, f_{rot} is the rotor frequency in Hz, s is the slip and p is the number of pole-pairs in the motor. These are depicted in Figures 8.2 and 8.3. It is to be noted that the ratio of number of poles to the rotor bars plays a significant role in the detectability of AGE fault in an induction motor [31]. The higher frequency harmonics, especially f_{AGE_2}, depend on this ratio. It was noticed that the same motor, with 39 rotor bars instead of 28, has shown significantly better magnitude of the fault signatures in the current spectrum. However, the f_{AGE_1} was certainly detectable in all cases. Besides, the fault magnitudes increase with increase in degree of severity e and the load as well. The fault signatures are significantly higher in magnitude with loads closer to the rated loads. The BRB fault increases with an increase in number of BRBs and severity of fault. For instance, in Figure 8.4, the current spectrum corresponding

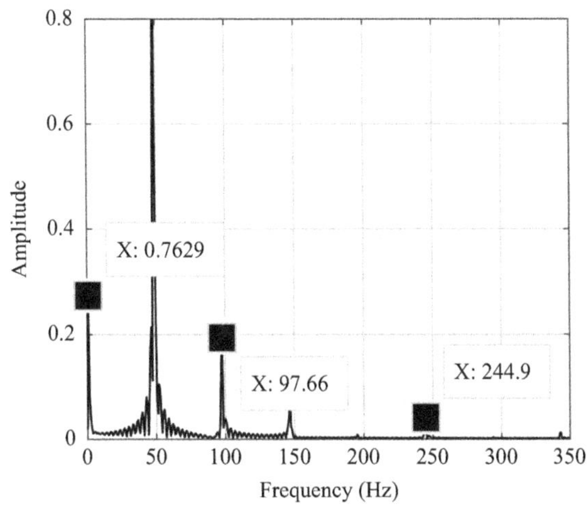

Figure 8.2 Fault frequencies at f_{AGE_1}

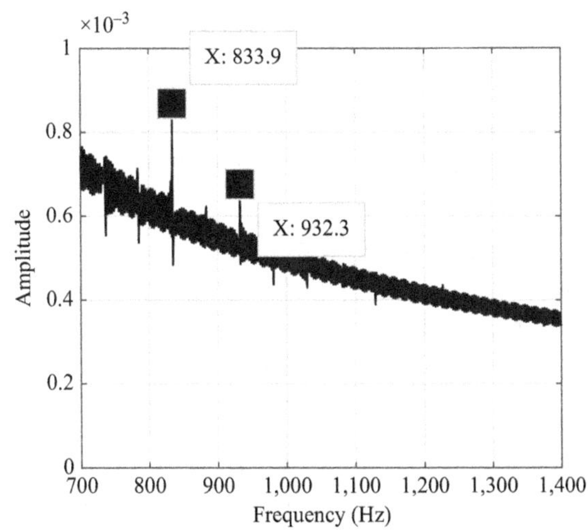

Figure 8.3 Fault frequencies at f_{AGE_2}

to two BRBs is shown. The supply frequency $f_s = 50$ Hz is filtered using a notch filter. The resistance of the two bars was increased by 50 times the normal resistance and the fault frequencies appear at

$$f_{BRB} = |f_s(1 \pm 2s)| \tag{8.21}$$

The faults increase in magnitude with an increase in bar resistance or load. It was noticed that the fault is undetectable at no-load, understandably so, because there will be almost no induced current in the rotor bars at negligible slip.

Bearing faults in the case of bearings on the rotor shafts are more likely to be outer race faults due to the fact that the centrifugal forces are always directed outwards as the shaft is rotating and the outer race is rigidly fixed. Therefore, only outer race

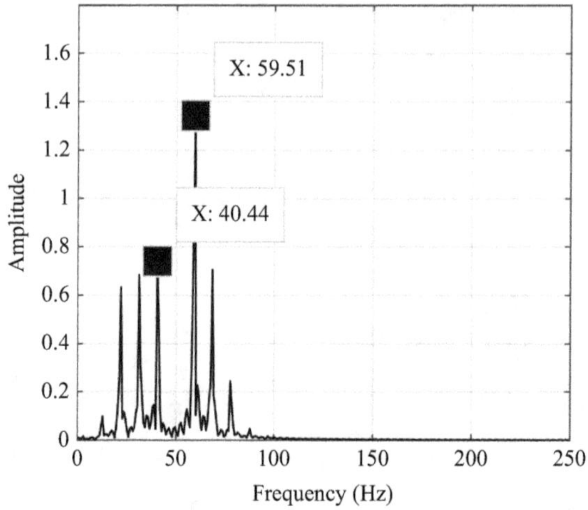

Figure 8.4 BRB fault spectrum

fault is simulated, and it can be seen that the bearing fault frequencies are detected in the single phase current spectrum as described by

$$f_{BRG} = |f_s \pm kf_c|, \quad k = 1, 2, 3\ldots \tag{8.22}$$

where f_c is the outer raceway fault frequency given by

$$f_c \approx 0.4N_b f_{rot} \tag{8.23}$$

for 6 to 12 rolling elements [32], where N_b is the number of rolling elements in the bearing. It was noticed that the bearing fault frequencies increase in magnitude with increase in degree of severity *e*. Besides, with increase in load, the spectrum shows increase in spectral content across the frequencies and decreases the detectability of fault signature. The bearing fault signature at 5 N m load is shown in Figure 8.5.

In case of the STFs, the single phase current spectrum did not show the fault at the principal slot harmonics as described in [33]. However, it was observed that the EPVA has shown promising results [34]. The *dq*-current components according to EPVA are described by

$$\begin{aligned} i_d &= \sqrt{\tfrac{2}{3}}i_A - \sqrt{\tfrac{1}{6}}i_B - \sqrt{\tfrac{1}{6}}i_C \\ i_q &= \sqrt{\tfrac{1}{2}}i_B - \sqrt{\tfrac{1}{2}}i_C \\ i_P &= |\sqrt{i_d^2 + i_q^2}| \end{aligned} \tag{8.24}$$

The i_d and i_q currents have shown a fault component at $3f_s$ that increases with increase in fault magnitude. However, the i_P currents show a $2f_s \approx 100$ Hz fault frequency in current signature that appears to be a reliable metric in load and speed variations. Besides, as described in [21], the loci of the plot (i_d, i_q) is circular in the case of healthy machine and takes an elliptical pattern in case of the stator winding faults. The stator winding faults signature at $3f_s$ and the ellipticity of *dq* loci increase with

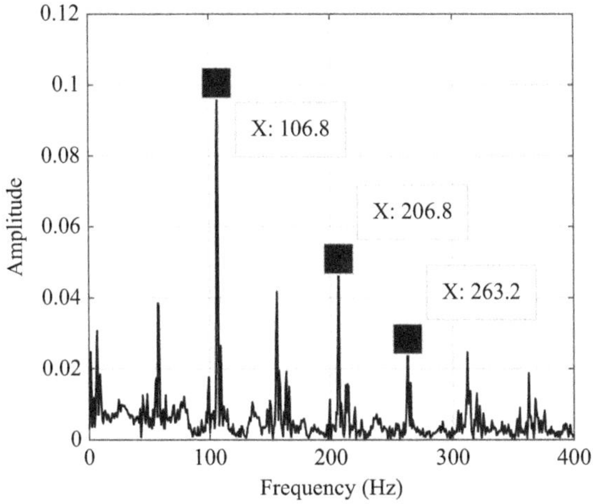

Figure 8.5 BRG fault spectrum

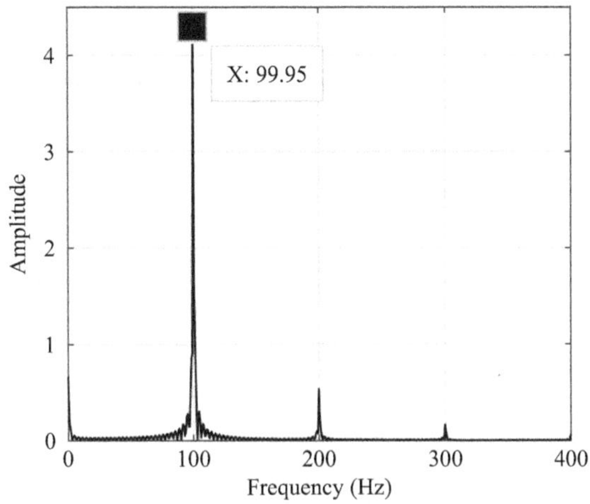

Figure 8.6 i_P Current spectrum stator fault

increase in the number of faulty windings. However, it was noticed that load has understandably, little effect on the fault signature. These results were described in Figures 8.6 and 8.7. The frequency signatures as calculated by the formulae for AGE, BRB, BRG described by (8.19)–(8.22) are shown in Table 8.1, which show good agreement with the frequency spectrum obtained from simulation of faulty models.

Insofar, all the described faults of induction motor are physically simulated at supply frequency, and the effect of fault severity and variation of load is studied. However, in pitch systems, the motor operates in closed-loop with a controller and runs at lower speeds with heavy loads. Further, the closed-loop behaviour of the induction motor shall be evaluated.

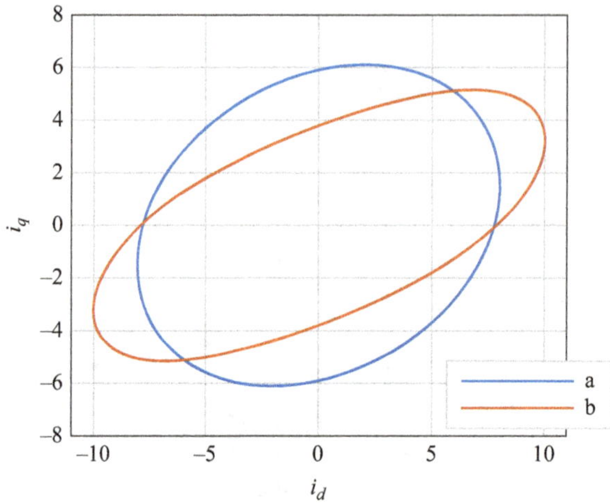

Figure 8.7 dq-Current loci stator fault (a) 5% fault and (b) 15% fault

Table 8.1 Calculated fault frequencies in supply-fed motor

f_{BRG} (Hz)	f_{BRB} (Hz)	f_{AGE_1} (Hz)	f_{AGE_2} (Hz)
106.5	59.97	99.11	1,250.3
206.5	40.03	0.88	1,154
263		148.63	1,202
		246.32	1,302

8.4 Faulty machine behaviour in closed-loop

In order to simulate pitch operations, vector control for induction motors has been adopted as the control strategy as described in Figure 8.9. An inverse half-Γ model of induction motor, as described in Figure 8.8, is developed as it is suitable for control applications. The necessary parameters for this model are derived based on the physical parameters of the induction motor as described in Appendix A. The system of equations representing the dynamics of the lumped model in stator coordinates are

$$\begin{aligned}\{V_s\}_{(3\times1)} &= [R_s]_{(3\times3)}\{i_s\}_{(3\times1)} + \tfrac{d}{dt}\{\psi_s^s\}_{(3\times1)} \\ 0_{(3\times1)} &= [R_R]_{(3\times3)}\{i_R^s\}_{(3\times1)} + \tfrac{d}{dt}\{\psi_R^s\}_{(3\times1)} - j\omega_r\{\psi_R^s\}_{(3\times1)}\end{aligned} \tag{8.25}$$

where $\psi_s^s = L_\sigma i_s^s + L_M i_M^s$, $\psi_R^s = L_M i_M^s$, $L_M = L_m^2/L_r$, $L_\sigma = L_s - L_M$, $R_R = (L_m/L_r)^2 R_r$, $L_s = L_{s\sigma} + L_m$ and $L_r = L_{r\sigma} + L_m$.

The principal idea behind vector control strategy is to make the control of induction motor similar to that of a dc machine. The field-oriented control is beneficial from the control point of view as the control inputs (V_{s_d}, V_{s_q}) directly affect the

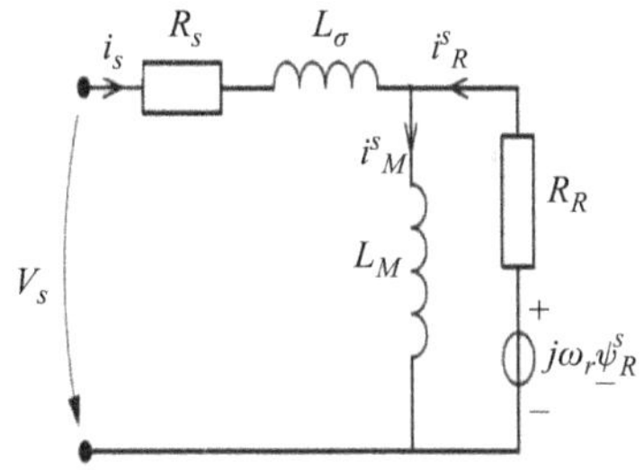

Figure 8.8 Inverse half-Γ model (source: [3])

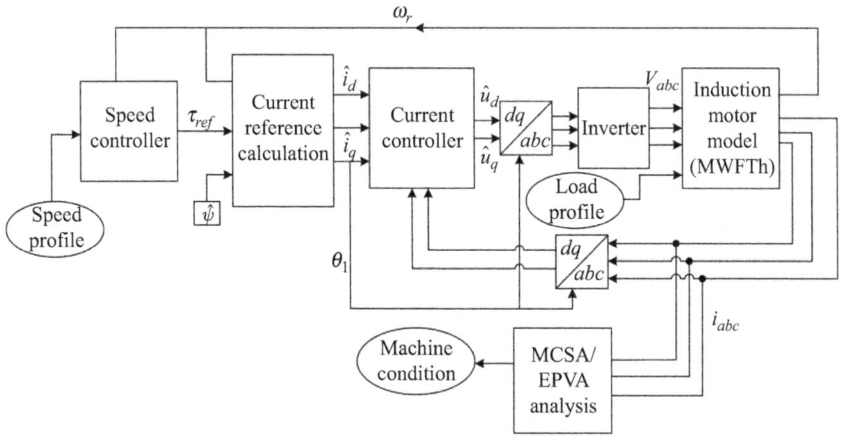

Figure 8.9 A vector control scheme (source: [3])

current dynamics only (i_{s_d}, i_{s_q})[1]. In this implementation, proportional-integrator (PI) controllers are used for both current and speed control applications [35]. As shown in Figure 8.9, the output of the PI speed controller is the torque set point reference τ_{ref} for the current controller. The output of the current controller is the voltage in *dq*0 frame, which is then mapped back to *abc* frame with a pulse width modulation inverter in order to include the inverter harmonics in the supply. The three-phase voltage output from the inverter is then fed to induction machine. The normalised speed profiles to the speed controller are supplied from the FAST simulations as described in Section 8.2. The load profiles from the FAST analysis are applied to the induction motor.

Once the controllers are designed and tested with the lumped model, they are simulated in closed-loop with the models incorporating faults, described in Section 8.3. The stator currents from the fault models are then captured and analysed using FFT. In this way, the effectiveness of current signature-based analyses is studied in the pitch system environment.

[1]Here, the subscripts s_d and s_q refer to the stator quantities in *dq*0 frame.

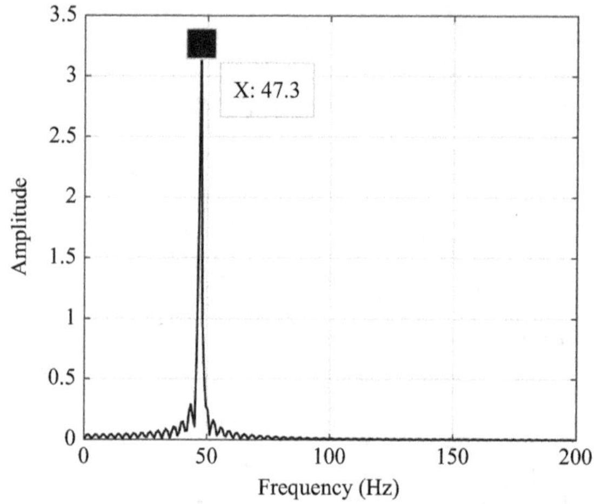

Figure 8.10 Healthy motor in closed-loop

The data obtained from FAST simulation described in Section 8.2 is used to simulate the developed closed-loop system. Since the detailed induction motor model is nonlinear, a very low sample time in the order of 1 μs is necessary to run the model accurately. In order to reduce computational burden, the 160 s of simulated data is cut into snapshots of 10 s each. The corresponding normalised load profile is applied. As the MCSA assumes fixed supply frequency, regions of relatively steady operation (with speed variations less than 10 rpm) are chosen for evaluation. One such case is presented here, wherein the motor is running steady at 2,762 rpm for about 3.5 s. The three phase currents stator currents i_s are collected for this period and analysed using standard FFT technique. The mean speed is obtained from the model, and the supply frequency is evaluated from the voltage supplied by the controller to be 46.54 Hz. First, the healthy motor is simulated, and the current spectrum as well as the *dq* loci are shown in Figures 8.10 and 8.11, respectively. Further, the AGE fault is simulated with a degree of eccentricity of about 3% variation in air gap similar to the case in open-loop. The results are shown in Figures 8.12 and 8.13. It can be seen that although the harmonics as described by (8.19) at the supply frequency are present, the higher frequency harmonics as described by (8.20) are almost invisible. This is due to the fact that the load is significantly varying in this region. The supply frequency of 46.54 Hz was filtered from the current signal using a notch filter.

The BRG fault is also similarly visible with about 4 μm variation in the air gap due to outer race crack. The results are shown in Figure 8.14. However, the BRG signature appears to be feeble, and especially in the presence of inverter harmonics, ambient noise and varying load conditions could be difficult to detect as the noise floor rises along with increasing load. The BRB fault was detectable by increasing the resistance of *three* rotor bars up to 50 times their normal values, similar to the open-loop. The spectrum for BRB fault is shown in Figure 8.15. However, it was observed that the BRB fault is undetectable in the case of low load conditions.

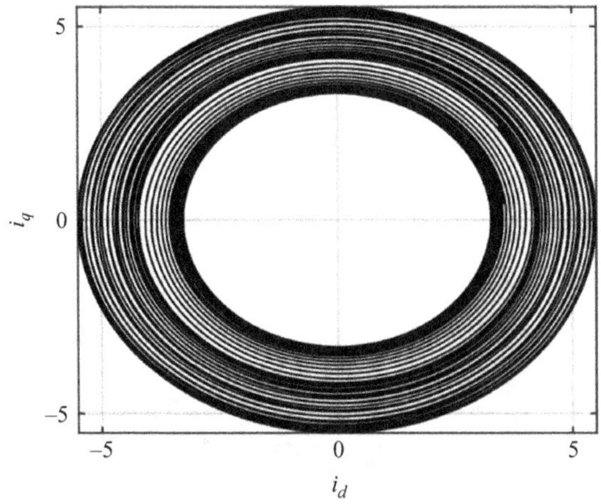

Figure 8.11 Healthy motor dq-current loci

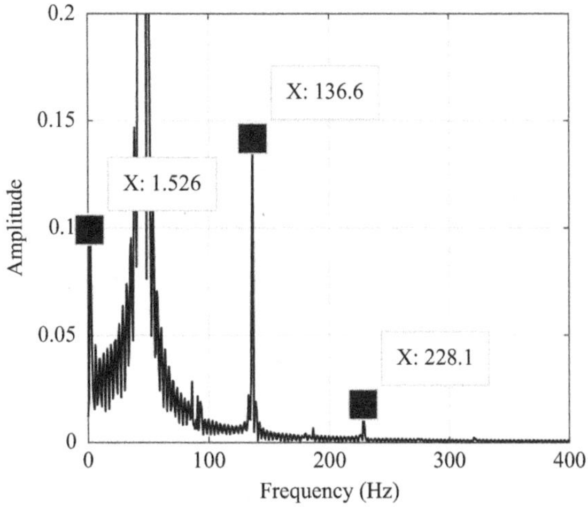

Figure 8.12 Fault frequencies at f_{AGE_1}

Finally, the STF is tested with 15% windings shorted in the phase A. The i_P current spectrum as well as the *dq* current loci in faulty condition are depicted in Figures 8.16 and 8.17, respectively. It can be noticed that the $2f_s \approx 93$ Hz frequency for the supply frequency $f_s = 46.54$ Hz is clearly noticeable in the spectrum. The *dq* loci is also elliptical, due to the presence of the fault as opposed to being circular in the case of healthy machine as shown in Figure 8.11.

The fault frequencies in all the fault cases are calculated based on (8.19)–(8.22). These are described in Table 8.2. These are in good agreement with the results obtained from the simulations.

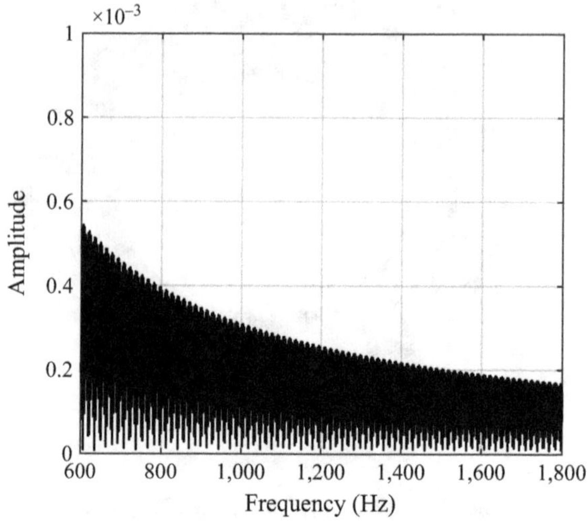

Figure 8.13 Fault frequencies at f_{AGE_2}

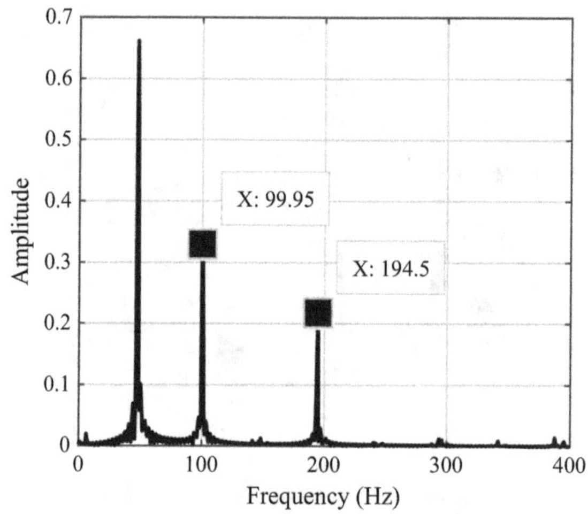

Figure 8.14 BRG fault spectrum

8.5 Conclusion

The MCSA based on FFT is tested in wind turbine pitch system scenario for detecting pitch motor faults. The tests are performed on a detailed model of induction motor, based on the MWFTh. Each fault condition is modelled capturing the physics of fault, and the current signatures are tested. It was noticed that the speed of motor, the duration of operation, fault severity and loading conditions affect the effectiveness of

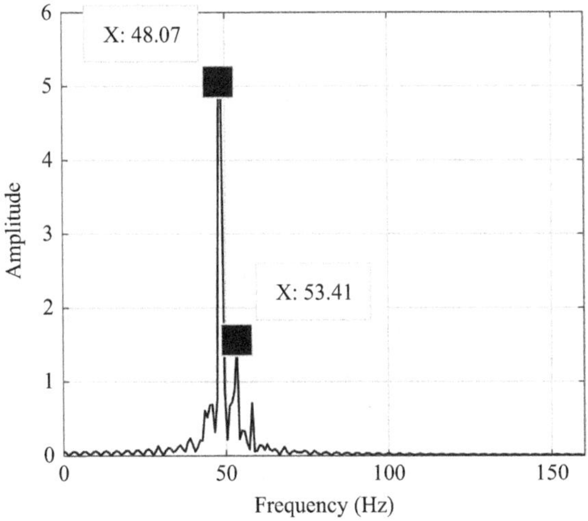

Figure 8.15 BRB fault spectrum

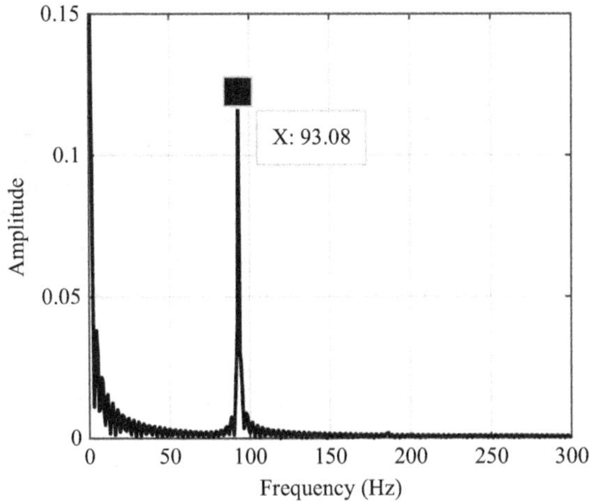

Figure 8.16 i_P-Current spectrum

FFT-based diagnostics. It has been observed that as long as the speed of the motor is relatively constant even for a short duration, there appears to be good chance in detecting the faults. If the recommended practice to rotate the pitch bearing by a large amplitude periodically (once per day) to ensure even spread of lubrication in pitch bearing [36] is followed, then the assumptions on availability of constant speed operation data may be reasonable. Further, a speed sensor or drive's internal speed measurement can be used as a basis to detect the short windows of constant

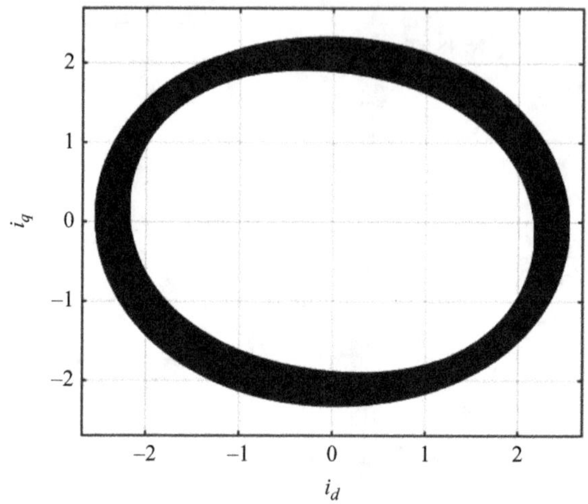

Figure 8.17 dq-Current loci

Table 8.2 Calculated fault frequencies under closed-loop operation

f_{BRG} (Hz)	f_{BRB} (Hz)	f_{AGE_1} (Hz)	f_{AGE_2} (Hz)
100.07	48.36	99.66	783.8
194.15	45.71	1.58	1,014.4
341.7		136.7	1,106.7
		229.10	1,245.9

speed for diagnostics. The diagnostics need not necessarily be continuous and can utilise these short windows of about 100 rotations as demonstrated here. Besides, it is also beneficial to do so as this would require relatively less computational resources and data transfers to maintenance personnel. The stator winding faults, bearing faults and AGE faults appear detectable in relatively steady operating speeds, despite load variation within very short time windows, using conventional MCSA. However, if the speed is inconstant, and very quick start-stop operations are to be considered, then advanced methods based on time-frequency analysis like short-time Fourier transforms, wavelets [37] or other techniques such as the so-called Vienna monitoring method [38] are to be evaluated. Further, if the MCSA technique is found to be reliable, it is well suited for farm-level implementation of pitch system diagnostics for two reasons. One, it can be integrated into the controller as current and speed measurements are usually included in motor drives. If this is not possible, it only requires installation of current sensors, which are inexpensive and clip on to the supply cables with minimal modifications to the existing setup. And two, FFT-based methods are advantageous as they require relatively less computational power. Therefore, this study appears to be a promising start for further analysis and development of reliable diagnostics of electrically operated pitch systems.

Acknowledgements

This work has been funded by Norwegian Centre for Offshore Wind Energy (NORCOWE) under Grant 193821/S60 from Research Council of Norway (RCN). NORCOWE is a consortium with partners from industry and science, hosted by Christian Michelsen Research.

Appendix A

The healthy motor parameters [25] used for modelling the induction motor are detailed in Table A.1.

Table A.1 Motor parameters used for simulation

Parameter	Units	Values
Rotor (effective) axial length l	m	120×10^{-3}
Rotor radius r	m	70×10^{-3}
Air gap length g	m	0.28×10^{-3}
Number of stator turns N_t		156
Number of rotor bars N_r		28
Rotor inertia J	kg/m^2	0.002
Number of poles p		2
Stator resistance r_s	Ω	1.5
Stator leakage inductance L_{ls}	ℍ	7×10^{-3}
Rotor bar resistance r_b	Ω	96×10^{-6}
Rotor bar leakage inductance L_b	ℍ	0.28×10^{-6}
Rotor end ring resistance r_e	Ω	5×10^{-6}
Rotor end ring inductance L_e	ℍ	0.036×10^{-6}

References

[1] S. T. Kandukuri, K. G. Robbersmyr, and H. R. Karimi, "Towards farm-level health management of offshore wind farms for maintenance improvements," *The International Journal of Advanced Manufacturing Technology*, vol. 83, pp. 1557–67, Apr 2016.

[2] C. J. Crabtree, D. Zappalá, and P. J. Tavner, "Survey of commercially available condition monitoring systems for wind turbines," Technical Report, SUPERGEN, Durham University, May 2014.

[3] S. T. Kandukuri, V. K. Huynh, H. R. Karimi, and K. G. Robbersmyr, "Fault diagnostics for electrically operated pitch systems in offshore wind turbines," *Journal of Physics: Conference Series*, vol. 753, p. 052005, Sep 2016.

[4] Douglas-Westwood Limited & BVGAssociates, "Doing business with wind turbine manufacturers: becoming part of their supply chain," Technical Report, Douglas-Westwood Limited & BVG Associates, Jul 2006. Available at https://bvgassociates.com/publications/

[5] J. B. Gayo, "Final publishable summary of results of project ReliaWind," Technical Report, FP7-Energy-2007-1-RTD, 2011.

[6] Y. Lin, L. Tu, H. Liu, and W. Li, "Fault analysis of wind turbines in China," *Renewable and Sustainable Energy Reviews*, vol. 55, pp. 482–90, Mar 2016.

[7] E. A. Bossanyi, "Wind turbine control for load reduction," *Wind Energy*, vol. 6, no. 3, pp. 229–44, 2003.

[8] P. F. Odgaard, J. Stoustrup, and M. Kinnaert, "Fault-tolerant control of wind turbines: a benchmark model," *IEEE Transactions on Control Systems Technology*, vol. 21, no. 4, pp. 1168–82, 2013.

[9] S. Yin, G. Wang, and H. R. Karimi, "Data-driven design of robust fault detection system for wind turbines," *Mechatronics*, vol. 24, no. 4, pp. 298–306, 2014.

[10] J. Jonkman, S. Butterfield, W. Musial, and G. Scott, "Definition of a 5-MW reference wind turbine for offshore system development," Technical Report NREL/TP-500-38060, National Renewable Energy Laboratory, 2009.

[11] B. Jonkman, "Turbsim user's guide: Version 1.50," Tech. Rep. NREL/TP-500-46198, National Renewable Energy Laboratory, 2009.

[12] LM Wind Power, *Blade Summary*, Available at www.lmwindpower.com/Rotor-Blades/

[13] Rothe Erde Thyssen Krupp, Dortmund, Germany, *Slewing Bearings (Catalogue)*, 2007.

[14] D. Townsend, C. Allen, and E. Zaretsky, "Study of ball bearing torque under elastohydrodynamic lubrication," Technical Memorandum NASA TM X-68271, National Aeronautics and Space Administration, 1973.

[15] *Slewing Drives Catalogue*, Brevini Power Transmission S.p.A., 2014. Available at www.brevini.de.

[16] T. Bakka and H. R. Karimi, "H_∞ static output-feedback control design with constrained information for offshore wind turbine system," *Journal of the Franklin Institute*, vol. 350, no. 8, pp. 2244–60, 2013.

[17] *IEC 61400-1 Wind turbine generator systems – Part 1: Design requirements, III Edition*. International Electrotechnical Commission (IEC), 2005.

[18] P. Albrecht, J. Appiarius, R. McCoy, E. Owen, and D. Sharma, "Assessment of the reliability of motors in utility applications – updated," *IEEE Transactions on Energy Conversion*, vol. EC-1, pp. 39–46, Mar 1986.

[19] S. Nandi, H. A. Toliyat, and L. Xiaodong, "Condition monitoring and fault diagnosis of electrical motors – a review," *IEEE Transactions on Energy Conversion*, vol. 20, no. 4, pp. 719–29, 2005.

[20] W. Thomson and M. Fenger, "Case histories of current signature analysis to detect faults in induction motor drives," in *IEEE International Electric Machines and Drives Conference, 2003. IEMDC'03.*, vol. 3, pp. 1459–65, IEEE, 2003.

[21] A. Marques Cardoso, S. Cruz, and D. Fonseca, "Inter-turn stator winding fault diagnosis in three-phase induction motors, by Park's vector approach," *IEEE Transactions on Energy Conversion*, vol. 14, no. 3, pp. 595–98, 1999.

[22] A. Bellini, F. Filippetti, G. Franceschini, and C. Tassoni, "Closed-loop control impact on the diagnosis of induction motors faults," *IEEE Transactions on Industry Applications*, vol. 36, no. 5, pp. 1318–29, 2000.

[23] P. Krause and C. Thomas, "Simulation of symmetrical induction machinery," *IEEE Transactions on Power Apparatus and Systems*, vol. 84, pp. 1038–53, Nov 1965.

[24] T. A. Lipo, *Analysis of Synchronous Machines.* Taylor & Francis Group, Florida, US: CRC Press, 2nd ed., 2012.

[25] A. K. Ibrahim, M. I. Marei, H. S. El-Gohary, and S. A. M. Shehata, "Modeling of induction motor based on winding function theory to study motor under stator/rotor internal faults," in *Proceedings of the 14th International Middle East Power Systems Conference (MEPCON'10)*, pp. 494–500, Dec 2010.

[26] H. Toliyat, S. Nandi, S. Choi, and H. Meshgin-Kelk, *Electric Machines Modeling Condition Monitoring and Fault Diagnosis.* Taylor & Francis Group, LLC, Boca Raton, FL, USA.: CRC Press, 2013.

[27] S. Nandi, R. Bharadwaj, and H. Toliyat, "Performance analysis of a three-phase induction motor under mixed eccentricity condition," *IEEE Transactions on Energy Conversion*, vol. 17, pp. 392–99, Sep 2002.

[28] M. Blodt, P. Granjon, B. Raison, and G. Rostaing, "Models for bearing damage detection in induction motors using stator current monitoring," *IEEE Transactions on Industrial Electronics*, vol. 55, pp. 1813–22, Apr 2008.

[29] A. R. Munoz and T. A. Lipo, "Complex vector model of the squirrel-cage induction machine including instantaneous rotor bar currents," *IEEE Transactions on Industry Applications*, vol. 35, no. 6, pp. 1332–40, 1999.

[30] M. Sahraoui, A. Ghoggal, S. Zouzou, and M. Benbouzid, "Dynamic eccentricity in squirrel cage induction motors – simulation and analytical study of its spectral signatures on stator currents," *Simulation Modelling Practice and Theory*, vol. 16, pp. 1503–13, Oct 2008.

[31] S. Nandi, S. Ahmed, and H. Toliyat, "Detection of rotor slot and other eccentricity related harmonics in a three phase induction motor with different rotor cages," *IEEE Transactions on Energy Conversion*, vol. 16, no. 3, pp. 253–60, 2001.

[32] R. Schiltz, "Forcing frequency identification of rolling element bearings," *Journal of Sound and Vibration*, vol. 24, no. 5, pp. 16–19, 1990.

[33] W. T. Thomson and M. Fenger, "Current signature analysis to detect induction motor faults," *IEEE Industry Applications Magazine*, vol. 7, no. 4, pp. 26–34, 2001.

[34] S. Cruz and A. Cardoso, "Stator winding fault diagnosis in three-phase synchronous and asynchronous motors, by the extended Park's vector approach," *IEEE Transactions on Industry Applications*, vol. 37, no. 5, pp. 1227–33, 2001.

[35] L. Harnefors, *Control of Variable-Speed Drives*. Mälarden University, Sweden, 2003.

[36] T. Harris, J. H. Rumbarger, and C. P. Butterfield, "Wind turbine design guideline DG03: yaw and pitch rolling bearing life," Technical Report NREL/TP-500-42362, National Renewable Energy Laboratory (NREL), 2009.

[37] K. Teotrakool, M. J. Devaney, and L. Eren, "Adjustable speed drive bearing fault detection via wavelet packet decomposition," in *IEEE Instrumentation and Measurement Technology Conference Proceedings*, pp. 22–25, IEEE, Apr 2006.

[38] R. Wieser, M. Schagginger, C. Kral, and F. Pirker, "The integration of machine fault detection into an indirect field oriented induction machine drive control scheme – the Vienna Monitoring Method," in *IEEE Industry Applications Conference. Thirty-Third IAS Annual Meeting*, vol. 1, pp. 278–85, 1998.

Chapter 9

Magnetic bearing for wind turbine power generator shaft: an emulator prototype design for vibration control

Francisco Palacios-Quiñonero[1], Leonardo Acho[1], Josep M. Rossell[1], and Hamid Reza Karimi[2]

9.1 Introduction

Energy is a fundamental element of human societies. In industrialized and urban areas, a huge amount of energy is required by the productive processes, building, commercial and residential activities, transportation, mobility and communications. In rural environments, the energy needs are smaller but equally relevant. Specifically, energy is demanded in agricultural and farming activities, small-scale commerce, cottage industries, domestic uses, transportation, communications, schools and clinics. Currently, about 87% of the global energy demand is covered by fossil fuels that are subject to a limited availability and produce serious environmental issues. These facts have motivated an increasing interest in cleaner and renewable energy sources, such as hydropower, photovoltaic and thermal solar energy, wind energy, tidal energy and geothermal energy [1–3].

Wind energy is playing an increasingly relevant role among renewable energy sources. Fostered by international policies to reduce greenhouse gas (GHG) emissions, an impressive technological development has been carried out in the field over the last 20 years. As a result, big inland and offshore parks of megawatt-scale wind turbines have been extensively deployed, and, nowadays, wind energy has grown into a real alternative to conventional power generation systems in some countries [4–7]. By the end of 2015, wind energy reached a share of about 7% of the global power generation, with a total capacity of 435 GW. The main part of this capacity (97.0%) corresponds to large inland wind turbines, with 422 GW and 210,000 units, while large offshore wind turbines have a contribution of around 2.7%, with 12 GW and 4,000 units. The contribution of small wind turbines (SWTs) is around 0.03%, with an overall capacity of about 1 GW and more than 900,000 units [1,8]. Current average capacity ratings of large-scale wind turbines are about 2 MW, and the trend points towards

[1]Department of Mathematics, Universitat Politècnica de Catalunya (UPC), Spain
[2]Department of Mechanical Engineering, Politecnico di Milano, Italy

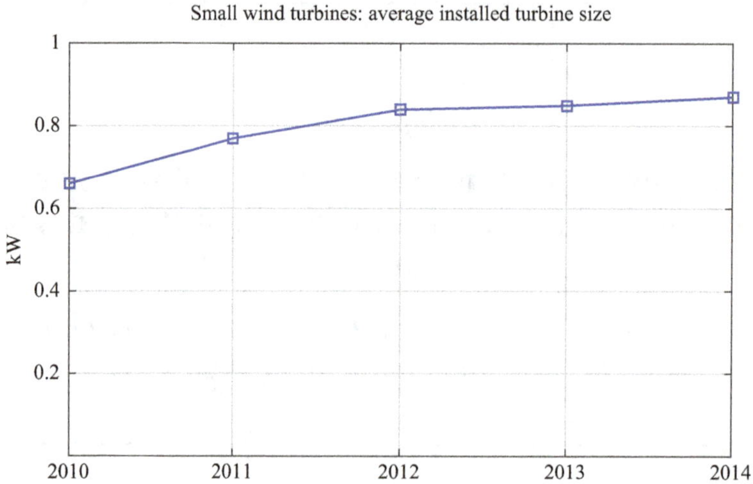

Figure 9.1 Average capacity rating for small wind turbines corresponding to the period 2010–14 (data from Small Wind World Report, WWEA 2016)

more powerful and larger machines with capacities of 4–8 MW [9]. In contrast, the average capacity rating of SWTs is inferior to 1 kW. As displayed in Figure 9.1, the average size of SWTs has slightly increased from 0.66 kW in 2010 to 0.87 kW in 2014. According to the WWEA [8], there were about 945,000 SWTs installed all over the world by the end of 2014. China was by large the most important small-wind country with 689,000 units (73%), followed by the United States of America (USA) with 159,300 units (17%). In addition to the differences in the number of installed units, there are also remarkable differences in the SWTs sizes among countries. As it can be observed in Figure 9.2, Chinese SWTs have an average size of 0.5 kW, while the average capacity rating of SWTs is around three times higher in the USA (1.4 kW) and about ten times larger in the United Kingdom (4.7 kW).

Considering these big numbers, it might seem that SWTs are practically irrelevant as a power generation source. However, it must be highlighted that a large part of the world population is living without access to electricity. Indeed, some estimates point out that one-fifth of the global population has no access to electricity, and the proportion increases in developing countries of Asia and Africa, particularly in rural areas [10–12]. In this context, electrical energy is a critical element in a number of basic technological systems such as illumination, water purification, desalination and cold storage. Moreover, electrical energy is also a key element of modernization, economical development and education, providing access to computers, modern communication systems and distance learning [10,13].

In rural and remote areas, effective and sustainable power generation systems can be designed by means of a proper combination of SWTs, solar photovoltaic cells, small hydroturbines and biodiesel generators. This hybrid approach can provide a suitable solution to the limited capacity and intermittent characteristics of renewable

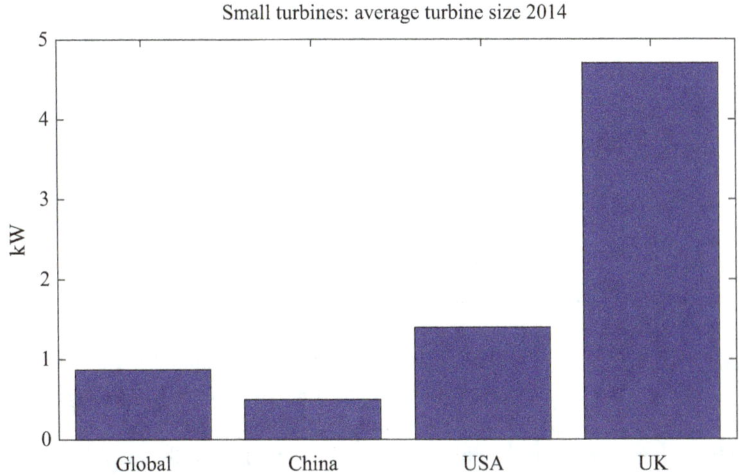

Figure 9.2 Average capacity rating for small wind turbines in 2014 corresponding to China, USA and UK (data from Small Wind World Report, WWEA 2016)

energy resources. It is certainly an unfortunate fact that, despite the favourable wind conditions in wide regions of Latin America, Africa and Asia, the small wind industry has very little impact outside China [14]. Clearly, small wind generation can produce great social and economic benefits in countries with a poor electric grid. A key factor that limits this development is the lack of a local industry that can provide effective and low-cost technological solutions for the different wind and climate conditions and the efficient integration with other energy sources [11,15].

According to the previous discussion, it becomes apparent that SWTs can make a positive impact in the living conditions and development of an important number of human communities worldwide. The objective of this chapter is to demonstrate that SWTs and, in particular, micro-sized wind turbines (μSWTs) can also play a major role in research and education. To this end, we present an experimental platform for active magnetic bearing (AMB) vibration control of a rotary shaft. This particular experimental setup includes an interesting combination of electric, mechanical, electronic, magnetic and control elements and, at the same time, makes it possible to conduct advanced full-scale experimental studies of some fundamental aspects of μSWTs at a very low cost and with very limited resources.

The rest of the chapter is organized as follows: In Section 9.2, SWTs are classified into three more homogeneous subcategories: μSWTs, small-sized wind turbines and medium-sized wind turbines, and their main characteristics and common usages are briefly reviewed. In Section 9.3, a more detailed discussion of the challenges and opportunities associated to μSWTs is presented, paying particular attention to the ecological, social, economical, technological, scientific and educational aspects. In Section 9.4, a minimal background on AMB vibration control is provided, and the main elements of the experimental platform are discussed. In Section 9.5, some final remarks and conclusions are presented. Finally, a readable assembler code for the microcontroller implementation of the proposed control strategy is included in the Appendix.

9.2 Small wind turbines characteristics

Quite surprisingly, a good number of contradictory statements about the main characteristics of SWTs can be found in the literature as, for example, the prevalence of on-grid or off-grid operation, the relevance of the initial installation costs or the financial viability of SWTs. These apparent inconsistencies are mainly motivated by the fact that SWT is an imprecise concept that includes a wide range of designs with very different characteristics, functionalities and applications. The term SWT is defined by opposition to large-scale wind turbines, and, broadly speaking, it includes machines with a maximum rated capacity of 100 kW [13]. A more consistent and meaningful discussion can be performed by introducing a proper classification of SWTs into more homogeneous subcategories [1,2,16–20]. In this work, the following three kinds of SWTs will be considered:

- *Micro-sized wind turbines* (μSWTs), with rated capacity up to 1.5 kW.
- *Small-sized wind turbines* (SSWTs), with rated capacity of 1.5–10 kW.
- *Medium-sized wind turbines* (MSWTs), with rated capacity of 10–100 kW.

Typically, μSWTs operate off-grid, admit an easy combination with photovoltaic panels and small diesel generators and can use batteries for energy storage. In fact, most commercial μSWTs are equipped with an AC/DC converter to transform the output current into direct current suitable for battery charging [21]. A μSWT requires a very modest initial economic investment and can supply a significant part of the electricity demand of an average Chinese household, which is estimated in about 1 kW. An important share of the SWT market in China is related to μSWTs. Thus, for example, from a total number of 165,000 units sold in the Chinese small-wind market in 2011, around a 62% corresponded to the 300 and 600 W models, with 71,200 and 31,000 units, respectively [13]. Clearly, μSWTs have a great potential in rural areas of developing countries. They can also provide effective solutions for small isolated systems in developed countries such as wireless monitoring sensors, road signalization, stand-alone illumination poles or advertising panels [14].

SSWTs can cover the electricity demand of small farms, commerce, cottage industries or small rural communities in developing countries. They can also satisfy the power needs of an average European or US household, which are estimated in 4 and 10 kW, respectively [13]. In developing countries and remote areas, SSWTs are typically operated off-grid or forming part of a local microgrid. In this case, to improve the power supply stability, SSWTs have to be combined with other power generation sources as photovoltaic panels, small hydroturbines or diesel generators. In developed countries, SSWTs are mainly used as complementary on-grid power sources in small farms, commerce and residential applications. Also, in addition

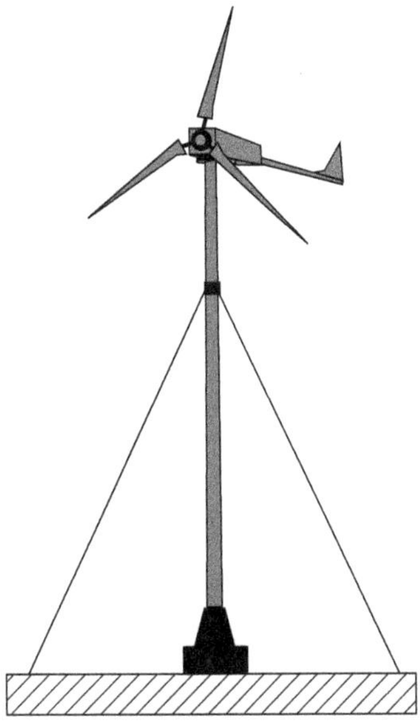

Figure 9.3 Upwind horizontal axis three-blade small wind turbine with a guyed pole mounting

to the typical rural installations, there is an increasing interest in the integration of SSWTs in urban areas [22]. Both, off-grid hybrid power generation systems and grid connected SSWTs are technically complex and require an important financial investment [1,14].

MSWTs can cover the electricity demand of a small community or a medium-sized company in a developed country. In recent years, MSWTs have been installed in some European countries in response to governmental energy policies on GHG reduction. MSWTs require a large financial investment, careful financial and technical feasibility studies and a very specialized installation. Moreover, the effectiveness of MSWTs in reducing GHG emissions and their financial viability are both unclear [6,15].

As it happens with large-scale wind turbines, the most usual layout for small turbines is the horizontal-axis three-blade design schematically displayed in Figure 9.3. This horizontal-axis wind turbine (HAWT) design provides a well-tested technology with low cost and high efficiency. For SSWTs and μSWTs, however, there exists a good number of less common designs, which include different kinds of HAWTs and, also, a wide variety of vertical-axis wind turbines (VAWTs) [14]. Two usual rotor configurations for VAWTs are displayed in Figure 9.4. This kind of rotors are omnidirectional and have a good response to changes in wind direction. Moreover, due to their lower visual impact and reduced noise level, VAWTs are gaining increasing relevance in urban applications [7,14,23]. An urban rooftop mounting and a street pole mounting of a Darrieus μSWT are schematically displayed in Figure 9.5.

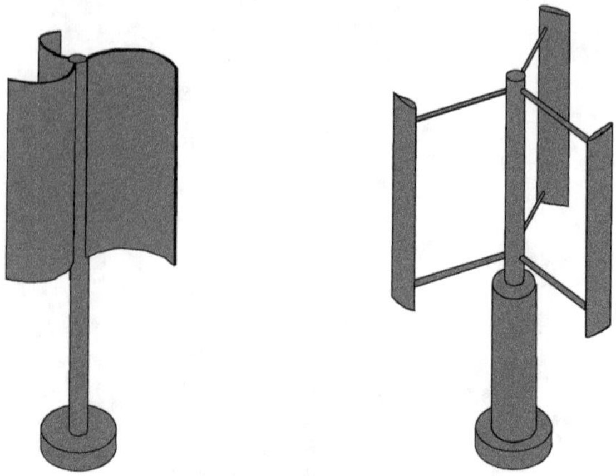

(a) (b)

Figure 9.4 Vertical rotor configurations. (a) Savonius rotor and (b) Darrieus H-rotor

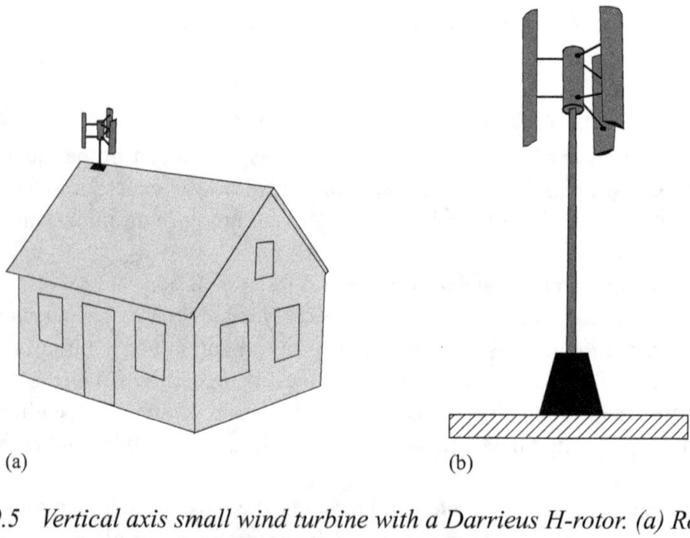

(a) (b)

Figure 9.5 Vertical axis small wind turbine with a Darrieus H-rotor. (a) Rooftop mounting and (b) stand-alone pole mounting

9.3 Micro-sized wind turbines: challenges and opportunities

Most of the open literature on SWTs is focused on small or medium sized turbines. However, μSWTs have some distinctive features that make them objects of singular interest in a large number of fields, including ecology, economy, town planning, industrial design, social and technological development, research and education. In all these fields, μSWTs introduce major theoretical and practical challenges and provide excellent opportunities of developing innovative technological solutions, which can have a remarkable impact in local communities of developing countries.

From an ecological point of view, a combination of μSWTs, small solar panels and biodiesel generators may be a good option to obtain efficient, well dimensioned and more reliable hybrid energy systems. Certainly, μSWTs cannot make a significant contribution in reducing the global GHG emissions. However, this kind of small hybrid generation systems can help to reduce environmental degradation and pollution in areas without grid power supply. They can also help to promote the use of renewable energy sources and develop a culture of sustainable energy [11,15].

Based on the low efficiency, poor wind conditions and high financial costs, the lack of economic viability is one of the main objections to SWTs. These concerns are probably well-founded for SSWTs and MSWTs installed in developed countries with a reliable electric grid that can supply energy at a more competitive price. However, it is quite clear that this kind of economic viability assessment can hardly be applied to μSWTs in areas with no access to grid power supply and, frequently, with a weak and unstable political structure, a defective transportation system and without oil, gas and hydropower local resources. It is worth noting that, far from being a rare exception, the described situation can well be applied to a large number of human communities worldwide [10–12,15].

In urban and residential areas, μSWTs can be installed in rooftop mountings and stand-alone street poles as depicted in Figure 9.5. There are also more advanced placements such as inter-building mountings and ducted areas inside of tall buildings. In this context, the economical drawbacks can be overcome by the positive effect on air pollution, the willingness of extending eco-awareness among the urban population and the possibility of integration in advanced urban design. In this case, it should be highlighted the positive visual impact of the elegant vertical axis designs and the relevance of minimizing the noise emissions, which can certainly be a major issue in residential areas [3,17,18,22,24].

In addition to an extreme attention to cost optimization, there is a number of other important factors that make μSWTs an interesting and challenging field for industrial design. These factors include enhanced efficiency requirements, reliability, robustness, durability, low noise and visual impact, size and weight optimization, modularity and portability. Also relevant is the ease of installation, maintenance, energy storage and combination with other power generation sources [16,25]. Providing cost-effective and practical solutions to these design challenges can certainly constitute a source of local technological development. It is worth highlighting that "local" is the keyword in this point, and that the efforts should probably be focused on adapting well-tested technologies to the local conditions, such as the local wind characteristics, local labour force, local climate, local row materials and local availability of other power sources [10,16–18,21,26–28].

Certainly, the development of a local industry and the availability of skilled manpower are key elements for a successful deployment of μSWTs in developing countries. Particularly relevant is the availability of spear parts and well-qualified local technicians that can perform a proper installation and, if necessary, adapt the turbines to the local conditions. From this point of view, μSWTs constitute a good opportunity for social and economical development [10,11]. It has also to be noted that *local* should by no means be understood as *small*. Thus, for example, the present discussion could be applied to the adaptation of a Chinese or European μSWT design to the particular local conditions in a wide geographical area in Africa [11,15].

From a scientific and technological point of view, μSWTs are a perfect instance of mechatronical systems that require the integrated action of advanced electrical, magnetic, mechanical, electronic and control principles. For example, some fundamental topics are advanced vibration control and the integration with other power generation sources. From this perspective, it is clear that μSWTs can be a proper field to develop theoretical and applied research in a wide variety of advanced topics. Certainly, some of these characteristics can also be found in other fields as, for example, large-scale wind turbines. However, μSWTs have two very singular and distinctive features: (i) the possibility of carrying out full-scale physical experimentation at a very low cost and with very limited resources, and (ii) the possibility of making a direct impact in the local society and local industry of developing countries.

Finally, μSWTs can also be a very important element in technological and scientific educational programmes, providing an endless number of exciting challenges and practical applications suitable for all education levels, from the very basic to the advanced post-graduated level. For instance, at a basic level, μSWTs can be an excellent tool to illustrate and motivate the fundamental concepts in physics, mechanics, electricity or electronics, and also to provide an early glimpse of the mechatronical multidomain approach. At a higher level, μSWTs can be equally effective in introducing advanced problems in nonlinear dynamics, constrained optimization, multidomain and stochastic modelling, control design, sound and vibration analysis, fault identification and heath monitoring. In all these fields, μSWTs bring the possibility of performing a huge variety of full-scale experimental projects at a very low cost and with a clear real-life impact. Some examples of practical topics that could be suitable for this kind of experimental projects are the following: study, modelling and characterization of the local wind conditions, construction of open-source wind turbine designs using local raw materials, study and design of improved starting mechanisms, assessment of the noise impact and design of noise mitigation strategies, development of optimized designs for maintenance, development of low-cost and well-adapted mounting systems, design of hybrid wind–solar microsystems, etc.

9.4 Experimental platform for AMB vibration control

To provide a practical demonstration of the ideas previously discussed, we present in this section an experimental platform for AMB vibration control of a rotary shaft. The design of this particular experimental setup provides a clear illustration of the mechatronical multidomain approach, involving the combined action of basic concepts and elements of electricity, mechanics, electronics, electromagnetism and control and, at the same time, makes it possible to conduct advanced full-scale experimental studies of some fundamental aspects of μSWTs, such as vibration control, fault detection or health monitoring. In order to be consistent with the low-cost constraints and the usage of limited resources, the concepts of cost-efficiency and reusing have been applied at two different levels. At a primary level, inexpensive parts have been selected, and available parts and materials have been reused as much as possible, making the necessary adaptations. At a secondary level, the whole experimental platform has also been designed as a reusable system, which can be used in different and more complex experiments with minimal modifications. To this end, a digital control approach has been selected. This approach allows introducing assembler programming and implementing different control strategies with minor electronic modifications. It also eases the computational scalability, making it possible an easy implementation of more complex control strategies by means of a simple microprocessor change. Thus, although the presentation in this paper has been focussed on active vibration control, it is clear that the proposed platform can also be a useful experimental tool in other advanced applications as, for instance, fault detection and health monitoring, design and optimization of the electromagnets and sensors, and design and implementation of advanced control configurations.

To facilitate a clearer presentation, the section has been divided into two parts. In the first part, a minimal background on AMBs and vibration control is provided. Next, in the second part, the main elements of the proposed experimental platform are discussed. Additionally, the assembler code for the microcontroller implementation of the control procedure is presented in the Appendix.

9.4.1 Active magnetic bearings and vibration control

Rotating machinery plays a central role in practically all the industrial sectors. Common instances include motors, turbines, compressors, generators and machining tools. Rotational bearings are machine elements that support and constrain the motion of rotational parts, such as axles and shafts. A simple type of rotational bearing is the journal bearing schematically depicted in Figure 9.6. In this case, the shaft (or journal) slides over the bearing surface, which incorporates some kind of lubrication system to reduce friction losses and prevent mechanical wear.

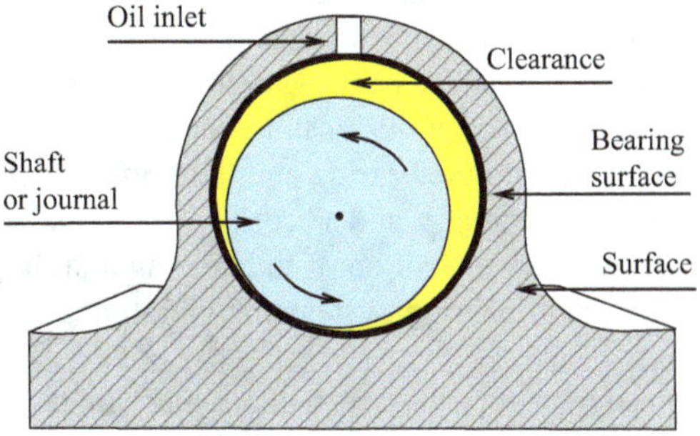

Figure 9.6 Journal bearing with oil lubrication

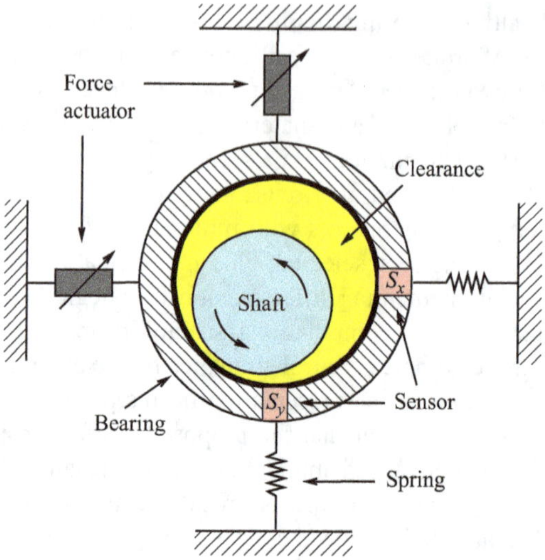

Figure 9.7 Schematic representation of an active journal bearing system with mechanical actuators

Mechanical vibrations produced by mass imbalance and the action of external disturbances are a major issue in rotating systems. The objective of vibration control is to mitigate the damaging effect of mechanical vibrations by keeping the vibration amplitude within safe operational limits [29–32]. To meet the demanding performance requirements of modern high-speed rotational applications, a variety of sophisticated passive bearing strategies have been developed, such as high-speed ball bearings, static air bearings, dynamic air bearings, gas foil bearings and floating ring bearings. These advanced passive bearings can effectively reduce the friction power losses, but they have a limited capacity of mitigating the mechanical vibrations.

An alternative approach consists in considering an actively controlled bearing system that combines bearing elements, actuation devices and sensors to produce an intelligent bearing system with improved vibration control capabilities [32–35]. A schematic representation of a mechanically actuated active bearing system is presented in Figure 9.7. In this case, the control forces are exerted by piezoelectric or hydraulic actuators, and preloaded springs can be included to simplify the actuation scheme. The interaction between the bearing and the shaft is performed through the lubricant fluid that fills the clearance, and a proper set of sensors is implemented to obtain information of the relative position of the bearing and the shaft. The generic control structure of an active bearing system for vibration control is displayed in Figure 9.8.

AMBs are able to support rotary shafts without mechanical contact by means of properly controlled magnetic forces. AMBs operate without lubricant, can eliminate wear and friction and allow implementing advanced vibration control strategies. Due to their unique characteristics, AMBs are used in advanced technical and industrial

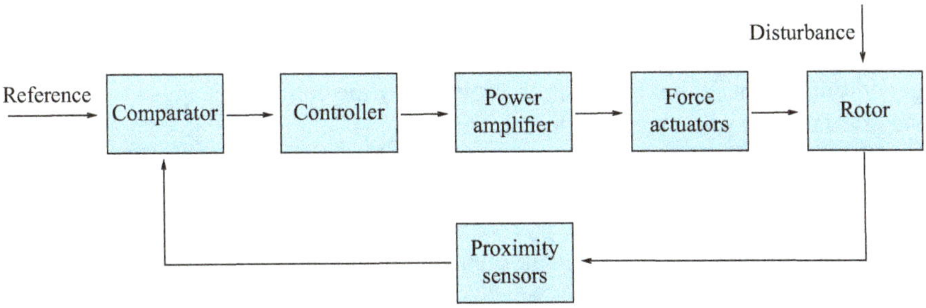

Figure 9.8 General structure of an active bearing system

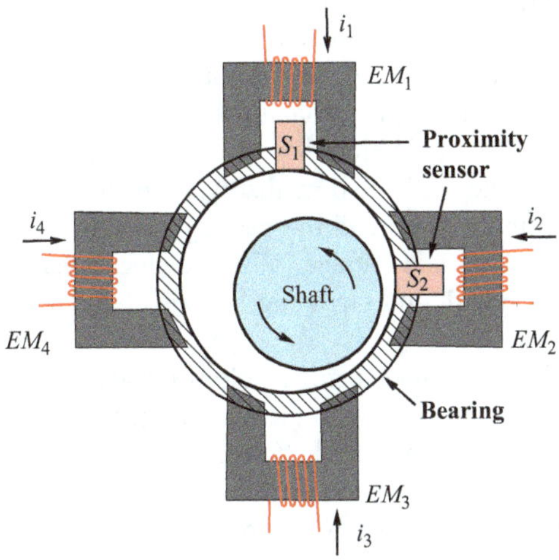

Figure 9.9 Schematic representation of an active magnetic bearing

applications such as power generators, high-speed machine tools, turbo molecular pumps, power storage flywheels and rotary gyros [36–39]. As schematically shown in Figure 9.9, the actuation forces in AMBs are produced by a set of electromagnets that support the rotor in magnetic levitation. The control system is an essential element of AMBs, which are open-loop unstable. In addition to the electromagnetic actuators, a typical AMB digital control system includes multiple proximity sensors, an embedded controller for fast real-time operation and a power amplifier (see Figure 9.8) [40–42].

In wind turbines, bearings are used to support the rotor shaft, the gearbox, the generator shaft and the yaw system. The cyclic loads generated by the rotor introduce bending effects and misalignments on the drive train, producing undesirable vibrations. Additional vibrations are generated by wind speed variations and turbulences. Bearings are a relevant element in wind turbines reliability. For example, it has been estimated that more than 50% of the gearbox faults are generated by bearing failures [43]. Active bearing systems, and in particular AMBs, can be used in wind turbines to simultaneously improve performance and implement vibration control. As an additional benefit, the information provided by the control system sensors can be used for fault detection and health monitoring [44–47].

9.4.2 Experimental platform design

In this section, we present the main elements of the proposed experimental platform for AMB vibration control. Specifically, the following elements are briefly discussed: electromagnetic actuators, rotor shaft, proximity sensors, comparators, microcontroller, power drivers, velocity sensor, motor driver and digital control strategy. A schematic representation and an overall view of the platform are displayed in Figures 9.10 and 9.11, respectively. The *electromagnetic coils* have been handmade using a C-form ferromagnetic core and 0.8 mm ∅ varnished wire. A total amount of 2.0 kg of copper wire has used to make the four electromagnets, which have been mounted in opposed vertical and horizontal pairs, as it can be seen in the front view presented in Figure 9.13(a). A lateral view of the coils EM_1 and EM_2 can be appreciated in Figures 9.11 and 9.12. Two methacrylate plates have been used to make the frame that supports the coils and the bearing housing [see Figure 9.13(b)]. To facilitate the transportation and storage, all the elements have been assembled on a 70×70 cm plywood board, and clamping brackets have been used to secure the mounting supports.

The *rotor shaft* has been made using a hollow plastic tube with an iron sleeve in the front end and a flexible coupling in the rear end [see Figures 9.10 and 9.13(b)], which has been linked to the motor shaft with adhesive tape. To generate a rotor imbalance, a small mass has been attached to the rotor shaft (see Figures 9.10 and 9.12). The rotor is driven by a small AC motor (0.55 A, 26 W).

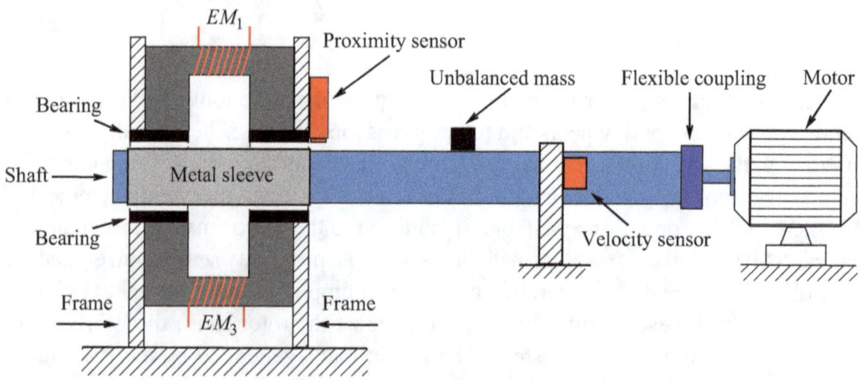

Figure 9.10 Schematic representation of the experimental platform for active magnetic bearing vibration control

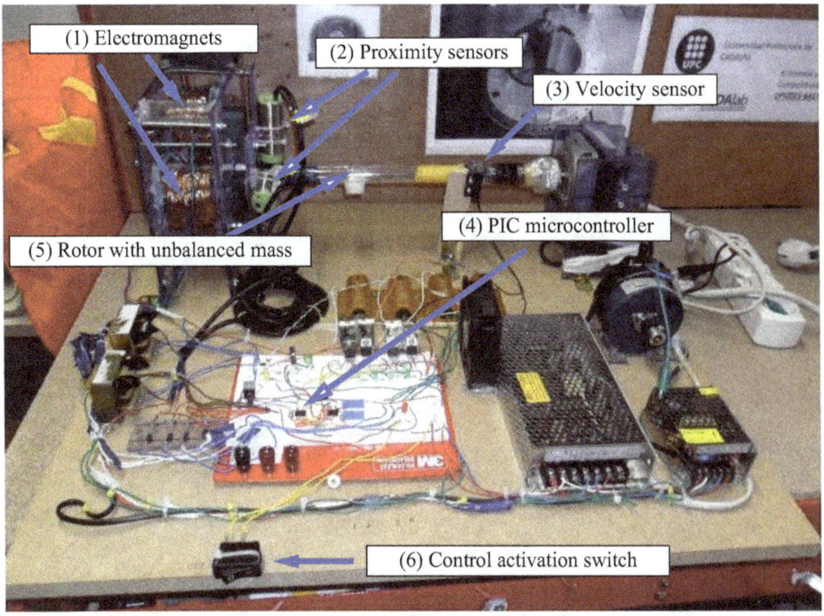

Figure 9.11 *Overall view of the experimental platform for active magnetic bearing vibrational control and the main elements*

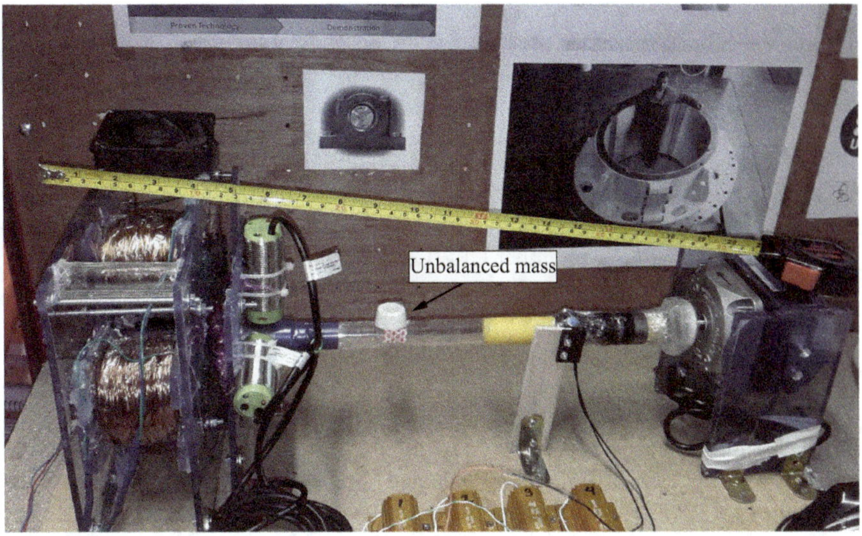

Figure 9.12 *Dimensional perspective and closer view of the electromagnet coils, proximity sensors, rotor shaft and unbalanced mass*

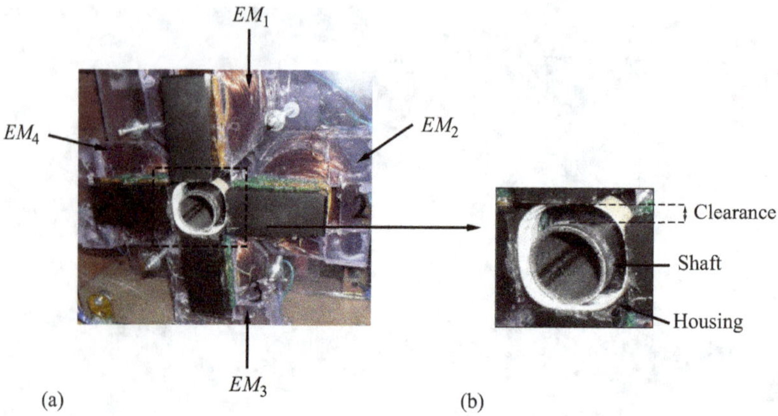

Figure 9.13 Frontal view of the active magnetic bearing. (a) Electromagnetic actuators (coils) and (b) close-up view of the rotor shaft and the bearing housing

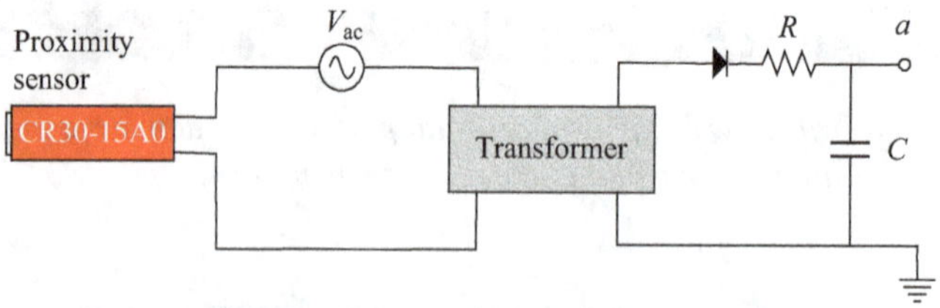

Figure 9.14 Block schematic diagram for the instrumentation design of the proximity sensors

To detect the vertical and horizontal displacements of the shaft, two capacitive type *proximity sensors* have been implemented in a perpendicular layout (see Figures 9.9 and 9.11). These sensors are of the series CR30 and type AC2 manufactured by Autonics with part number CR30-15A0. The electronic instrumentation of each proximity sensor is depicted in Figure 9.14, where the transformer performs a reduction from 220 V (AC) to 15 V (AC), the resistor and capacitor have the values $R = 330\ \Omega$ and $C = 47\ \mu\text{F}$, respectively, and the point a denotes the sensor output.

Two differential *comparators* are used to tune the digital controller. These comparators are implemented with the popular IC LM293p, taking as inputs the proximity signals S_1 and S_2 supplied by the proximity sensors and the reference voltages V_{ref1} and V_{ref2}, which are manually adjusted through mechanical potentiometers (see Figure 9.15). The binary proximity outputs P_1 and P_2 are activated when the corresponding proximity signals exceed the threshold indicated by the reference voltages.

To implement the digital vibration control, we have selected the 8-bit *microcontroller* PIC12F508 by Microchip, which provides high-performance and flash

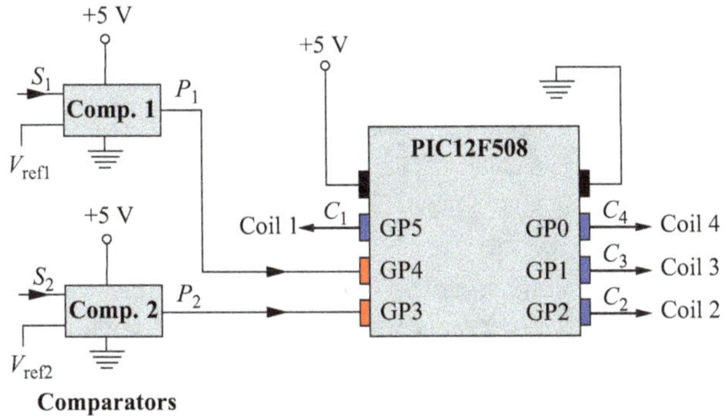

Figure 9.15 Schematic diagram of the general PIC microcontroller wiring and the comparator system. The inputs S_1 and S_2 denote the outputs of the corresponding proximity sensors; P_1 and P_2 are the binary proximity signals and C_j, $j = 1,\ldots,4$, denotes the digital control signal for the jth electromagnet

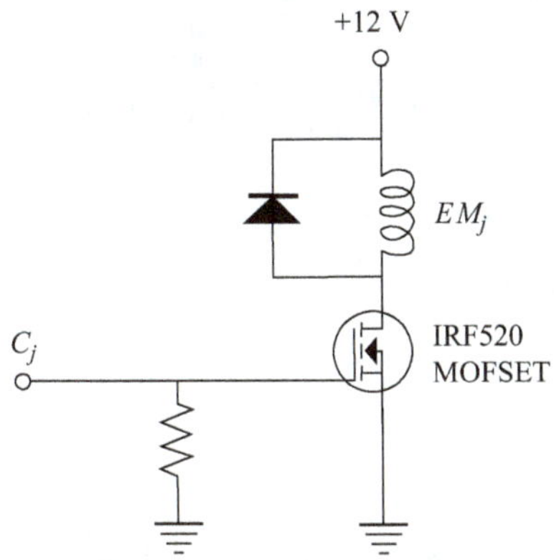

Figure 9.16 Power electronic driver for the jth electromagnetic coil. C_j denotes the corresponding control signal generated by the microcontroller

programming flexibility at a very low cost and also has a proper number of I/O ports to process the sensors and actuation signals. Specifically, the binary proximity signals P_1 and P_2 have been allocated in the input ports $GP4$ and $GP3$, respectively, and the binary activation signals for the electromagnetic coils have been allocated in the output ports $GP0$, $GP1$, $GP2$ and $GP5$ as indicated in Figure 9.15. The *power drivers* have been implemented using IRF520 Power MOFSETs, which allow a cost-effective and fast digital manipulation of the electromagnetic coils (see Figure 9.16).

To capture the rotational angular speed of the rotor shaft, an economical realization of a magnetic *velocity sensor* has been performed [48]. The selected design

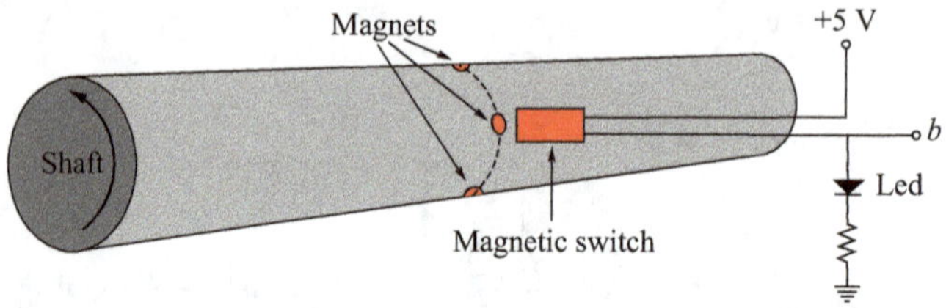

Figure 9.17 Angular velocity magnetic sensor based on a reed-switch position sensor excited by permanent magnets. Point b is the sensor output

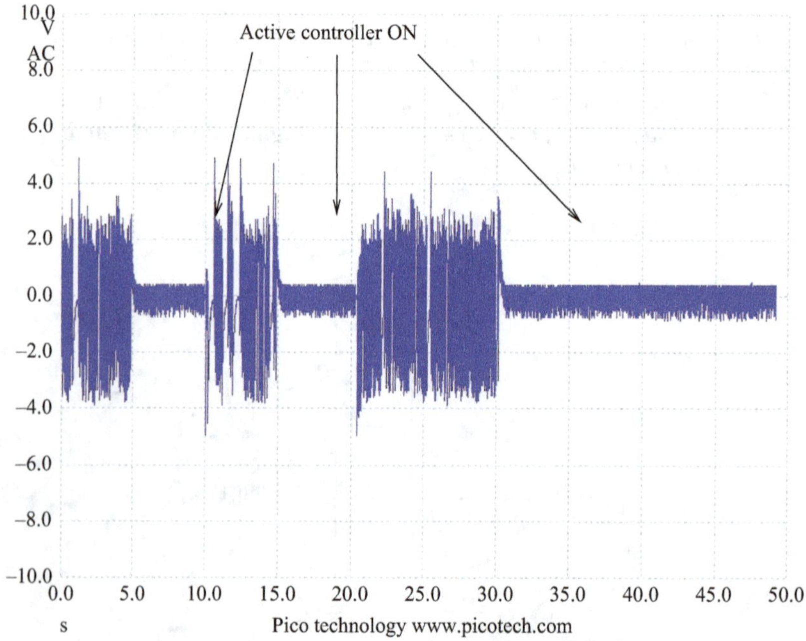

Figure 9.18 Output of the vertical proximity sensor. Signal obtained with a PicoScope data acquisition card in AC mode operation. The AMB vibration control system is manually activated at the time instants 5, 15 and 30 s and deactivated at the time instants 0, 10 and 20 s

is based on a magnetic reed-switch Cherry MP201901 excited by four permanent magnets attached to the shaft as indicated in Figure 9.17. Point *b* is the sensor output.

The adopted control strategy is very simple and intuitive. The electromagnets EM_1 and EM_3 are associated to the vertical sensor S_1 and the electromagnets EM_2

and EM_4 are associated to the horizontal sensor S_2 (see Figure 9.9). Each coil is deactivated when the associated sensor indicates that the rotor is close to the electromagnet and activated otherwise. The state of the four coils is sequentially updated and, at the end of each cycle, all the magnets are deactivated to keep the coils temperature within safety levels. A readable assembler code [49] to implement this simple control strategy in the PIC12F508 microcontroller has been included in the Appendix. The overall control system is activated/deactivated by means of a manual switch [element number (6) in Figure 9.11].

The plot in Figure 9.18 illustrates the behaviour of the proposed AMB vibration control system. The displayed signal has been obtained at the output of the vertical proximity sensor (point *a* in Figure 9.14) with a PicoScope 2205A data acquisition card in AC mode operation. As indicated in the figure, the control system is switched on at the time instants 5, 15 and 30 s and switched off at the time instants 0, 10 and 20 s. Looking at the plot, it can be clearly appreciated that, despite its simplicity, the proposed control system exhibits a well-balanced and uniform closed loop response and attains a remarkable reduction of the rotor vibration. Moreover, it is also worth mentioning that, according to the data provided by the velocity sensor, the effect of the vibration control system on the rotor angular velocity is practically negligible.

9.5 Conclusions and final remarks

In this chapter, the relevance and potential of micro-sized wind turbines (μSWTs) have been discussed and the following facts have been stated:

- In wide areas of developing countries with a poor grid power supply, μSWTs can be an important factor of social, economic and technological development.
- From an educational point of view, μSWTs can be a valuable source of motivation, helping to provide a practical presentation of basic and advanced mechatronical principles. Moreover, μSWTs can also be an excellent field for cooperative and project-based learning, allowing to define a wide variety of experimental projects with a direct real-life application.
- From a research perspective, μSWTs constitute a unique opportunity to perform full-scale physical experimentation on advanced research topics at a very low cost and with very limited resources.

To provide a practical demonstration of the μSWTs potential in academic and research experimentation, a low-cost platform for active magnetic bearing vibration control has been presented.

Acknowledgements

This work was partially supported by the Spanish Ministry of Economy and Competitiveness under Grant DPI2015-64170-R (MINECO/FEDER).

Appendix

Assembler program for the proposed AMB vibration control.

```
list p=12F508 ;list directive to define the processor
#include "p12f508.inc"
_CONFIG _OSC_IntRC & _WDT_OFF & _CP_OFF & _MCLRE_OFF
ORG   0000H
;port configuration
    MOVLW b'00011000'
    TRIS 6
;timer configuration
    MOVLW b'11010111'
    OPTION
INI NOP
;MAGNETIC ACTUATOR 1
    BTFSS  GPIO,04H
    BSF    GPIO,05H
    BTFSC  GPIO,04H
    BCF    GPIO,05H
;MAGNETIC ACTUATOR 3
    BTFSS  GPIO,04H
    BCF    GPIO,01H
    BTFSC  GPIO,04H
    BSF    GPIO,01H
;MAGNETIC ACTUATOR 2
    BTFSS  GPIO,03H
    BSF    GPIO,02H
    BTFSC  GPIO,03H
    BCF    GPIO,02H
;MAGNETIC ACTUATOR 4
    BTFSS  GPIO,03H
    BCF    GPIO,00H
    BTFSC  GPIO,03H
    BSF    GPIO,00H
;ZEROS
        BCF    GPIO,05H
        BCF    GPIO,01H
    BCF    GPIO,02H
    BCF    GPIO,00H
GOTO INI
END
```

References

[1] World Energy Resources 2016. World Energy Council. 2016.

[2] Satpathy AS, Kastha D, Kishore NK. Modeling of small wind turbine characteristics. In: IEEE 6th International Conference on Power Systems (ICPS), New Delhi, India. 2016; 1–6.

[3] Cooney C, Byrne R, Lyons W, O'Rourke F. Performance characterisation of a commercial-scale wind turbine operating in an urban environment, using real data. Energy for Sustainable Development. 2017; 36:44–54.

[4] Njiri JG, Söffker D. State-of-the-art in wind turbine control: trends and challenges. Renewable and Sustainable Energy Reviews. 2016; 60:377–393.

[5] McKenna R, Ostman vd Leye P, Fichtner W. Key challenges and prospects for large wind turbines. Renewable and Sustainable Energy Reviews. 2016; 53:1212–1221.

[6] Bortolini M, Gamberi M, Graziani A, Manzini R, Pilati F. Performance and viability analysis of small wind turbines in the European Union. Renewable Energy. 2014; 62:629–639.

[7] Alaimo A, Esposito A, Messineo A, Orlando C, Tumino D. 3D CFD analysis of a vertical axis wind turbine. Energies. 2015; 8:3013–3033.

[8] Pitteloud JD, Gsänger S. 2016 Small Wind World Report. World Wind Energy Association. 2016.

[9] Renewables 2016 Global Status Report. REN21. 2016.

[10] Sen R, Bhattacharyya SC. Off-grid electricity generation with renewable energy technologies in India: an application of HOMER. Renewable Energy. 2014; 62:388–398.

[11] Kamp LM, Vanheule LFI. Review of the small wind turbine sector in Kenya: status and bottlenecks for growth. Renewable and Sustainable Energy Reviews. 2015; 49:470–480.

[12] Bhandari B, Poudel SR, Lee KT, Ahn SH. Mathematical modeling of hybrid renewable energy system: a review on small hydro-solar-wind power generation. International Journal of Precision Engineering and Manufacturing-Green Technology. 2014; 1:157–173.

[13] Small Wind World Report 2013. World Wind Energy Association. 2013.

[14] Bukala J, Damaziak K, Kroszczynski K, *et al.* Small wind turbines: specification, design, and economic evaluation. In: Wind turbines – design, control and applications. IntechOpen, London, 2016; 73–93.

[15] Kealy T. Stakeholder outcomes in a wind turbine investment; is the Irish energy policy effective in reducing GHG emissions by promoting small-scale embedded turbines in SME's? Renewable Energy. 2017; 101:1157–1168.

[16] Kishore RA, Priya S. Design and experimental verification of a high efficiency small wind energy portable turbine (SWEPT). Journal of Wind Engineering and Industrial Aerodynamics. 2013; 118:12–19.

[17] Ying P, Chen YK, Xu YG. An aerodynamic analysis of a novel small wind turbine based on impulse turbine principles. Renewable Energy. 2015; 75: 37–43.

[18] Bukala J. Analytical-theoretical study of SWTs in terms of pre-selection of the aero layouts best suited for further optimization. Part 1. STOW Project Technical Report. 2014.

[19] De Santoli L, Albo A, Garcia DA, Bruschi D, Cumo F. A preliminary energy and environmental assessment of a micro wind turbine prototype in natural protected areas. Sustainable Energy Technologies and Assessments. 2014; 8:42–56.

[20] Nguyen TH, Al Hosani K, Al Sayari N. Grid integration improvement for single-phase inverters of small wind turbines under distorted voltage conditions. International Journal of Electrical Power and Energy Systems. 2017; 87:144–153.

[21] Breeze P. Small wind turbines. In: Wind power generation. Academic Press – Elsevier, London, 2016; 67–73.

[22] Tabrizi AB, Whale J, Lyons T, Urmee T. Rooftop wind monitoring campaigns for small wind turbine applications: effect of sampling rate and averaging period. Renewable Energy. 2015; 77:320–330.

[23] Mabrouk IB, El Hami A, Walha L, Zghal B, Haddar M. Dynamic vibrations in wind energy systems: application to vertical axis wind turbine. Mechanical Systems and Signal Processing. 2017; 85:396–414.

[24] Jafari SAH, Hassanli S, Kwok KCS. Performance analysis of a small wind turbine mounted inside a tall building. In: 18th Australasian Wind Engineering Society Workshop. McLaren Vale, South Australia. 2016; 6–9.

[25] Bukala J, Damaziak K, Karimi HR, *et al.* Modern small wind turbine design solutions comparison in terms of estimated cost to energy output ratio. Renewable Energy. 2015; 83:1166–1173.

[26] Pourrajabian A, Mirzaei M, Ebrahimi R, Wood D. Effect of air density on the performance of a small wind turbine blade: a case study in Iran. Journal of Wind Engineering and Industrial Aerodynamics. 2014; 126:1–10.

[27] Pourrajabian A, Afshar PAN, Ahmadizadeh M, Wood D. Aero-structural design and optimization of a small wind turbine blade. Renewable Energy. 2016; 87:837–848.

[28] Shu L, Liang J, Hu Q, *et al.* Study on small wind turbine icing and its performance. Cold Regions Science and Technology. 2017; 134:11–19.

[29] Zhou S, Shi J. Active balancing and vibration control of rotating machinery: a survey. The Shock and Vibration Digest. 2001; 33:361–371.

[30] Krysinski T, Malburet F. Mechanical vibrations: active and passive control. Wiley-ISTE, London, 2007.

[31] Allaire PE, Lewis DW, Knight JD. Active vibration control of a single mass rotor on flexible supports. Journal of the Franklin Institute. 1983; 315:211–222.

[32] Arias-Montiel M, Silva-Navarro G, Antonio-García A. Active vibration control in a rotor system by an active suspension with linear actuators. Journal of Applied Research and Technology. 2014; 12:898–907.

[33] Rajasekhar M, Srinivas J. Active vibration control in engine rotors using electromagnetic actuator system. Journal of Mechanical Design and Vibration. 2014; 2:25–30.

[34] Looser A, Tüysüz A, Zwyssig C, Kolar JW. Active magnetic damper for ultrahigh-speed permanent-magnet machines with gas bearings. IEEE Transactions on Industrial Electronics. 2017; 64:2982–2991.
[35] Chen WJ, Gunter EJ. Dynamics of rotor-bearing systems. Trafford Publishing, Canada. 2010.
[36] Toh CS, Chen SL. Design, modeling and control of magnetic bearings for a ring-type flywheel energy storage system. Energies. 2016; 9:1–19.
[37] Zhu H, Ding S, Jv J. Modeling for three-pole radial hybrid magnetic bearing considering edge effect. Energies. 2016; 9:1–15.
[38] Koehler B, Denk J, Van Maanen G, Lang M. Applying standard industrial components for active magnetic bearings. Actuators. 2017; 6:1–9.
[39] Han B, Zheng S, Li H, Liu Q. Weight-reduction design based on integrated radial-axial magnetic bearing of a large-scale MSCMG for space station application. IEEE Transactions on Industrial Electronics. 2017; 64:2205–2214.
[40] Humphris RR, Kelm RD, Lewis DW, Allaire PE. Effect of control algorithms on magnetic journal bearing properties. Journal of Engineering for Gas Turbines and Power. 1986; 108:624–632.
[41] El-Shafei A, Dimitri AS. Controlling journal bearing instability using active magnetic bearings. Journal of Engineering for Gas Turbines and Power. 2009; 132:1–9.
[42] Knospe CR, Fedigan SJ, Hope RW, Williams RD. A multitasking DSP implementation of adaptive magnetic bearing control. IEEE Transactions on Control Systems Technology. 1997; 5:230–238.
[43] Machado HD, Araújo AM, Bouchonneau N. A review of wind turbine bearing condition monitoring: state of the art and challenges. Renewable and Sustainable Energy Reviews. 2016; 56:368–379.
[44] Ouyang H, Liu F, Zhang G, *et al.* Vibration suppression for rotor system of magnetic suspended wind turbines using cross-feedback-based sliding mode control. In: IEEE/SICE International Symposium on System Integration. 2015; 112–115.
[45] Wymore ML, Van Dam JE, Ceylan H, Qiao D. A survey of health monitoring systems for wind turbines. Renewable and Sustainable Energy Reviews. 2015; 52:976–990.
[46] Luo N, Vidal Y, Acho L. Wind turbine control and monitoring. Springer, Cham, Switzerland, 2014.
[47] Saidi L, Ali JB, Bechhoefer E, Benbouzid M. Wind turbine high-speed shaft bearings health prognosis through a spectral Kurtosis-derived indices and SVR. Applied Acoustics. 2017; 120:1–8.
[48] Gamazo-Real JC, Vázquez-Sánchez E, Gómez-Gil J. Position and speed control of brushless DC motors using sensorless techniques and application trends. Sensors. 2010; 10:6901–6947.
[49] Predko M. Programming and customizing the PIC microcontroller. McGraw-Hill. 2008.

Chapter 10

Condition monitoring and diagnostics of wind turbine power train

Steven Chatterton[1] *and Paolo Pennacchi*[1]

10.1 Background

Wind and solar energies are playing a leading role for achieving the climate goals given by the reduction of the global CO_2 emissions. The landmark was given by climate negotiations at the UNFCCC's COP21 Climate Conference, held in Paris in 2015, where a long-term target has been set for a 100 per cent emission-free power sector by 2050.

Wind generation represents the source of renewable energy that showed the highest growing in the last years. The annual reports of the World Wind Energy Association [1] and Global Wind Energy Council (GWEC) [2] show that the worldwide wind capacity reached about 450 GW in 2016, representing an annual installed capacity growth rate of about 22 per cent with respect to 2015 (+63 GW). In relative terms, the wind power contributes to only the 4.7 per cent of the world's electricity demand.

In total, more than 100 countries are today using wind power on commercial basis, and more than 300k large wind turbines are spinning around the world at the end of 2015.

By the end of 2015, eight countries had more than 1 GW of installed capacity including, in order, China (145 GW with an annual energy production of 186 TW h), USA (75 GW, annual energy production of 190 TW h), Germany (45 GW), India (25 GW), Spain (23 GW), United Kingdom (14 GW), Canada (11 GW) and France (10 GW).

The same countries show the highest installed capacity growth rate in 2015: China (+30 GW, share of 49 per cent), USA (+9 GW – 14 per cent), Germany (+6 GW – 10 per cent), Brazil (+3 GW – 4 per cent) and India (+3 GW – 4 per cent).

Onshore is already the cheapest option of wind power generation, with costs of about 30–40 EUR/MW h instead of 100 EUR/MW h for offshore installations.

Offshore option takes advantages by higher wind speeds and less difficulties in finding a proper installation site. This notwithstanding, the total offshore installed capacity represents only the 3 per cent of total wind turbine installations (about

[1] Department of Mechanical Engineering, Politecnico di Milano, Italy

12 GW), where more than 90 per cent of all offshore installations are in waters off the coasts of European countries at the average distance of about 43 km from the shore. The remaining installations are in China, Japan and South Korea.

The average size of offshore wind turbines is 4.2 MW. Higher costs and installation complexity compared to onshore are the two main drawbacks that can be mainly reduced by developing larger turbines.

A special case is represented by the small wind turbine world market given by turbine size less than 100 kW. At the end of 2014, the cumulative total of less than 1 million small wind turbines was installed all over the world, for a total installed capacity of about 830 MW. In terms of installed capacity, China accounts for 41 per cent of the global capacity, USA for 30 per cent and United Kingdom for 15 per cent [3].

According to the 2016s GWEC report, the growth of the whole wind energy sector will continue. At the end of the year 2020, the expected installed power will reach 800 GW globally, with an average annual installed capacity growth rate of about 6 per cent. The increase of installed capacity is also achieved by building larger and larger wind turbines up to 10 MW, able to achieve the highest wind energy harvest. Large turbines have higher failure rates and require more maintenance than small turbines.

The popularity of wind energy increased the interest of companies and research centres about the technical problems related to wind turbines.

10.2 Failure analysis

Horizontal-axis wind turbine is the most common configuration, which is mainly composed of a tower and a nacelle. The nacelle can follow the direction of the wind by means of a yaw system. The main horizontal rotor in the nacelle connects the hub of the turbine blades to the generator. Turbine blades can rotate along their longitudinal axis, by means of the pitch system, increasing the wind energy harvest.

Several configurations are available, depending on the presence of the active pitch system, on constant or variable rotational speed and on the generator type (squirrel-cage induction gen., permanent magnet synchronous gen., double fed induction gen., electrically excited synchronous gen.) [4].

Two main architectures can be identified:

- Geared wind turbines, with a gearbox that connects the blade hub to the high-speed generator.
- Direct drive wind turbines, with a gearless drive train equipped with low-speed generator with many poles.

The most typical installed configuration employs turbines equipped with three blades, active pitch control, double-fed induction generators and running in variable speed conditions [4]. The life for a new wind turbine is around 20 years, where the highly cost components are, in descending order, tower (about 30 per cent of the total cost), blades (25 per cent), gearbox (15 per cent) and generator/converter (10 per cent).

The failures for a wind turbine follow the well-known bathtub curve, whose three phases can be modelled as a Power Law Process, with a failure intensity function [5] given by

$$\lambda(t) = \rho\beta t^{\beta-1} \tag{10.1}$$

High rates of failures can be observed in the initial phases of operation ($\beta < 1$) and in the end of life ($\beta > 1$) of the machine. A minimum and constant failure rate can be observed in the period of useful life ($\beta = 1$). When $\beta = 1$, the process is a homogeneous Poisson process, in which the times between failures are identically distributed exponential random variables.

Several studies have been conducted to investigate the failures and downtimes of wind turbine components.

Faulstich *et al.* in [6] analysed the failure rates and downtimes of about 1,500 onshore wind turbines in Europe in the period from 1989 to 2006. Data had been extracted by a collection of 64,000 maintenance and repair reports. The survey showed that onshore wind turbines in Europe achieved a high availability of 95–99 per cent. The study revealed that electrical and electronic components fail more frequently (in average every 2–2.5 years) than mechanical system. Conversely, the replacement of mechanical parts required longer downtimes. By considering the drive train, the annual failure rate is about 0.05, whereas the average downtime is about 5.7 days. The authors also highlighted a correlation of failure rate to the wind turbine location. Turbines located near the coast and in the highlands suffered higher failure rates, due to the weather conditions (humidity, temperature and saline concentration). They concluded that similar faults can be expected as offshore wind turbine directly derived from onshore turbines, but the downtimes would be expected higher due to the limited accessibility.

Spinato *et al.* in [7] investigated the reliability of more than 6,000 wind turbines of assorted sizes and manufacturers in Denmark and Germany, over more than 10 years from 1994 to 2004. Their analysis is based on available statistical data (WindStats) about failures of European wind turbines. They concluded that the components with the highest failure rate are (in descending order) electrical system, blades and hub, converter, generator, hydraulics and gearbox. Larger wind turbines are complex system and, therefore, have lower reliability than smaller wind turbines. The conclusion about the reliability of direct drive and geared turbines is interesting. The reliability of wind turbine gearboxes is the same of equivalent size industrial gearboxes, but they exhibit higher downtime. In particular, gearboxes are operating in the wear phase of the bathtub curve ($\beta = 1.2 - 1.8$). On the contrary, generators and converters of direct drive turbines show higher failure rates, especially in the initial phases of operation, but lower downtime than gearboxes in geared turbines. Furthermore, the failure rates of electrical equipment, in direct drive turbines, are greater than equivalent generators/converters in other industries [7].

Ribrant and Bertling in [8] investigated the failures of wind turbines installed in Sweden, Finland and Germany occurred during 1997–2005. They concluded that the components with the highest failure rate are similar to the results obtained in [7] and are, in descending order, electrical system, sensors, blades/pitch, control system, hydraulic and gears. Also in this case, they highlighted that the gearbox is the component that exhibits the highest downtime. In particular, they analysed the failures occurred in the gearboxes, discovering that the highest failure rate affects the bearings.

From a mechanical point of view, gearbox gained increased attention from industries and academia due to its long downtime. McNiff in [9] showed that the most frequent failure is due to the bearings in planetary gear trains. Highest failure rates are achieved by the planet bearings, the intermediate shaft bearing and the high-speed shaft bearings. Low-speed shaft bearings showed the lowest failure rates. The same author participates to a wide experimental project, described in [10], where a lot of tests had been performed on the artificially damaged gearbox by means of a suitable test rig.

Bearing failures are also the main cause of generator faults, about 40 per cent for induction generators as highlighted by Popa *et al.* in [11]. Other faults in generators occurred in the stator and rotor windings. In this case, vibration can occur due to unbalances of the air-gap flux and phase current and to torque pulsation.

10.3 Maintenance policies

The studies by Morthorst and Jacobsen [12] and Amicarelli and Tresca [13] showed that the design and installation of wind turbines are quite expensive, especially offshore. In particular, operational and maintenance (O&M) costs are significantly higher than other energy sources, and they represent a big share of the life cycle cost of wind farms. O&M costs have been estimated in the range of 20–25 per cent/kW h produced, a level comparable to nuclear power, and far above the 7–17 per cent of the fossil fuels (coal and gas) and the 10 per cent of photovoltaic. Other statistical examinations showed that O&M cost may be even higher than expected, since unpredicted failures in wind turbine components may cause dramatically high replacement costs and production losses. An SKF study [14] underlined that an early identification of faults is important for the consequent damages that a severe failure would otherwise cause. For instance, they reported an unnoticed defect in a $1,500 bearing could lead to $100,000 of gearbox replacement costs, $50,000 of generator rewinding costs and further $70,000 just for accessing the failed components. The correct choice of the maintenance strategy and the related tools is, therefore, crucial to reduce the overall maintenance cost and increases the availability of the wind turbine as well. Run-to-failure maintenance (RFM) and scheduled maintenance (SM) are the two common maintenance approaches. RFM is generally the most expensive maintenance strategy for complex system. SM can lead to over-maintenance, that is excessive maintenance activities, resulting in high maintenance costs. Conversely, under-maintenance in SM approach can result in unexpected failures.

The condition-based maintenance (CBM) is a more efficient approach than traditional maintenance techniques (RFM and SM). If properly and effectively implemented (see e.g. Jardine *et al.* [15]), CBM can significantly reduce maintenance costs by reducing the number of unnecessary SM operations. It consists in monitoring the state of a machine and detecting any deterioration of the components, based on parameters obtained by measurements. Conversely, the CBM approach is more complex and requires knowledge, expertise and a set of sensor and data acquisition systems. The CBM process requires three fundamental phases: (i) data acquisition, where the information of the component is collected and stored; (ii) data processing, where the collected information is processed to obtain a suitable fault indicator; and (iii) fault detection and isolation, where the proper decision-making is executed to achieve fault diagnosis and plan efficient maintenance policies.

10.4 Condition monitoring and diagnostics

The efficiency of a CBM approach relies on the continuous condition monitoring of the physical state of the components. Condition monitoring data for wind turbine power train can be successfully obtained by three technologies that still dominate the field of wind turbines: vibration, electric and acoustic emission (AE) measurements.

Vibration-based condition monitoring (VCM) is the most common technique especially for rotating machinery and components, such as gears and bearings. Vibration data can be acquired by different sensors as a function of the characteristic fault frequency of the component to be monitored. Proximity sensors are employed in the monitoring of the low-frequency shaft vibrations, when the fault frequencies (unbalance, misalignment, crack among all) are typically a multiple of the rotating speed of the shaft. Accelerometers can be successfully employed for the monitoring of vibrations of rolling element bearings or gears, when the characteristic fault frequency is higher enough than the frequency resolution of the acquired signal. Typically, the characteristic fault frequency of bearings is a fraction of the rotational speed of the bearing components. Accelerometers are installed close to the housing of the bearings or on the case of the gearbox. AE methods can be successfully used for the diagnostics of bearings located on the slow-speed shaft, allowing the early detection of cracking and pitting of the rolling elements of the bearings. For instance, Swiszcz *et al.* in [16] described a typical sensor installation for a 660 kW wind turbine. Seven accelerometers are mounted on the drive train (generator, gearbox and main bearing, in axial and radial directions) and two laser meters measure displacements of the low-speed shaft, in vertical and horizontal direction.

Vibrations strongly depend on the operating conditions, such as load, speed and temperature. Therefore, as shown in the works of Wakui and Yokoyama [17], Kusiak *et al.* [18] and Zimroz *et al.* [19], auxiliary measurements are needed for an optimal VCM; therefore, a series of additional sensors are installed, such as strain gauges, power and rotational speed transducers. In particular, variation of the fault indicators in the time domain needs to be analysed under the same operating conditions.

Vibration analysis is mainly aimed at diagnosing the health status of bearings. Studies in [20–24] report examples of experimental tests on the gearbox's bearings, while the papers [25,26] are focused on the bearings of the generator. The bearings of the main shaft have been investigated in the paper of An *et al.* [27]. The VCM has also been proposed for monitoring gearbox outer case, by Liu *et al.* [28], and for diagnosing rotor imbalance and misalignment, by Li *et al.* [29].

Electric measurements, such as current, voltage and power, represent the second most common technique for wind turbine condition monitoring being simpler and cheaper than vibration analysis. Generators and accumulators of wind turbines can be continuously monitored by using cheap current/voltage sensors installed on the generator. Al Ahmar *et al.* [30] showed that dangerous situations such as air-gap eccentricity, broken rotor and bearing damages are identified by analysing rotor and stator currents measured in the generator. Also, Yazidi *et al.* [31] showed a case history in which electrical asymmetry of a wind turbine generator rotor has been successfully diagnosed by monitoring both stator and rotor currents. In the studies of Xiang *et al.* [32] and Amirat *et al.* [33], stator currents have been successfully used to detect bearing faults of direct-drive wind turbines. Other studies [34,35] confirmed that signatures of wind turbine generator bearing faults are successfully extracted using stator current measurements.

AE is another condition-monitoring tool, mostly used in CBM as a structural non-destructive testing, which applies piezoelectric sensors in the very high-frequency range (ultrasonic). This technique is mainly used for detecting cracks in structures and blades. In particular, changes in amplitude/velocity of the waves can be monitored by measuring the frequency or phase characteristics in order to detect an incipient crack and its propagation. The literature provides some examples of AE both for the diagnostics of wind turbine bearings and gears, as in the work by Li *et al.* in [36]. An interesting example of combined AE and VCM on gearbox and generator shafts during service has been reported by Soua *et al.* in [37].

10.5 Signal processing and fault identification

The techniques based on vibration analysis, electric measurements and AEs produce a large amount of data, consisting of long digital signals, mainly due to the high sampling frequencies required. Therefore, digital signal processing algorithms must be employed to extract diagnostic information and highlight potential symptoms of faults. Basic statistical indexes, such as statistical moments and maximum/minimum values, are often used to set alarms to avoid catastrophic failures. In the study of Zheng and Xu [38] on the diagnostics of wind turbine bearings, a list of the most commonly used indexes has been provided: the root mean square value (rms), the peak-to-peak amplitude, the crest factor, the skewness and the kurtosis. However, this approach is insufficient to extract detailed diagnostic information to identify early stage faults. For this reason, different techniques have been developed and combined in complex algorithms to process the signal. The signal may be processed by mainly three steps: (i) order tracking (OT), (ii) signal enhancement (mainly demodulation) and (iii) spectral transformation.

OT consists of using a reference revolution signal (i.e. tachometer/keyphasor or encoder) to transform the signal from the time domain to the angular domain of the shaft. The use of OT allows performing diagnostics also in case of speed fluctuations or operating transients.

OT allows splitting the signal in two different components: those synchronous with the shaft harmonics and those not directly depending on them. This is a first very effective way to separate damage symptoms originating from distinct parts of the drive train.

OT is performed by resampling the signal using a fixed angular step. This is obtained by interpolating the signal sampled at a constant sample rate. This procedure is called computed OT. Other possibilities to perform fast online OT are the Velocity Synchronous Discrete Fourier Transform, proposed by Borghesani *et al.* [39], and the Vold–Kalman filter, described by Vold *et al.* [40].

Signal enhancement allows highlighting the diagnostic information and attenuating/removing the components not related to the health status of the machine. The most common signal-enhancement techniques belong to the families of auto-selective filtering and amplitude and frequency/phase demodulation.

Auto-selective filtering consists of a filter bank and an index to evaluate the potential of each spectral band to contain useful diagnostic information. Xingjia *et al.* [41] used a Morlet-wavelet filter bank and kurtosis index to select the best band for the analysis of a signal of a cracked tooth gear of a wind turbine. The kurtosis, as an index of impulsiveness, is commonly used for the selection of bands in case of bearing and gear damages. Conversely, the band kurtosis index $\kappa_{l,h}$, that is the kurtosis index of a filtered signal in the frequency band $f_l - f_h$, can be successfully used for bearing diagnostics.

Anyway, the amplitude demodulation remains the most used approach for the diagnostics of bearings.

By considering a local defect in one of the rolling elements of the bearing, such as inner ring, outer ring, balls/rollers or cage, the vibrations originate from the periodic contact, in an ideal case, between the rolling elements of the bearing with the damage when the shaft rotates. After each impact, the system vibrates at its natural frequencies.

Therefore, the signal of a damaged rolling element bearing consists of a repetition of system response to the impulses given by the train of impacts. The typical ideal vibration signal for a defect in the outer ring of a rolling element bearing is shown in Figure 10.1.

The period and the amplitude of the repetitions can provide information about the type and the level of the damage.

The theoretical period or the frequency of repetition of impacts, called characteristics damage frequency, can be easily obtained from the geometry of the bearing (D_P = pitch diameter, D_R = roller diameter, Z = number of roller, α = contact angle) and the frequency of revolution of the inner (f_I) and outer (f_O) rings.

For a ball bearing, the damaged frequencies are

$$BPFI = \frac{1}{2} Z(f_O - f_I) \left(1 + \frac{D_R}{D_P} \cos\alpha\right) \tag{10.2}$$

$$BPFO = \frac{1}{2} Z(f_O - f_I) \left(1 - \frac{D_R}{D_P} \cos\alpha\right) \tag{10.3}$$

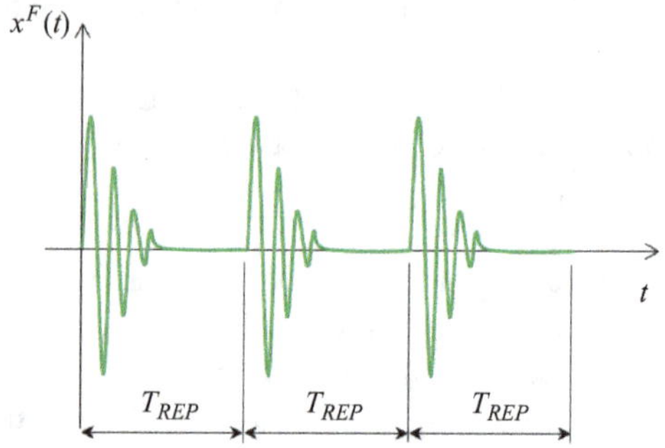

Figure 10.1 Typical vibration signal for a defect in the outer ring of a rolling element bearing (ideal signal)

$$FTF = \frac{1}{2} f_I \left(1 - \frac{D_R}{D_P}\cos\alpha\right) + \frac{1}{2} f_O \left(1 + \frac{D_R}{D_P}\cos\alpha\right) \tag{10.4}$$

$$BSF = \frac{1}{2}(f_O - f_I)\frac{D_P}{D_R}\left(1 - \left(\frac{D_R}{D_P}\cos\alpha\right)^2\right) \tag{10.5}$$

where BPFI is the ball pass frequency for a defect on the inner race, BPFO is the ball pass frequency for a defect on the outer race, FTF is the fundamental train frequency, that is the frequency of the cage and BSF is the ball spin frequency. The presence of a defect on the roller/ball is given by twice the BSF due to the two consecutive contacts of the defect on the outer and inner races for a revolution of the roller/ball.

The signal in Figure 10.1 can be analytically described by the convolution of the impulse train of the excitation $\delta(t - iT)$ due to the damage (impact of rolling elements to the defect) with the impulse response of the structure $h(t)$:

$$x(t) = \sum_i A_0 \cdot \delta(t - iT) \otimes h(t) = \sum_i A_0 \cdot h(t - iT) \tag{10.6}$$

In real bearings, several phenomena can occur, such as fluctuation of load, roller–cage interactions, presence of loaded zone and unloaded zone, slipping, variation of oil-film thickness and deformations. The result is a not constant period of the damaging phenomenon (Figure 10.2), where the delay between two consecutive impacts is called as jitter τ. The corresponding model is given by

$$x(t) = \sum_i A_0 \cdot h(t - iT - \tau_i) + n(t) \tag{10.7}$$

The vibration signal in Figure 10.1 can be approximated by an amplitude modulation of the resonant vibrations of the structure [carrier signal $C(t)$ in Figure 10.3(a)] by the cyclic force caused by the fault [$A_m(t)$ in Figure 10.3(b)]:

$$x(t) = A_m(t) \cdot C(t) \tag{10.8}$$

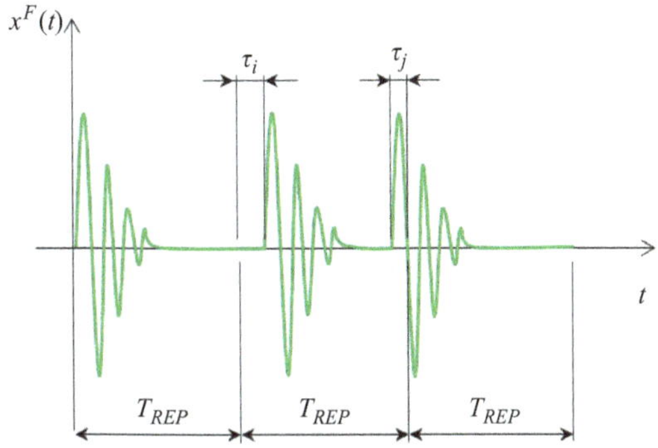

Figure 10.2 Typical vibration signal for a defect in the outer ring of a rolling element bearing (more realistic signal)

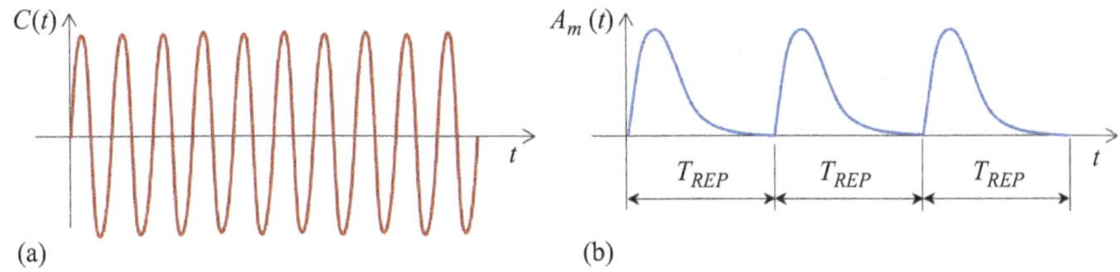

Figure 10.3 Carrier component (a) and modulating signal (b)

The plain application of Fourier transform to a faulty bearing signal rarely provides clear symptoms of faults. The spectrum may appear dominated by the natural frequencies of the system, without any clear symptom related to the bearing.

To extract the information regarding the modulating signal (the one strictly related to the damage), the Hilbert transform is generally used to obtain the so-called envelope signal, which corresponds to the instantaneous amplitude (power) of the signal. The envelope analysis consists of the following steps:

1. Preprocessing the raw signal to remove the deterministic part (i.e. the part that is synchronous with the shaft rotation) and enhance the signal-to-noise ratio. The residual signal is obtained by subtracting the deterministic component (the set of data obtained after applying the synchronous averaging) from the raw signal
2. Filter the residual signal $x(t)$ by means of a band-pass filter. Filtering is the most critical part of the entire process. At the medium and low frequencies, the signal shows a lot of modulations due to other mechanical components and mechanical

noise. Therefore, it is necessary to filter the signal in the neighbourhood of the frequency of the carrier f_c (resonance frequency of the bearing system):

$$x(t)\xrightarrow{\Im}X(f)\overset{Filtering}{\longrightarrow}X^F(f)\xrightarrow{\Im^{-1}}x^F(t) \tag{10.9}$$

3. Evaluating the analytic signal (of the filtered signal) $x_a^F(t)$. An analytic signal is a complex-valued function that has no negative frequency components. The real part is given by the signal itself and the imaginary part can be obtained by the Hilbert transform:

$$x_a^F(t) = x^F(t) + \mathrm{i}\cdot\tilde{x}^F(t) \tag{10.10}$$

4. Evaluating the envelope of the filtered signal that is the absolute value of the analytic filtered signal:

$$env\left[x^F(t)\right] = \left|x_a^F(t)\right| = \sqrt{\left(x^F(t)\right)^2 + \left(\tilde{x}^F(t)\right)^2} \tag{10.11}$$

Actually, instead of the envelope, it is preferred to use the square of the envelope:

$$env_2\left[x^F(t)\right] = \left|x_a^F(t)\right|^2 = \left(x^F(t)\right)^2 + \left(\tilde{x}^F(t)\right)^2 \tag{10.12}$$

5. Performing the Fourier transforms the square of the envelope to obtain the envelope spectrum (ES):

$$ES = \Im\left[env_2\left[x^F(t)\right]\right] = \Im\left[\left|x_a^F(t)\right|^2\right] \tag{10.13}$$

Also in this case, it is preferred to use the squared ES (SES):

$$SES = \left|\Im\left[\left|x_a^F(t)\right|^2\right]\right|^2 \tag{10.14}$$

In the case of a damaged bearing, it is possible to highlight the clear presence of frequency components close to the theoretical damage frequencies of the bearing in the SES. In this way, it is possible to diagnose the type of the defect. Conversely, an estimation of the level of the damage can be obtained by suitable elaboration of the amplitude of the actual frequency component of the defect.

The main critical point of the envelope analysis is the selection of the frequency band for the demodulation of the vibration signal. Usually, the definition of the frequency band based on the band-Kurtosis index is performed by mean of the Fast Kurtogram diagram [42], or by the Protrugram [43], in which the value of BK is reported as a function of the central frequency for a fixed bandwidth. The constancy in the time of this optimal frequency band is also critical. External phenomena, such as the temperature gradient of the bearings, can change the optimal bands suggested by the two diagrams as can be highlighted by the KurtoMap diagram in [44].

Regarding the severity of the damage, being the rms of the signal equal to the zero-frequency component of the SES, Borghesani *et al.* demonstrated in [45] the relationship between the band kurtosis indicator and the SES:

$$\kappa_{l,h} = \frac{\sum_{k=0}^{h-l} SES_{l,h}[k]}{SES_{l,h}[0]} \tag{10.15}$$

Another useful indicator is given by the ratio of content component (RCC) [45], associated to the cyclic band q, p:

$$RCC_{l,h}^{p,q} = \frac{\sum_{k=p}^{q} SES_{l,h}[k]}{\sum_{k=o}^{h-l} SES_{l,h}[k]} \tag{10.16}$$

The peaks in the SES can be normalized by the statistical threshold $TH_{SES}[\bar{f}_{damage}]$, in correspondence of the damage frequency, obtaining the STR index (SES to threshold ratio):

$$STR[\bar{f}_{damage}] = \frac{SES[\bar{f}_{damage}]}{TH_{SES}[\bar{f}_{damage}]} \tag{10.17}$$

The statistical threshold is defined in the domain of the cyclic frequencies [46] and represents the frequency distribution of coloured noise signal of healthy bearing with the probability p:

$$TH_{SES}(\alpha) = \chi^{-1}(p, 2) \cdot \left\{ \frac{1}{2} \cdot \left| R_x\left(\left| \Im\left(|x_a(t)|^2 \right) \right|^2 \right) \right| \right\} \tag{10.18}$$

where χ is a 2 dof Chi-square distribution.

Similar thresholds can be defined for the band kurtosis indicator ($\kappa = 2$ for a healthy bearing) and RCC index.

Eventually, the presence of damage of rolling element bearings can be detected by the deviation of the fault indicator during the bearing life.

Conversely, frequency demodulation (FD) is commonly used for gear diagnostics, where damaged teeth cause an irregularity of the gear contact. This, in turn, generates a phase variation of the signals, which show a short-term acceleration/deceleration profile when the damaged tooth is transferring power. Different FD and mixed AD/FD techniques have been proposed in the past. In the field of wind turbines, empirical mode decomposition (EMD) has been successfully applied by Feng *et al.* [47], to detect two types of sun gear damage (worn gear and chipped tooth) in different operating conditions. EMD decomposes the signal in non-predetermined variable frequency components, with different spectral content, called intrinsic mode functions. In the study by Liu *et al.* [48], a gear box cracked gear was successfully diagnosed by using the local mean decomposition technique. This method decomposes the signal into product functions and each product function corresponds to a product of an envelope signal and a frequency-modulated signal [49]. Yang *et al.* [50] applied bivariate empirical mode decomposition to current measurements for the diagnosis of the rotor mechanical unbalance in wind turbine.

References

[1] WEA. *World Wind Energy Association Report* 2016. http://www.wwindea.org. 2016.

[2] GWEC. *Global Wind Energy Council Report* 2015. http://www.gwec.net. 2015.

[3] WWEA. *Small Wind World Report* 2016. http://www.wweindea.org. 2016.

[4] Pinar Pérez J.M., García Márquez F.P., Tobias A., Papaelias M. 'Wind turbine reliability analysis'. *Renewable Sustainable Energy Reviews* 2013;23: 463–472.

[5] Tavner P.J., Xiang J., Spinato F. 'Reliability analysis for wind turbines'. *Wind Energy* 2007;10:1–18.

[6] Faulstich S., Hahn B., Tavner P.J. 'Wind turbine downtime and its importance for offshore deployment'. *Wind Energy* 2011;14(3):327–337.

[7] Spinato F., Tavner P.J., Van Bussel G.J.W., Koutoulakos E. 'Reliability of wind turbine subassemblies'. *IET Renewable Power Generation* 2009;3(4): 387–401.

[8] Ribrant J., Bertling L.M. 'Survey of failures in wind power systems with focus on Swedish wind power plants during 1997–2005'. *IEEE Transaction on Energy Conversion* 2007;22(1):167–173.

[9] McNiff B. 'The Gearbox reliability collaborative'. *Reliability Workshop Proc. 2nd Sandia National Laboratories Wind Turbine*; 2007.

[10] Link H., LaCava W., van Dam J., McNiff B., Sheng S., Wallen R., McDade M., Lambert S., Butterfield S., Oyague F. 'Gearbox Reliability Collaborative Project Report: Findings from Phase 1 and Phase 2 Testing'. *Technical Report NREL/TP-5000-51885* 2011.

[11] Popa L.M., Jensen B.B., Ritchie E., 'Condition monitoring of wind generators'. *Proc. IAS Annual Meeting*; 2003.

[12] Morthorst P.E., Jacobsen H, 'EWEA: wind energy – the facts – costs & price', *EWEA Executive Summary*, 2004;2:94–110.

[13] Amicarelli V., Tresca F.A. 'Considerazioni economiche sulla produzione di energia eolica', *Energia Ambiente e Innovazione* 2011;6:41–48.

[14] SKF, 'Managing the wind: reducing kilowatt-hour costs with condition monitoring'. *Refocus* 2005;6(3):48–51.

[15] Jardine A.K.S., Lin D., Banjevic D. 'A review on machinery diagnostics and prognostics implementing condition-based maintenance'. *Mechanical Systems and Signal Processing* 2006;20(7):1483–1510.

[16] Swiszcz G., Cruden A., Booth C., Leithead W. 'A data acquisition platform for the development of a wind turbine condition monitoring system'. *Proceedings of 2008 International Conference on Condition Monitoring and Diagnosis*, CMD 2008; 2007.

[17] Wakui T., Yokoyama R. 'Wind speed sensorless performance monitoring based on operating behavior for stand-alone vertical axis wind turbine'. *Renew Energy* 2013;53:49–59.

[18] Kusiak A., Zhang Z., Verma A. 'Prediction, operations, and condition monitoring in wind energy'. *Energy* 2013;60:1–12.

[19] Zimroz R., Bartelmus W., Barszcz T., Urbanek J. 'Diagnostics of bearings in presence of strong operating conditions non-stationarity – a procedure of load-dependent features processing with application to wind turbine bearings'. *Mechanical System and Signal Processing* 2014;46(1):16–27.

[20] Yang W., Little C., Court R. 'S-Transform and its contribution to wind turbine condition monitoring'. *Renew Energy* 2014;62:137–146.

[21] Zhou F. 'Research on online monitoring and diagnosis of bearing fault of wind turbine gearbox based on undecimated wavelet transformation'. *Proceedings – 2010 IEEE Youth Conference on Information, Computing and Telecommunications, YC-ICT 2010*; 2010.

[22] Shenggang Y., Xiaoli L., Ming L. 'Bearing fault diagnosis of a wind turbine using maximum likelihood detection'. *Proceedings of 2011 International Conference on Electronic and Mechanical Engineering and Information Technology, EMEIT 2011*; 2011.

[23] Shenggang Y., Xiaoli L., Ming L. 'Bearing fault detection and diagnosis of a wind turbine'. *2011 International Conference on Consumer Electronics, Communications and Networks, CECNet 2011 – Proceedings*; 2011.

[24] Zhang X., Yang S., Li X. 'Wind turbine bearing condition monitoring based on high frequency resonance method'. *2011 International Conference on Electronics, Communications and Control, ICECC 2011 – Proceedings*; 2011.

[25] Urbanek J., Barszcz T., Antoni J. 'Time-frequency approach to extraction of selected second-order cyclostationary vibration components for varying operational conditions'. *Measurement* 2013;46(4):1454–1463.

[26] Sun H., Zi Y., He Z. 'Wind turbine fault detection using multiwavelet denoising with the data-driven block threshold'. *Applied Acoustics* 2014;77:122–129.

[27] An X., Jiang D., Li S. Zhao M. 'Application of the ensemble empirical mode decomposition and Hilbert transform to pedestal looseness study of direct-drive wind turbine'. *Energy* 2011;36(9):5508–5520.

[28] Liu W.Y., Zhang W.H., Han J.G., Wang G.F. 'A new wind turbine fault diagnosis method based on the local mean decomposition'. *Renewable Energy*, 2012;48:411–415.

[29] Li J., Chen X., Du Z., Fang Z., He. Z. 'A new noise-controlled second-order enhanced stochastic resonance method with its application in wind turbine drivetrain fault diagnosis', *Renewable Energy* 2013;60:7–19.

[30] Al Ahmar E., Choqueuse V., Benbouzid M. E. H., Amirat Y., El Assad J., Karam R., Farah S. 'Advanced signal processing techniques for fault detection and diagnosis in a wind turbine induction generator drive train: a comparative study'. *Energy Conversion Congress and Exposition (ECCE), Atlanta, GA, Sep. 12–16 2010*: 3576–3581.

[31] Yazidi A., Capolino G.A., Filippetti F., Casadei D., 'A new monitoring system for wind turbines with doubly-fed induction generators', *in 2006 IEEE Mediterranean Electrotechnical Conference (MELECON), May 16–19*. 2006; 1142–1145.

[32] Xiang G., Wei Q. 'Current-based online bearing fault diagnosis for direct-drive wind turbines via spectrum analysis and impulse detection', *in 2012 IEEE Power Electronics and Machines in Wind Applications (PEMWA), Denver, CO*, 2012;1–6.

[33] Amirat Y., Choqueuse V., Benbouzid M.E.H., Charpentier J.F. 'Bearing fault detection in DFIG-based wind turbines using the first Intrinsic Mode Function', *in XIX International Conference on Electrical Machines (ICEM)*, Rome, Italy, 2010; 1–6.

[34] Amirat Y., Choqueuse V., Benbouzid M.E.H. 'Condition monitoring of wind turbines based on amplitude demodulation', *in 2010 IEEE Energy Conversion Congress and Exposition (ECCE), Atlanta, GA*, 2010:2417–2421.
[35] Xiang G., Wei Q. 'Bearing fault detection for direct-drive wind turbines via stator current spectrum analysis', *in 2011 IEEE Energy Conversion Congress and Exposition (ECCE), Phoenix, AZ*, 2011;313–318.
[36] Li L., Wenxiu L., Fulei C. 'Application of AE techniques for the detection of wind turbine using Hilbert-Huang transform' *in PHM '10 Prognostics and Health Management Conference, Macao, China*, 2010;1–7.
[37] Soua S., Van Lieshout P., Perera A., Gan T., Bridge B. 'Determination of the combined vibrational and acoustic emission signature of a wind turbine gearbox and generator shaft in service as a pre-requisite for effective condition monitoring', *Renewable Energy* 2010;51(0):175–181.
[38] Zheng X., Xu H. 'Fault diagnosis of wind turbine rolling bearing based on wavelet and Hilbert transforms', *in 31st Chinese Control Conference (CCC), Hefei, China*, 2012:5290–5293.
[39] Borghesani P., Pennacchi P., Chatterton S., Ricci R. 'The velocity synchronous discrete Fourier transform for order tracking in the field of rotating machinery', *Mechanical Systems and Signal Processing* 2014;44(1–2):118–133.
[40] Vold H., Herlufsen H., Mains M., Corwin-Renner D. 'Multi axle order tracking with the Vold–Kalman tracking filter', *Sound and Vibration* 1997;31(5):30–35.
[41] Xingjia Y., Guo C., Zhong M., Li Y., Shan G., Zhang Y. 'Wind turbine gearbox fault diagnosis using adaptive Morlet wavelet spectrum', *in Second International Conference on Intelligent Computation Technology and Automation (ICICTA09), Changsha, Hunan, China*, 2009:580–583.
[42] Antoni J. 'Fast computation of the Kurtogram for the detection of transient faults'. *Mechanical Systems and Signal Processing* 2007;21:108–124.
[43] Barszcz T., Jabłooski A. 'A novel method for the optimal band selection for vibration signal demodulation and comparison with the Kurtogram'. *Mechanical Systems and Signal Processing* 2011;25(1):431–451.
[44] Chatterton S., Borghesani P., Pennacchi P., Vania A. 'Experimental evidences in the monitoring of rolling element bearings', *Journal of Vibrational Engineering and Technologies* 2016;4(5):395–402.
[45] Borghesani P., Pennacchi P., Chatterton S. 'The relationship between kurtosis- and envelope-based indexes for the diagnostic of rolling element bearings'. *Mechanical Systems and Signal Processing* 2014;43(1–2):25–43.
[46] Borghesani P., Pennacchi P., Ricci R., Chatterton S. 'Testing second order cyclostationarity in the squared envelope spectrum of non-white vibration signals', *Mechanical Systems and Signal Processing* 2013;40(1):38–55.
[47] Feng Z., Liang M., Zhang Y., Hou S. 'Fault diagnosis for wind turbine planetary gearboxes via demodulation analysis based on ensemble empirical mode decomposition and energy separation', *Renewable Energy* 2002;47:112–126.
[48] Liu W.Y., Zhang W.H., Han J.G., Wang G.F. 'A new wind turbine fault diagnosis method based on the local mean decomposition', *Renewable Energy* 2012;48:411–415.

[49] Li Z., Dai W. 'Local mean decomposition combined with SVD and application in telemetry vibration signal processing', *Applied Mechanics and Materials*, 2013: 854–858.

[50] Yang W., Court R., Tavner P.J., Crabtreeb C.J. 'Bivariate empirical mode decomposition and its contribution to wind turbine condition monitoring', *Journal of Sound and Vibration* 2011;330(15):3766–3782.

Chapter 11

Robust fuzzy fault tolerant control wind energy system subject to actuator and sensor faults

Abdel Aitouche[1], Elkhatib Kamal[2], and Maher Kharrat[3]

11.1 Introduction

Nonlinearities and system uncertainties are the most important difficulties in designing controllers that ensure stability and acceptable closed-loop performance. Many significant results on the stability and robust control of uncertain nonlinear systems using Takagi–Sugeno (TS) fuzzy model have been reported [1–12] over the past decades, and considerable advances have been made. However, as stated in [11], many approaches for stability and robust control of uncertain systems are often characterized by conservatism when dealing with uncertainties. It has been well known that the TS fuzzy model is a very effective representation of complex nonlinear systems. In TS fuzzy model, the state space of a nonlinear system is divided into different fuzzy regions with a local linear model being used in each region. The overall model output is obtained by defuzzification using the Center of Gravity method. Once the fuzzy model is obtained, control design can be carried out via the so-called parallel distributed compensation (PDC) approach, which employs multiple linear controllers corresponding to the locally linear plant models [13]. This class of systems is described as a weighted sum of some simple linear subsystems and thus is easily analyzable.

In practice, many control systems are subjected to faults which can be caused by sensors, actuators or system faults. Therefore, it is an important issue in control system design as to how the system is kept stable and acceptable performance levels maintained when a failure occurs. Generally speaking, fault tolerant control (FTC)

[1]YNCREA-HEI, Lille Cedex Research Laboratory: CRIStAL UMR CNRS 9189 University of Lille, Science and Technology, France
[2]Department of Industrial Electronics and Control Engineering, Faculty of Electronic Engineering, Menofia University, Egypt
[3]National School of Electronics and Telecommunications of Sfax, (ENET'COM), Laboratory of Sciences and Techniques of Automatic & Computer Engineering (Lab-STA), National Engineering School of Sfax, University of Sfax, Tunisia

can be achieved either passively or actively. In passive FTC, the formed may be viewed as robust control. It requires a priori knowledge of possible faults which may affect the system, and its controller is based on treating all possible faults as uncertainties which are taken into account for the design of tolerant control by using different techniques such as in [14,15]. The interest of this approach lies in the fact that no online information is needed and the structure of the control law remains unchanged. Generally, the structure of the uncertainties (faults) is not taken into account in order to lead to a convex optimization problem. Furthermore, the class of the faults considered is limited, and it then becomes risky to use only passive FTC [16]. While in active FTC, the controller is designed to be reconfigurable for occurring faults according to the fault detection and estimation performed by an observer to allow the faulty system to accomplish its mission. Indeed, the active FTC has been introduced to overcome the passive control drawbacks. In addition, the active FTC can usually get better control performances, it has attracted more attentions in recent years. The knowledge of some information about these is required and obtained from a fault detection and diagnosis block. Different ideas are developed in the literature, e.g., control law rescheduling [17–19]. This approach requires a very robust fault detection and isolation block, which constitutes its major disadvantage. Indeed, a false alarm or an undetected fault may lead to degraded performance or even to instability. In the linear framework, the FTC problem is widely treated [20–22]. However in practice, most of physical systems are nonlinear; hence, it is primary to consider the FTC design for nonlinear systems. Some approaches dealing with this problem are proposed by [23,24]. The proposed methods in [25] are confined to the actuator failures and did not consider the fuzzy systems with sensor failures. Afterwards, several FTC methods have been extended to the nonlinear systems with sensor faults [26]. In the above studies, the considered faults affecting the system behavior are modeled by a constant function. However, in practice, the faults are often time variant and as we know; in practice, many control systems are subjected to faults which can be caused by actuators, sensors or systems faults and parameter uncertainties. In addition, the problem of control in the maximization of power generation in variable speed (VS)-wind energy system (WES) has been greatly studied, and such control still remains an active research area [27–35].

However, in this chapter, our goal is to present three different proposed approaches, namely, robust fuzzy FTC (RFFTC), robust fuzzy scheduled FTC (RFSFTC) and robust dynamic fuzzy FTC (RDFFTC) for TS models subject to time-varying actuator faults, sensor faults and parameter uncertainties and maximize the output power from the wind turbine (WT).

The chapter is organized as follows: the RFFTC control for the WES subject to time-varying actuator faults, time-varying sensor faults, and parameter uncertainties are presented in Section 11.2. The RFSFTC control for the WES subject to actuator faults, sensor faults with time varying and parameter uncertainties are presented in Section 11.3. Intelligent RDFFTC of WES in the presence the parameter uncertainties, sensor fault and actuator faults are studied in Section 11.4. Chapter conclusion will be given in Section 11.5.

11.2 RFFTC of WES with DFIG

This section presents a new method for RFFTC of nonlinear systems described by TS fuzzy systems subject to sensor faults, parameter uncertainties and time-varying actuator faults. The algorithm based on reconfiguration mechanism is then investigated for detection, isolation and accommodation of faults. The idea is to use a fuzzy-dedicated observers (FDOS), fuzzy proportional-integral estimation observer (FPIEO) and design a new control law to minimize the state deviation between a healthy observer and the eventually faulty actual model. This scheme requires the knowledge of the system states and the occurring faults. These signals are estimated from FDOS and FPIEO. TS fuzzy systems are classified into three families based on the input matrices and a design RFFTC for each family. In each family, the FTC law is designed by using the Lyapunov method to obtain conditions which are given in linear matrix inequalities (LMIs) formulation. The effectiveness of the proposed controller design methodology is finally demonstrated through a WES with DFIG to illustrate the effectiveness of the proposed method.

This section is organized as follows: Section 11.2.1 provides the proposed RFFTC scheme, TS fuzzy model FDOS and FPIEO. The proposed algorithm and the augmented system are presented in Section 11.2.2. Section 11.2.3 shows the stability and robustness conditions for the proposed algorithm followed by the calculation of state RFFTC, FDOS and FPIEO gains. Section 11.2.4 shows WES model system with DFIG. Simulation and results are shown in Section 11.2.5.

11.2.1 TS fuzzy model with parameter uncertainties and fuzzy observer

In this subsection, we will present the TS fuzzy plat model subject to sensor faults, parametric uncertainties and time-varying actuator faults. In addition, we will present the FDOS and FPIEO and the RFFTC scheme design.

11.2.1.1 TS fuzzy plant model with parameter uncertainties, sensor faults and actuator faults

This subsection is dedicated to a brief presentation of TS models. Consider a nonlinear system described by

$$\begin{cases} \dot{x}(t) = f(x(t), u(t)) \\ y(t) = f(x(t), u(t)) \end{cases} \tag{11.1}$$

where $x(t)$ and $y(t)$ are the state and output vector, respectively, $u(t)$ is the control input and $f(x(t), u(t))$ is the nonlinear function. A fuzzy dynamic model proposed by TS is often used to represent a nonlinear system (11.1) by the interpolation of a set of linear submodels. The TS fuzzy model is a piecewise interpolation of several linear models through membership functions. The fuzzy model is described by fuzzy if–then rules and will be employed here to deal with the control design problem for the nonlinear system. The TS fuzzy systems can be classified into three families based on the diversity of their input matrices B_i. These input matrices can be defined as follows:

$$B_1/\alpha_1 = B_2/\alpha_2 = \cdots = B_p/\alpha_p = B \tag{11.2}$$

where $\alpha_1, \ldots, \alpha_p$ are positive values and $B \in \kappa^{n\times m}$. Let p be the number of fuzzy rules. The ith rule of the fuzzy model for the nonlinear system is given by [36]:
Plant Rule i: IF $q_1(x(t))$ is M_{1i} AND … AND $q_\varepsilon(x(t))$ is $M_{\varepsilon i}$
Then

$$\begin{cases} \dot{x}(t) = (A_i + \Delta A_i)\, x(t) + \alpha_i(B + \Delta B)u(t) & i = 1, 2, \ldots, p \\ y(t) = C_i\, x(t) & i = 1, 2, \ldots, p \end{cases} \tag{11.3}$$

Here, $M_{\ni i}$ is the fuzzy set ($i = 1, 2, \ldots, p$; $\ni = 1, 2, \ldots, \varepsilon$), p is the number of the rules, $x_i(t) \in \kappa^{n\times 1}$ is the state vector, $u(t) \in \kappa^{m\times 1}$ is the input vector, $y(t) \in \kappa^{g\times 1}$ is the output vector, $A_i \in \kappa^{n\times n}$ is the system matrix, $C_i \in \kappa^{g\times n}$ is the output matrix, $q_1(x(t)), \ldots, q_\varepsilon(x(t))$ are known premise variables and p is the number of if–then rules. Each linear consequent equation represented by $A_i x(t) + B_i u(t)$ is called a subsystem and $\Delta A_i \in \kappa^{n\times n}$ and $\Delta B_i \in \kappa^{n\times m}$ are the uncertainties of the constant system matrices $A_i \in \kappa^{n\times n}$ and $B_i \in \kappa^{n\times m}$, respectively, which are nontime-varying parametric uncertainties in the plant model. These uncertainties are admissibly norm-bounded and structured. The main feature of a TS fuzzy model is to express the local dynamics of each fuzzy implication (rule) by a linear system model. The overall fuzzy model of the system is achieved by fuzzy "blending" of the linear system models. Each submodel contributes to the global behavior of the nonlinear system through a weighting function $\mu_i(q(x(t)))$. The TS structure is given by

$$\begin{aligned} \dot{x}(t) &= \frac{\sum_{i=1}^{p} \varpi(q(x(t)))[(A_i + \Delta A_i)\, x(t) + \alpha_i(B + \Delta B)u(t)]}{\sum_{i=1}^{p} \varpi_i(q(x(t)))} \\ &= \sum_{i=1}^{p} \mu_i(q(t))[(A_i + \Delta A_i)\, x(t) + \alpha_i(B + \Delta B)u(t)\,] \end{aligned} \tag{11.4}$$

$$y(t) = \frac{\sum_{i=1}^{p} \varpi_i(q(x(t)))C_i\, x(t)}{\sum_{i=1}^{p} \varpi_i(q(x(t)))}$$

$$y(t) = \sum_{i=1}^{p} \mu_i(q(x(t)))C_i\, x(t) \tag{11.5}$$

where $q(x(t)) = [q_1(x_1(t))\; q_2(x_2(t))\; \cdots\; q_\varepsilon(x(t))]$

$$\varpi_i(q(x(t))) = \prod_{\ni=1}^{\varepsilon} M_{\varepsilon i}(q_\varepsilon(x(t))), \mu_i(q(x(t))) = \frac{\varpi_i(q(x(t)))}{\sum_{i=1}^{p} \varpi_i(q(x(t)))},$$

for all t, the term $M_{\varepsilon i} q_\varepsilon(x(t))$ is the grade of membership of $q_\varepsilon(x(t))$ in $M_{\varepsilon i}$. Since

$$\begin{cases} \sum_{i=1}^{p} \varpi_i(q(x(t))) > 0 \\ \varpi_i(q(x(t))) \geq 0 \qquad i = 1, 2, \ldots, p \end{cases} \tag{11.6}$$

It is clear that

$$\begin{cases} \mu_i(q(x(t))) \geq 0 \\ \sum_{i=1}^{p} \mu_i(q(x(t))) = 1 \qquad i = 1, 2, \ldots, p \end{cases} \tag{11.7}$$

For the first family, we consider the input matrices on a one-dimensional cone, therefore the $\alpha_1, \ldots, \alpha_p > 0$. Therefore, the plant dynamics is then described by (11.4) and (11.5).

For the second family, we consider the input matrices are not all the same; therefore, from (11.2) $\alpha_1 \neq \alpha_2 \cdots \neq \alpha_i$, thus $B_i = \alpha_i B$ and the state equation of the TS fuzzy systems can be described by

$$\dot{x}(t) = \sum_{i=1}^{p} \mu_i(q(x(t)))[(A_i + \Delta A_i)\, x(t) + (B_i + \Delta B_i)u(t)] \tag{11.8}$$

For the third family, to consider the input matrices are common input matrix, from (11.2), $\alpha_1 = \alpha_2 \cdots = \alpha_i = 1$, so the plant dynamics are then described by

$$\dot{x}(t) = \sum_{i=1}^{p} \mu_i(q(x(t)))[(A_i + \Delta A_i)\, x(t) + (B + \Delta B)u(t)] \tag{11.9}$$

The contribution of this *work* is to *consider* the occurrence of the actuator and sensor faults. For the first family, if we take the nontime-varying the parameter uncertainties in consideration, we write the plant dynamics (11.4) as the following [36,37]:

$$\dot{x}(t) = \sum_{i=1}^{p} \mu_i(q(x(t)))[A_i\, x(t) + \alpha_i B u(t) + \bar{D}_i f_a(t)] + \sum_{i=1}^{p} \mu_i(q(x(t)))\Delta A_i x(t)$$

$$y(t) = \sum_{i=1}^{p} \mu_i(q(x(t)))[(I + F_s)\, C_i x(t)] \tag{11.10}$$

where $\bar{D}_i \in \kappa^{n\times k}$ are known matrices of actuator faults and $f_a(t) \in \kappa^{k\times 1}$ is actuator faults time-varying signal ($k < n$), $\bar{D}_i = B_i D_i$, $D_i \in \kappa^{m\times k}$ actuator faults which are known. F_s are faults which are modelled as proportional signals to sensor outputs, let

$$F_s = diag(\eta_1, \eta_2, \ldots, \eta_g), -0.1 \leq \eta_a \leq 0.1, a = 1, 2, \ldots, g;$$

which is the sensor fault vectors which is assumed to be bounded and $(I + F_s)$ is sensor gain. It is assumed that at any given time, only one sensor fails at most, I is the identity matrix, the premise variables $q_1(x(t)), \ldots, q_p(x(t))$ are also assumed measurable variables and do not depend on the faults

For the second family, the plant dynamics can be described by

$$\dot{x}(t) = \sum_{i=1}^{p} \mu_i(q(x(t)))[A_i x_i(t) + B_i u(t) + \bar{D}_i f_a(t)] + \sum_{i=1}^{p} \mu_i(q(x(t)))\Delta A_i x_i(t)$$

$$y(t) = \sum_{i=1}^{p} \mu_i(q(x(t)))[(I + F_s)\, C_i x(t)] \tag{11.11}$$

For the third family, the plant dynamics are then given by

$$\dot{x}(t) = \sum_{i=1}^{p} \mu_i(q(x(t)))[A_i x_i(t) + Bu(t) + \bar{D}_i f_a(t)] + \sum_{i=1}^{p} \mu_i(q(x(t)))\Delta A_i x_i(t)$$

$$y(t) = \sum_{i=1}^{p} \mu_i(q(x(t)))[(I + F_s)\, C_i x(t)] \tag{11.12}$$

11.2.1.2 TS fuzzy observer and the RFFTC scheme design

In this subsection, we presented the proposed RFFTC scheme design, FPIEO and FDOS.

- Nonlinear FPIEO and FDOS

This subsection presents FPIEO and FDOS design methodologies involving actuator faults estimation for TS fuzzy models. PDC structure is employed to achieve the following unknown fuzzy observer structures [36,38]. The structure of the FPIEO to estimate the actuator faults and the dynamic of the fault error estimation, ***for the first family*** can be written as follows:

$$\dot{\hat{x}}_u(t) = \sum_{i=1}^{p} \mu_i[A_i\, \hat{x}_u(t) + \alpha_i Bu(t) + K_i(y(t) - \hat{y}_u(t)) + \bar{D}_i \hat{f}_a(t)]$$

$$\dot{\hat{f}}_a(t) = \sum_{i=1}^{p} \mu_i L_i(y(t) - \hat{y}_u(t)) = \sum_{i=1}^{p} \mu_i L_i \tilde{y}(t)$$

$$\hat{y}_u(t) = \sum_{i=1}^{p} \mu_i C_i \hat{x}_u(t) \tag{11.13}$$

where $\hat{x}_u(t)$ is the estimated state vector by unknown fuzzy observer, $K_i\,(1, 2, \ldots, p)$ are observation error matrices, L_i are their corresponding integral gains to be determined, $y(t)$ is the output vector and $\hat{y}_u(t)$ is the final output of the unknown fuzzy observer, $\tilde{y}(t) = y(t) - \hat{y}_u(t)$ is the output estimation error and $\hat{f}_a(t)$ is the estimated actuator faults.

In the same manner and from (11.2) and (11.3), we can induce the inferred output of the FPIEO for the other two cases.

Also this subsection presents FDOS. The inferred modified FDOS states are governed by

$$\dot{\hat{x}}_o(t) = \sum_{i=1}^{p} \mu_i[A_i \hat{x}_o(t) + \alpha_i Bu(t) + N_i(y(t) - \hat{y}_o(t)) + \bar{D}_i \hat{f}_a(\mathrm{t})]$$

$$\hat{y}_o(t) = \sum_{i=1}^{p} \mu_i C_i \hat{x}_o(t) \tag{11.14}$$

where $\hat{x}_o(t)$ is the estimated state vector by the FDOS, $\hat{y}_o(t)$ is the final output of the FDOS and $N_i \in \kappa^{n \times g}$ are the FDOS gains. From (11.11) and (11.12), in the

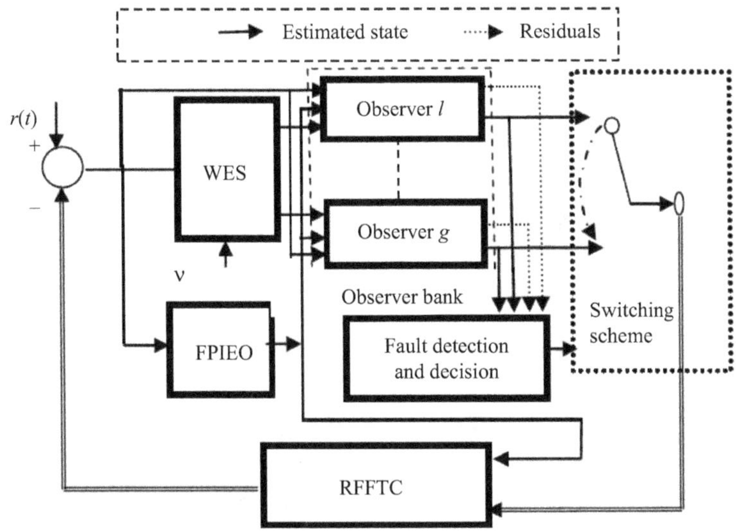

Figure 11.1 Block diagram of the proposed RFFTC scheme

same manner, we can induce the inferred output of the FDOS for the other two cases.

- The structure of the proposed RFFTC scheme

The proposed scheme, illustrated in Figure 11.1, which detect and isolate the sensor faults and an unknown input observer which estimates the actuator faults and reconstructs the state of the WES from a healthy estimate. Each one of the FDOS is driven by all sensed to generate residual signals. The estimated actuator faults from the FPIEO are fed to the FDOS. Through the decision and switcher mechanism, detecting and identifying the faulty sensor is possible. Finally, by using a switcher, selecting the healthy observer for reconstructing the controller input is enabled.

11.2.2 Proposed RFFTC based on FPIEO and FDOS

In this subsection, a unique RFFTC synthesis procedure is developed for each member of the TS family to deal with a wide range of uncertainties, sensor faults and actuator faults while maintaining the stability of the closed loop system.

11.2.2.1 Nonlinear RFFTC

For the fuzzy model (11.10), we construct the following RFFTC via the PDC. It is assumed that the fuzzy system (11.10) is locally controllable. A state-feedback with LMIs is used to design a controller for each subsystem.

For the first family, the modified RFFTC is defined as follows:
Rule j: IF $q_1(x(t))$ is M_{1i} AND … AND $q_\varepsilon(x(t))$ is $M_{\varepsilon i}$
Then

$$u(t) = -G_j x(t) - D_j \hat{f}_a(t) + r(t)/\alpha_j, \qquad j = 1, 2, \ldots, c \tag{11.15}$$

From (11.14), the estimated state by the healthy observer from FDOS is used, then the final output of the modified RFFTC becomes

$$u(t) = \frac{\sum_{j=1}^{c} \mu_j[-G_j\, \hat{x}_o(t) - D_j \hat{f}_a(t) + r(t)]}{\sum_{j=1}^{c} \alpha_j \mu_j} \tag{11.16}$$

For the second family, the overall output of the RFFTC is given by

$$\dot{u}(t) = z_k\, u(t) + \sum_{j=1}^{c} \mu_j[-G_j\, \hat{x}_o(t) - D_j \hat{f}_a(t) + r(t)] \tag{11.17}$$

For the last family, the RFFTC is given by

$$u(t) = \sum_{j=1}^{c} \mu_j[-G_j\, x_o(t) - D_j \hat{f}_a(t) + r(t)] \tag{11.18}$$

11.2.2.2 The augmented fuzzy control system

In order to carry out the analysis, the closed-loop fuzzy system should be obtained first by establishing the conditions for the asymptotic convergence of the observers (11.13) and (11.14).

For the first family, the fuzzy control system of the state and the errors can be obtained.

Let

$$e_1(t) = x(t) - \hat{x}_o(t) \tag{11.19}$$

$$\begin{aligned} \dot{x}(t) = & \sum_{i=1}^{p} \mu_i A_i\, x_i(t) + \left\{ \sum_{i=1}^{p} \alpha_i \mu_i \right\} Bu(t) + \sum_{i=1}^{p} \mu_i \bar{D}_i f_a(t) \\ & + \sum_{i=1}^{p} \mu_i \Delta A_i x_i(t) \end{aligned} \tag{11.20}$$

With the modified TS fuzzy fault tolerant controller (11.16) employed, the TS fuzzy system (11.19) has the following closed-loop:

$$\begin{aligned} \dot{x}(t) = & \sum_{i=1}^{p} \mu_i A_i x(t) + \sum_{i=1}^{p} {}_i\mu_i \bar{D}_i f_a(t) + \sum_{i=1}^{p} \mu_i \Delta A_i x_i(t) \\ & + \left\{ \sum_{i=1}^{p} \alpha_i \mu_i \right\} B \left\{ \sum_{j=1}^{p} \mu_j[-G_j\, \hat{x}_o(t) - D_j \hat{f}_a(t) + r(t)] \right\} / \left\{ \sum_{j=1}^{p} \alpha_j \mu_j \right\} \end{aligned} \tag{11.21}$$

Let

$$\tilde{f}_a(t) = f_a(t) - \hat{f}_a(t) \tag{11.22}$$

From (11.19) and (11.22), a TS fuzzy closed-loop can be observed:

$$\begin{aligned} \dot{x}(t) = & \sum_{i=1}^{p} \sum_{j=1}^{c} \mu_i \mu_j [(A_i - BG_j) x_i(t) + BG_j e_1(t) + \bar{D}_j \tilde{f}_a(t) + Br(t)] \\ & + \sum_{i=1}^{p} \mu_i \Delta A_i x_i(t) \end{aligned} \tag{11.23}$$

By taking the derivative of (11.10) and substituting from (11.10), (11.15) and (11.22), the following is obtained:

$$\dot{e}_1(t) = \sum_{i=1}^{p}\sum_{j=1}^{c} \mu_i\mu_j[(\Delta A_i - N_iF_sC_j)x_i(t) + (A_i - N_iC_j)e_1(t) + \bar{D}_i\tilde{f}_a(t)] \tag{11.24}$$

Using

$$e_2(t) = x(t) - \hat{x}_u(t) \tag{11.25}$$

then taking the derivative of (11.25) and substituting from (11.10), (11.13) and (11.22), the following is obtained:

$$\dot{e}_2(t) = \sum_{i=1}^{p}\sum_{j=1}^{c} \mu_i\mu_j[(\Delta A_i - K_iF_sC_j)x_i(t) + (A_i - K_iC_j)e_2(t) + \bar{D}_i\tilde{f}_a(t)] \tag{11.26}$$

we assume the actuator fault is not constant but time varying, the derivative of $\dot{\tilde{d}}(t)$ can be written as,

$$\dot{\tilde{f}}_a(t) = \dot{f}_a(t) - \dot{\hat{f}}_a(t) = \dot{f}_a(t) - \sum_{i=1}^{p} \mu_i[L_iC_ie_2(t) + L_iF_sC_ix(t)] \tag{11.27}$$

Equations (11.23), (11.24), (11.26) and (11.27) can be written:

$$\dot{X}(t) = \sum_{i=1}^{p}\sum_{j=1}^{c} \mu_i\mu_j(H_{ij} + \Delta H_{ij})X(t) + Sr(t) + \Psi_a\phi(t) \tag{11.28}$$

with $X(t) = \begin{bmatrix} x(t) \\ e_1(t) \\ e_2(t) \\ \tilde{f}_a(t) \end{bmatrix}$, $\phi(t) = [\dot{f}_a(t)]$, $S = \begin{bmatrix} B \\ 0 \\ 0 \\ 0 \end{bmatrix}$, $\Psi_a = \begin{bmatrix} 0 \\ 0 \\ 0 \\ I \end{bmatrix}$,

$$\Delta H_{ij} = \begin{bmatrix} \Delta A_i & 0 & 0 & 0 \\ \Delta A_i - N_iF_sC_j & 0 & 0 & 0 \\ \Delta A_i - K_iF_sC_j & 0 & 0 & 0 \\ L_iF_sC_i & 0 & 0 & 0 \end{bmatrix},$$

$$H_{ij} = \begin{bmatrix} (A_i - BG_j) & BG_j & 0 & \bar{D}_j \\ 0 & (A_i - N_iC_j) & 0 & \bar{D}_i \\ 0 & 0 & (A_i - K_iC_j) & \bar{D}_i \\ 0 & 0 & -L_iC_i & 0 \end{bmatrix}$$

For the second family, when the input matrices are different, using (11.11), (11.17), (11.24), (11.26) and (11.27), the augmented fuzzy system is given by

$$\dot{X}(t) = \sum_{i=1}^{p}\sum_{j=1}^{c} \mu_i\mu_j(H_{ij} + \Delta H_{ij})X(t) + S_ir(t) + \Psi_a\phi(t) \tag{11.29}$$

$$\text{with } X(t) = \begin{bmatrix} x(t) \\ u(t) \\ e_1(t) \\ e_2(t) \\ \tilde{f}_a(t) \end{bmatrix}, \phi(t) = \begin{bmatrix} \dot{f}_a(t) \\ \hat{f}_a(t) \end{bmatrix}, S_i = \begin{bmatrix} 0 \\ B_i \\ 0 \\ 0 \\ 0 \end{bmatrix}, \Psi_a = \begin{bmatrix} 0 & \bar{D}_i \\ 0 & 0 \\ 0 & 0 \\ 0 & 0 \\ I & 0 \end{bmatrix},$$

$$H_{ij} = \begin{bmatrix} A_i & B_i & 0 & 0_i & \bar{D}_j \\ G_j & Z_k & -G_j & 0 & \bar{D}_i \\ 0 & 0 & (A_i - N_iC_j) & 0 & \bar{D}_i \\ 0 & 0 & 0 & (A_i - K_iC_j) & \bar{D}_i \\ 0 & 0 & 0 & -L_iC_i & 0 \end{bmatrix},$$

$$\Delta H_{ij} = \begin{bmatrix} \Delta A_i & 0 & 0 & 0 & 0 \\ 0 & 0 & 0 & 0 & 0 \\ \Delta A_i - N_iF_sC_j & 0 & 0 & 0 & 0 \\ \Delta A_i - K_iF_sC_j & 0 & 0 & 0 & 0 \\ L_iF_sC_i & 0 & 0 & 0 & 0 \end{bmatrix}.$$

In the same manner, the augmented fuzzy system for the third family can be deduced.

11.2.3 Proposed RFFTC stability and robustness analysis

The analysis procedures are based on [39]. The main result for the global asymptotic stability of a TS fuzzy model with parameter uncertainties, sensor faults and actuator faults are summarized by the following theorem.

Theorem 11.1. *The fuzzy control system as given by* (11.28) *is stable if the controller and the observer gains are set to* $G_j = M_{a11}^{-1}Y_j$ *and* $N_i = P_{a22}{}^{-1}\, O_i$ *and* $\bar{E}_i = P_2^{-1}X_i$ *with the matrices* X_i, M_{a11}, Y_j *and* O_i *satisfying the following LMIs.*

$$M_{a11}A_i^T + A_iM_{a11} - (B_iY_j)^T - (B_iY_j) < 0 \tag{11.30}$$

$$A_i^TP_{a22} + P_{a22}A_i - (O_iC_j)^T - (O_iC_j) < 0 \tag{11.31}$$

$$H_{bij}^TP_2 + P_2H_{bij} - (X_i\bar{C}_j)^T - (X_i\bar{C}_j) < 0 \tag{11.32}$$

where,

$$\begin{gathered} H_{bij} = \begin{bmatrix} A_i & \bar{D}_i \\ 0 & 0 \end{bmatrix}, \quad \bar{E}_i = \begin{bmatrix} K_i \\ L_i \end{bmatrix}, \quad \bar{C}_j = \begin{bmatrix} C_j \\ 0 \end{bmatrix}^T, \\ P = \begin{bmatrix} P_1 & 0_{2x2} \\ 0_{2x2} & P_2 \end{bmatrix}, \quad P_1 = \begin{bmatrix} P_{a11} & 0 \\ 0 & P_{a22} \end{bmatrix}, \quad M_{a11} = P_{a11} \end{gathered} \tag{11.33}$$

11.2.4 WES with DFIG application

To show the effectiveness of the proposed controller design techniques, WES model system with DFIG [40–43] with parametric uncertainties is simulated.

11.2.4.1 The wind turbine characteristics and modelling

The mechanical output power at a given wind speed is drastically affected by the turbine's tip speed ratio (TSR), which is defined as the ratio of turbine rotor tip speed to the wind speed. At a given wind speed, the maximum turbine energy conversion efficiency occurs at an optimal TSR. Therefore, as wind speed changes, the turbine's rotor speed needs to change accordingly in order to maintain the optimal TSR and thus to extract the maximum power from the available wind resources [44]. The expression for aerodynamic power (P_m) captured by the WT is given by the nonlinear expression [45].

$$P_m = 0.5C_p(\lambda, \beta)\rho\pi R^2 v^3 \tag{11.34}$$

where ρ is the air density (kg/m^3), R is the rotor radius (m), v is the wind speed (m/s), β is the blade pitch angle of the WT, λ is TSR and C_p is the power coefficient defined by the following relation [46].

$$C_p = (0.44 - 0.0167\beta)\sin\left[\frac{\pi(\lambda - 3)}{15 - 0.3\beta}\right] - 0.00184(\lambda - 3)\beta \tag{11.35}$$

and TSR and is given by [46]:

$$\lambda = \frac{\Omega_r R}{u} \tag{11.36}$$

where Ω_r is the turbine rotational speed on the low-speed side of the gearbox. Referring to (11.30), optimal TSR λ_{opt} can be obtained as follows

$$\lambda_{opt} = \left(\frac{15 - 0.3\beta}{\pi}\right)\cos^{-1}\left[\frac{0.00184\beta(15 - 0.3\beta)}{\pi(0.44 - 0.167\beta)}\right] + 3 \tag{11.37}$$

Thus, the maximum power captured from the wind is given by

$$P_{m(\max)} = 0.5\, C_{p(\max)}(\lambda_{opt}, \beta)\rho\pi R^2 v^3 \tag{11.38}$$

A typical C_p–$\boldsymbol{\lambda}$ curve for different values of β is shown in Figure 11.2. It can be seen that there is a maximum power coefficient $C_{p(\max)}$. If $C_{p(\max)} \approx 0.48$, the maximum value of C_p is achieved for $\beta = 0°$ and $\boldsymbol{\lambda}_{opt}$. Normally, a VS WT follows the $C_{p(\max)}$ to capture the maximum power up to the rated speed by varying the rotor speed to keep the system at λ_{opt}, then it operates at the rated power with power control during the periods of high wind by the active control of the blade pitch angle or the passive regulation based on aerodynamic stall.

11.2.4.2 WES model system with DFIG in dq coordinates

This section is devoted to the modelling of the mechanical and electrical system of a 1.5 MW WT, and state space representation. The aerodynamic torque (T_r) on the WT rotor can be obtained using the following relationships [42]:

$$T_r = \frac{P_m}{\Omega_r} = \frac{0.5C_p(\lambda, \beta)\rho\pi R^2 v^3}{\Omega_r} \tag{11.39}$$

T_r at the optimal rotor speed $\Omega r_{(opt)}$ is then given by

$$T_r = K_{opt}\Omega_r^2 \tag{11.40}$$

where $K_{opt} = 0.5\rho\pi R^5 C_{p(\max)}/\lambda_{opt}^2$.

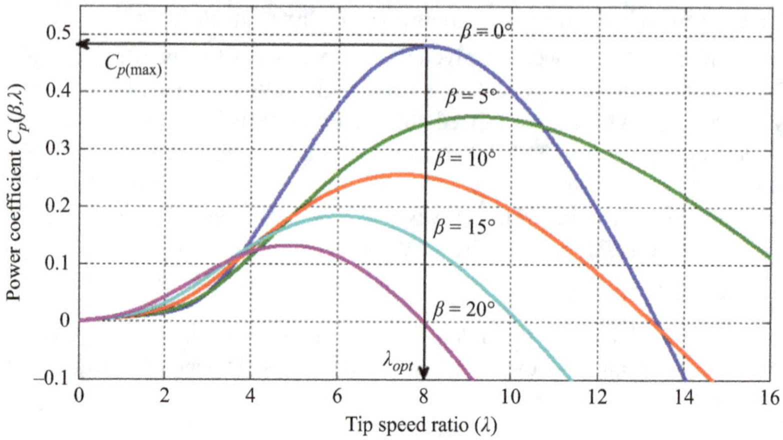

Figure 11.2 Power coefficient C_p versus λ

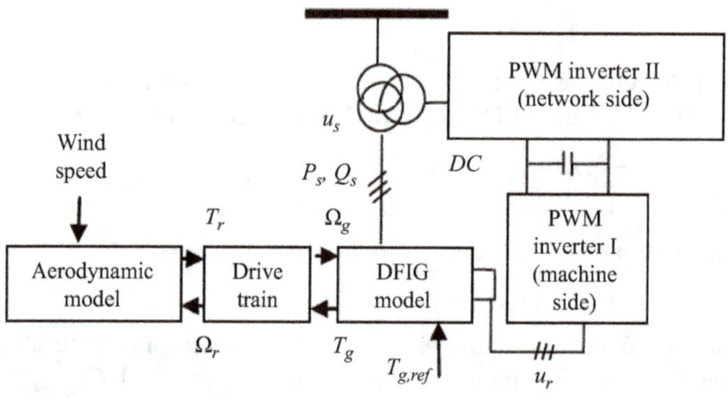

Figure 11.3 Structure of the DFIG-based wind turbine system

Since a commonly used model for induction generator (IG) converting power from the wind to serve the electric grid is shown in Figure 11.3 [40–43]. The stator of the wound rotor induction machine is connected to the low voltage balanced grid, and the rotor side is fed via the back-to-back voltage-source inverters with a common DC bus. These converters are voltage-sourced converters that use force commutated power electronic devices to synthesize an AC voltage from a DC source. A capacitor connected on the DC side acts as the DC voltage source. The WT catches the wind through its rotor blades and transfers it to the rotor hub. The rotor hub is attached to a low-speed shaft through a gear box. The high speed shaft drives an electric generator which converts the mechanical energy to electric energy and delivers it to the grid. As a summary, a doubly fed induction machine is a wound-rotor doubly fed electric machine and has several advantages over a conventional induction machine in wind power applications. First, as the rotor circuit is controlled by a power electronics converter, the IG is able to both import and export **reactive power**. This has important consequences for **power system stability** and allows the machine to support the grid during severe voltage disturbances. Second, the control of the rotor voltages and currents enables the induction machine to remain **synchronized** with the grid, while the WT speed varies. A VS WT utilizes the available wind resource more efficiently than a fixed speed WT, especially during light wind conditions. Third, the cost of the converter is low when compared with other VS solutions because only a fraction of the mechanical power, typically 25%–30%, is fed to the grid through the converter, the rest being fed to grid directly from the stator. The efficiency of the DFIG is very good for the same reason.

In VS WTs, the generator allows VS operation by providing an appropriate opposing torque T_g which follows the required value $T_{g,ref}$. However, modelling of the mechanical system of a 1.5 MW WT, electrical modelling, mechanical modelling in dq coordinates can be expressed as a function of different state variables [40–43].

$$
\begin{aligned}
\frac{di_{sd}}{dt} &= \frac{-R_s}{\xi L_s} i_{sd} + \left(\omega_s + \frac{1-\xi}{\xi} n_p \Omega_g\right) i_{sq} + \frac{R_r L_m}{\xi L_r L_s} i_{rd} + \frac{L_m}{\xi L_s} n_p \Omega_g i_{rq} \\
&\quad + \frac{L_m}{\xi L_r L_s} u_{rd} - \frac{1}{\xi L_s} u_{sd} \\
\frac{di_{sq}}{dt} &= \frac{-R_s}{\xi L_s} i_{sq} - \left(\omega_s + \frac{1-\xi}{\xi} n_p \Omega_g\right) i_{sd} + \frac{R_r L_m}{\xi L_r L_s} i_{rq} - \frac{L_m}{\xi L_s} n_p \Omega_g i_{rq} \\
&\quad + \frac{L_m}{\xi L_r L_s} u_{rq} - \frac{1}{\xi L_s} u_{sq} \\
\frac{di_{rd}}{dt} &= \frac{R_s L_m}{\xi L_r L_s} i_{sd} - \frac{L_m}{\xi L_r} n_p \Omega_g i_{sq} - \frac{R_r}{\xi L_r} i_{rd} + \left(\omega_s - \frac{1}{\xi} n_p \Omega_g\right) i_{rq} \\
&\quad - \frac{1}{\xi L_r} u_{rd} + \frac{L_m}{\xi L_r L_s} u_{sd} \\
\frac{di_{rq}}{dt} &= \frac{R_s L_m}{\xi L_r L_s} i_{sq} + \frac{L_m}{\xi L_r} n_p \Omega_g i_{sd} - \frac{R_r}{\xi L_r} i_{rq} - \left(\omega_s - \frac{1}{\xi} n_p \Omega_g\right) i_{rd} \\
&\quad - \frac{1}{\xi L_r} u_{rq} + \frac{L_m}{\xi L_r L_s} u_{sq} \\
\frac{d\Omega_r}{dt} &= \left(\frac{D_r}{J_r} + \frac{K_{opt}}{J_r}\Omega_r\right)\Omega_r - \frac{n_b}{J_r} T_h \\
\frac{d\Omega_g}{dt} &= -\frac{D_g}{J_g}\Omega_g + \frac{1}{J_g} T_h - \frac{1}{J_g} T_g \\
\frac{dT_h}{dt} &= \frac{1}{n_b}\left(K_{ls} - \frac{D_r D_{ls}}{J_r} + \frac{D_{lse} K_{opt}}{J_r}\Omega_r\right)\Omega_r - \frac{1}{n_b^2}\left(K_{ls} - \frac{D_g D_{ls}}{J_g}\right)\Omega_g \\
&\quad - D_{ls}\left(\frac{1}{J_r} + \frac{1}{n_b^2 J_g}\right) T_h + \frac{D_{ls}}{n_b^2 J_g} T_g \quad \frac{dT_g}{dt} = -\frac{1}{\tau_g} T_g + \frac{1}{\tau_g} T_{g,ref}
\end{aligned}
\tag{11.41}
$$

where $\xi = (1 - L_m{}^2)/L_sL_r$ is the leakage coefficient, $\omega_s = 2\pi f_s$, with f_s being the stator frequency, $\omega_r = n_p\Omega_g$, where ω_s and ω_r are the stator and rotor angular frequencies, respectively, Ω_g is the mechanical generator speed and n_p is the number of pole pairs. i_{sd} and i_{rd} denote the stator and rotor currents, respectively, in axis d, i_{sq} and i_{rq} are stator and rotor currents in axis q, R_s and R_r are stator and rotor resistance, L_s, L_r and L_m are stator, rotor leakage and magnetizing inductances, respectively, V_s is the stator voltage magnitude, τ_g is the time constant of the model, T_h is the high-speed shaft torque and n_b is the gearbox ratio. The gearbox is considered as a lossless device for this model. D_r, D_g and D_{ls} denote damping constants for the rotor, generator and the equivalent low-speed shaft, respectively, K_{ls} is the equivalent torsional stiffness of the low-speed shaft, J_r and J_g denote moments of inertia of the rotor and generator, respectively, and T_g and $T_{g,ref}$ are the generator torque and required generator torque, respectively. The used system parameters for the detailed generator and WT are shown in Tables 11.1 and 11.2 [40–43]. As far as the output of the system, the state will be assumed to be measurable. Moreover, other outputs of interest are the stator active power (P_s) and stator reactive power (Q_s). In the Park reference frame, they can be shown to have the following expression:

$$P_s = -\frac{L_m V_s}{L_s} i_{qr}, \quad Q_s = \frac{V_s^2}{\omega_s L_s} - \frac{L_m V_s}{L_s} i_{rd} \tag{11.42}$$

The state space of a 1.5 MW WT can be described by the following nonlinear equations from [40–43].

$$\dot{x}(t) = A(x)x(t) + Bu(t), y(t) = C(x)x(t) \tag{11.43}$$

where

$$x(t) = \left[i_{sd} \; i_{sq} \; i_{rd} \; i_{rq} \; \Omega_r \; \Omega_g \; T_h \; T_g \right]^T = [x_1(t) \; x_2(t) \; x_3(t) x_4(t) \; x_5(t) x_6(t) x_7(t) x_8(t)]^T$$

$$A(x) = \begin{bmatrix} -\frac{R_s}{\xi L_s} & \omega_s + \frac{(1-\xi)n_p\Omega_g}{\xi} & \frac{L_m R_r}{\xi L_r L_s} & \frac{L_m}{\xi L_s} n_p \Omega_g & 0 & 0 & 0 & 0 \\ \frac{-R_s}{\xi L_s} & -\left(\omega_s + \frac{1-\xi}{\xi} n_p \Omega_g\right) & \frac{R_r L_m}{\xi L_r L_s} & -\frac{L_m}{\xi L_r} n_p \Omega_g & 0 & 0 & 0 & 0 \\ \frac{R_s L_m}{\xi L_r L_s} & -\frac{L_m}{\xi L_r} n_p \Omega_g & -\frac{R_r}{\xi L_r} & \omega_s - \frac{1}{\xi} n_p \Omega_g & 0 & 0 & 0 & 0 \\ \frac{R_s L_m}{\xi L_r L_s} & \frac{L_m}{\xi L_r} n_p \Omega_g & -\frac{R_r}{\xi L_r} & -\omega_s + \frac{1}{\xi} n_p \Omega_g & 0 & 0 & 0 & 0 \\ 0 & 0 & 0 & 0 & \left(\frac{D_r}{J_r} + \frac{K_{opt}}{J_r}\Omega_r\right) & 0 & -\frac{n_b}{J_r} & 0 \\ 0 & 0 & 0 & 0 & 0 & \frac{-D_g}{J_g} & \frac{1}{J_g} & -\frac{1}{J_g} \\ 0 & 0 & 0 & 0 & a75 + \frac{D_{lse} K_{opt}}{n_b J_r}\Omega_r & a76 & a77 & \frac{D_{ls}}{n_b^2 J_g} \\ 0 & 0 & 0 & 0 & 0 & 0 & 0 & -\frac{1}{\tau_g} \end{bmatrix},$$

$$a_{75} = \frac{1}{n_b}\left(K_{ls} - \frac{D_r D_{ls}}{J_r}\right), a_{76} = -\frac{1}{n_b^2}\left(K_{ls} - \frac{D_g D_{ls}}{J_g}\right), a_{77} = -D_{ls}\left(\frac{1}{J_r} + \frac{1}{n_b^2 J_g}\right)$$

Table 11.1 DFIG parameters

Quantity	**CGS EMU to SI**
Rated power	1.5 [MVA]
Rated voltage (line to line)	575 [V]
Frequency/angular speed	$2\pi 60$ [rad/s]
Nominal system frequency	60 [Hz]
Stator resistance	0.0014 [Ω]
Stator leakage inductance	89.98 [μH]
Rotor resistance	0.99187 [mΩ]
Rotor leakage inductance	82.088 [μH]
Magnetizing inductance	1.526 [mH]
Inertia of the generator	53.036 [kg m^2]
Pole pairs	3

Table 11.2 Wind turbine parameters

Quantity	**CGS EMU to SI**
Wind turbine with a rotor diameter	70 [m]
Air density	0.55 [kg/m^3]
Cut-in wind speed	4 [m/s]
Cut-out wind speed	25 [m/s]
Rated wind speed	12 [m/s]
Rated rotor speed	19.7 [rpm]
Drivetrain torsion damper	1.0×10^7 [N m/s]
Drivetrain torsion spring	5.6×10^9 [N m/rad]
Inertia of the rotor	34.6×10^3 [kg m^2]

$$C(x) = \begin{bmatrix} 0 & 0 & 0 & 0 & 1 & 0 & 0 & 0 \\ 0 & 0 & 0 & 0 & 0 & 1 & 0 & 0 \\ 0 & 0 & 0 & -\frac{L_m V_s}{L_s} & 0 & 0 & 0 & 0 \\ 0 & 0 & \frac{V_s^2}{i_{rd}\omega_s L_s} & -\frac{L_m V_s}{L_s} & 0 & 0 & 0 & 0 & 0 \end{bmatrix}, y(t) = \begin{bmatrix} y_1 \\ y_2 \\ y_3 \\ y_4 \end{bmatrix} = \begin{bmatrix} \Omega_r \\ \Omega_g \\ P_s \\ Q_s \end{bmatrix},$$

$$u(t) = \begin{bmatrix} u_{sd} \\ u_{sq} \\ u_{rd} \\ u_{rq} \\ T_{g,ref} \end{bmatrix} = \begin{bmatrix} u_1 \\ u_2 \\ u_3 \\ u_4 \\ u_5 \end{bmatrix}, B = \begin{bmatrix} -\frac{1}{\xi L_s} & 0 & \frac{L_m}{\xi L_r L_s} & 0 & 0 \\ 0 & -\frac{1}{\xi L_s} & 0 & \frac{L_m}{\xi L_r L_s} & 0 \\ \frac{L_m}{\xi L_r L_s} & 0 & -\frac{1}{\xi L_r} & 0 & 0 \\ 0 & \frac{L_m}{\xi L_r L_s} & 0 & -\frac{1}{\xi L_r} & 0 \\ 0 & 0 & 0 & 0 & 0 \\ 0 & 0 & 0 & 0 & 0 \\ 0 & 0 & 0 & 0 & 0 \\ 0 & 0 & 0 & 0 & \frac{1}{\tau_g} \end{bmatrix},$$

- TS fuzzy WES description

The following steps should be taken to derive the TS fuzzy model of (11.43). Here Ω_r, Ω_g and $1/i_{rd}$ are nonlinear terms in the matrices $A(x(t))$ and $C(x(t))$ so we make them as our fuzzy variables. Generally, they are denoted as $q_1(x_5(t))$, $q_2(x_6(t))$ and $q_3(x_3(t))$ are known as premise variables that may be functions of state variables. Therefore, in our case, we have three nonlinear terms, $q_1(x_5(t)) = \Omega_r$, $q_2(x_6(t)) = \Omega_g$ and $q_3(x_3(t)) = 1/i_{rd}$. The first step for any kind of fuzzy modelling is to determine the fuzzy variables and fuzzy sets or so-called membership functions. Although there is no general procedure for this step and it can be done by various methods predominantly trial and error, in exact fuzzy modelling using sector nonlinearity, it is quite routine. It is assumed in this tutorial that the premise variables are just functions of the state variables for the sake of simplicity. To acquire membership functions, we should calculate the minimum and maximum values of $q_1(x_5(t))$, $q_2(x_6(t))$ and $q_3(x_3(t))$ which under

$$q_1(x_5(t)) \in [\min q_1(x_5(t)) \quad \max q_1(x_5(t))\,],$$

$$q_2(x_6(t)) \in [\min q_2(x_6(t)) \quad \max q_2(x_6(t))\,],$$

and $q_3(x_3(t)) \in [\min q_3(x_3(t)) \quad \max q_3(x_3(t))\,]$.

Therefore, $q_1(x_5(t))$, $q_2(x_6(t))$ and $q_3(x_3(t))$ can be represented by for membership functions M_1, M_2, D_1, D_2, N_1 and N_2 as follows:

$$\begin{aligned} q_1(x_5(t)) = \Omega_r &= M_1 q_1(x_5(t)) \cdot \max((q_1(x_5(t)))) \\ &\quad + M_2 q_1(x_5(t)) \cdot \min((q_1(x_5(t)))) \end{aligned} \tag{11.44}$$

$$\begin{aligned} q_2(x_6(t)) = \Omega_g &= D_1 q_2(x_6(t)) \cdot \max((q_2(x_6(t)))) \\ &\quad + D_2 q_2(x_6(t)) \cdot \min((q_2(x_6(t)))) \end{aligned} \tag{11.45}$$

$$\begin{aligned} q_3(x_3(t)) = \frac{1}{i_{rd}} &= N_1 q_3(x_3(t)) \cdot \max((q_3(x_3(t)))) \\ &\quad + N_2 q_3(x_3(t)) \cdot \min((q_3(x_3(t)))) \end{aligned} \tag{11.46}$$

and because M_1, M_2, D_1, D_2, N_1 and N_2 are actually fuzzy sets according to fuzzy mathematics

$$M_1(q_1(x_5(t))) + M_2(q_1(x_5(t))) = 1 \tag{11.47}$$

$$D_1(q_2(x_6(t))) + D_2(q_2(x_6(t))) = 1 \tag{11.48}$$

$$N_1(q_3(x_3(t))) + N_2(q_3(x_3(t))) = 1 \tag{11.49}$$

The degree of membership function for state $q_1(x_5(t))$ is depicted in Figure 11.4. Each membership function also represents model uncertainty for each sub-system. The degree of membership function for state $q_2(x_6(t))$ and $q_3(x_3(t))$ are implemented in the same manner.

The nonlinear systems are subject to large parameter uncertainties, actuator faults and sensor faults. The objective here is to conceive an actuator and sensor FTC for WES with parameters uncertainties within 35% of the nominal values and disturbance (20% of wind speed). We can generalize that the ith rules of the continuous TS fuzzy models including actuator faults and sensor faults are of the following forms:

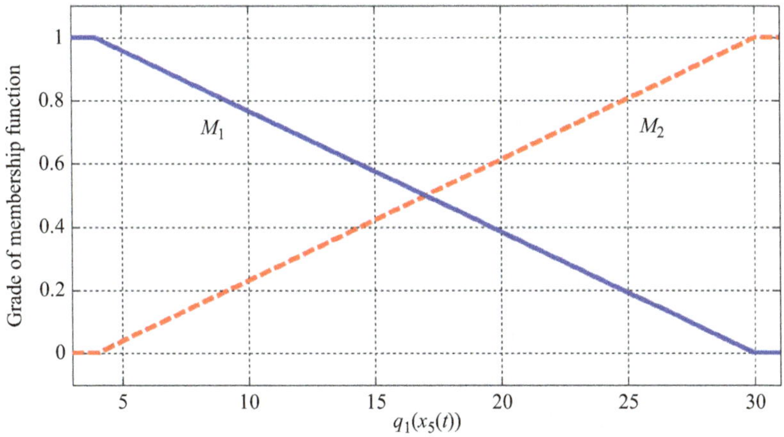

Figure 11.4 Membership functions $M_1\ (q_1(x_5(t)))$ and $M_2(q_1(x_5(t)))$

Rule i: IF $q_1(x_5(t))$ is $M_{\ni i}$ and $q_2(x_6(t))$ is $D_{\ni i}$ and $q_3(x_3(t))$ is $N_{\ni i}$
Then, $\dot{x}(t) = (A_i + \Delta A_i)x(t) + B_i u(t) + \bar{D}_i f_a(t)$

$$y(t) = (I + F_s)\, C_i x(t) \quad i = 1, 2, \ldots, 8; \ni = 1, 2 \tag{11.50}$$

where the subsystems are determined as $\bar{D}_i = \begin{bmatrix} 0 & 0 & 0 & 1 & 0 \\ 0 & 1 & 0 & 1 & 0 \end{bmatrix}^T$

$$A_i = \begin{bmatrix} -\frac{R_s}{\xi L_s} & \omega_s + \frac{(1-\xi)n_p q_{2i}}{\xi} & \frac{L_m R_r}{\xi L_r L_s} & \frac{L_m}{\xi L_s} n_p q_{2i} & 0 & 0 & 0 & 0 \\ \frac{-R_s}{\xi L_s} & -\left(\omega_s + \frac{1-\xi}{\xi} n_p q_{2i}\right) & \frac{R_r L_m}{\xi L_r L_s} & -\frac{L_m}{\xi L_r} n_p q_{2i} & 0 & 0 & 0 & 0 \\ \frac{R_s L_m}{\xi L_r L_s} & -\frac{L_m}{\xi L_r} n_p q_{2i} & -\frac{R_r}{\xi L_r} & \omega_s - \frac{1}{\xi} n_p q_{2i} & 0 & 0 & 0 & 0 \\ \frac{R_s L_m}{\xi L_r L_s} & \frac{L_m}{\xi L_r} n_p q_{2i} & -\frac{R_r}{\xi L_r} & -\omega_s + \frac{1}{\xi} n_p q_{2i} & 0 & 0 & 0 & 0 \\ 0 & 0 & 0 & 0 & \left(\frac{D_r}{J_r} + \frac{K_{opt}}{J_r} q_{1i}\right) & 0 & -\frac{n_b}{J_r} & 0 \\ 0 & 0 & 0 & 0 & 0 & \frac{-D_g}{J_g} & \frac{1}{J_g} & -\frac{1}{J_g} \\ 0 & 0 & 0 & 0 & a75 + \frac{D_{lse} K_{opt}}{n_b J_r} q_{1i} & a76 & a77 & \frac{D_{ls}}{n_b^2 J_g} \\ 0 & 0 & 0 & 0 & 0 & 0 & 0 & -\frac{1}{\tau_g} \end{bmatrix},$$

$$B_i = \begin{bmatrix} -\frac{1}{\xi L_s} & 0 & \frac{L_m}{\xi L_r L_s} & 0 & 0 \\ 0 & -\frac{1}{\xi L_s} & 0 & \frac{L_m}{\xi L_r L_s} & 0 \\ \frac{L_m}{\xi L_r L_s} & 0 & -\frac{1}{\xi L_r} & 0 & 0 \\ 0 & \frac{L_m}{\xi L_r L_s} & 0 & -\frac{1}{\xi L_r} & 0 \\ 0 & 0 & 0 & 0 & 0 \\ 0 & 0 & 0 & 0 & 0 \\ 0 & 0 & 0 & 0 & 0 \\ 0 & 0 & 0 & 0 & \frac{1}{\tau_g} \end{bmatrix},$$

$$C_i = \begin{bmatrix} 0 & 0 & 0 & 0 & 1 & 0 & 0 & 0 \\ 0 & 0 & 0 & 0 & 0 & 1 & 0 & 0 \\ 0 & 0 & 0 & -\frac{L_m V_s}{L_s} & 0 & 0 & 0 & 0 \\ 0 & 0 & \frac{V_s^2}{\omega_s L_s} q_{31} - \frac{L_m V_s}{L_s} & 0 & 0 & 0 & 0 & 0 \end{bmatrix},$$

where ΔA_i $(i = 1, 2, \ldots, 8)$ represents the nontime-varying system parameters uncertainties (J_r, J_g, R_s and R_r) but bounded, the elements of ΔA_i randomly achieve the values within 35% of their nominal values corresponding to A_i $(i = 1, 2, \ldots, 8)$; ΔB_i $(i = 1, 2, \ldots, 8) = 0$. The actuator faults $f_a(t) = [f_{a1}(t) f_{a2}(t)]^T$ are time varying and defined as follows:

$$f_{a1}(t) = \begin{cases} 0 & t < 120\,\text{s} \\ 4\sin(\pi t) & t \geq 120\,\text{s} \end{cases}, \quad f_{a2}(t) = \begin{cases} 0 & t < 120\,\text{s} \\ 2\sin(\pi t) & t \geq 120\,\text{s} \end{cases} \tag{11.51}$$

Faults are modelled as additive signals to generator speed sensor outputs.

According to the analysis above, the procedure for finding the proposed RFFTC with a nonlinear unknown input observer and the FDOS observer is summarized as follows:

1. Obtain the mathematical model of the WES to be controlled.
2. Obtain the fuzzy plant model for the system stated in step (11.1) by means of a fuzzy modelling method.
3. Solve LMIs (11.30)–(11.33) to obtain X_i, Y_j, P_{a22}, P_2, K_i, L_i, N_i, and $\bar{E}_i$ thus ($G_j = M_{a11}^{-1} Y_j$ and $N_i = P_{a22}^{-1} O_i$ and $\bar{E}_i = P_2^{-1} X_i$).
4. Construct fuzzy observers (11.13), (11.14) and the fuzzy controller (11.16)–(11.18).

11.2.5 Simulations and results

The proposed controller for the WES is tested for the same random variation of wind speed as shown in Figure 11.5 to illustrate the effectiveness of the proposed method. The control objective of this chapter is to design a RFFTC law for the system (11.43) to ensure that all signals in the closed-loop system are bounded. Figure 11.6 shows the time evolution of the time-varying actuator faults $f_a(t)$ and its estimate $\hat{f}_a(t)$, which occur after 120 s. Figure 11.7 shows the proportional signals that represent sensor failures which has been added to the output of sensor 6 between $t = 40$ s and $t = 80$ s. The magnitudes of faults are between ≈10% and ≈20% of the nominal values of the variables, and the parametric uncertainties J_r, J_g, R_s and R_r are considered within 35% of their nominal values.

The simulation results are given in Figures 11.8–11.11 with (left) and without (right) the RFFTC strategy, and all the simulations are realized on the nonlinear model with the presence of nontime-varying parametric uncertainties, sensor faults and time-varying actuator faults. In Figures 11.8–11.11 (left), the control law is based on one observer (observer 6) without using the switching block and the unknown observers. We can observe that between $t = 0$ s to $t \approx < 40$ s and $t > \approx 80$ s to $t \approx < 120$ s there are no faults, providing good tracking performance. However between $t = 40$ s to $t = 80$ s and after $t = 120$ s, we can see that the WES's performance is reduced right

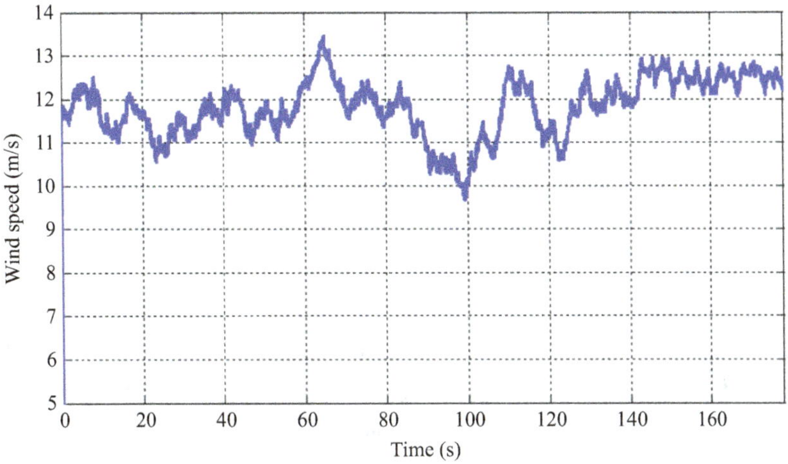

Figure 11.5 Wind speed profile

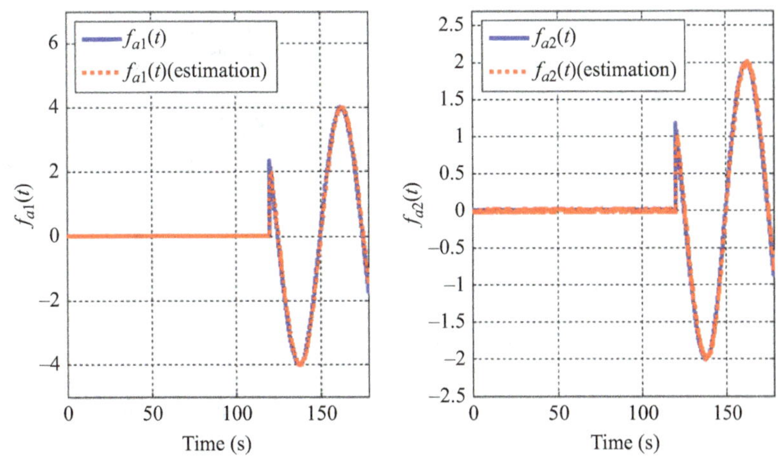

Figure 11.6 Actuator faults [$f_{a1}(t)$ and $f_{a2}(t)$] and their estimations

after the generator speed sensor and actuators became faulty. Figures 11.8 and 11.9 (right) shows WES state variables and their estimated signals, when the law control is based on the bank observers (observer 5) with the switch block and the unknown observers. We can note that the WES remains stable despite the presence of faults and parameter uncertainties, which shows the effectiveness of the proposed RFFTC strategy. The rotational speed of the WT and generator (dashed line), respectively, and their estimates (dotted lines) in the presence of parametric uncertainties, sensor faults and actuator faults, are shown in Figures 11.8 and 11.9. In order to obtain optimality, the $r(t) = \Omega_{gref} = \Omega_{opt} = \omega_s - n_b\lambda_{opt}V/R$ profile are chosen in such a way to follow the optimal TSR (λ_{opt}) (solid line).

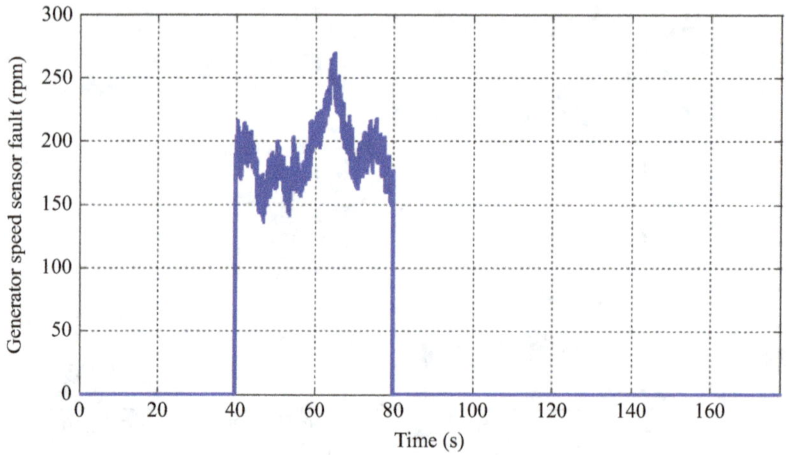

Figure 11.7 Proportional error on the generator speed sensor

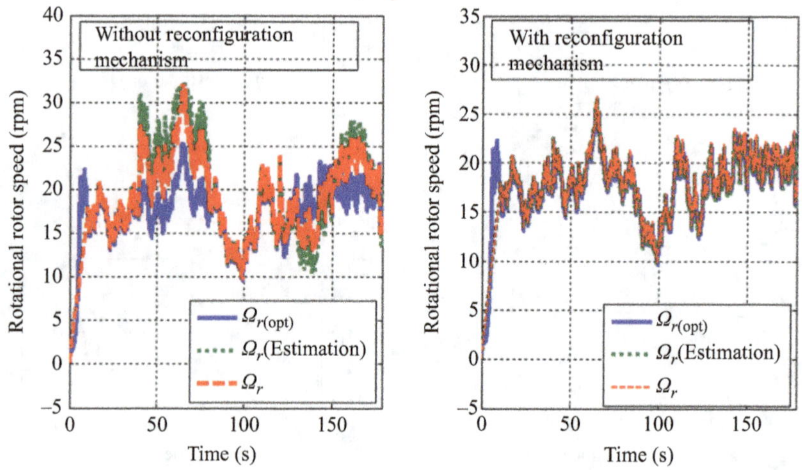

Figure 11.8 The trajectories of ω_r and its estimate without (left) and with (right) RFFTC strategy

The switching from observer 6 to observer 5 is visualized clearly at $t \approx 40$ s [Figures 11.8–11.11 (right)]. We notice that switching observers is carried out without loss of control of the system state.

From the simulation, it can be seen that without the reconfiguration mechanism, the WES lost its performance after the generator speed sensor and the actuator became faulty. Whereas for the same reference input and by using the RFFTC scheme strategy proposed, the WES remains stable in the presence of sensor faults, parameter uncertainties and actuator faults. This demonstrates the effectiveness of the proposed RFFTC strategy.

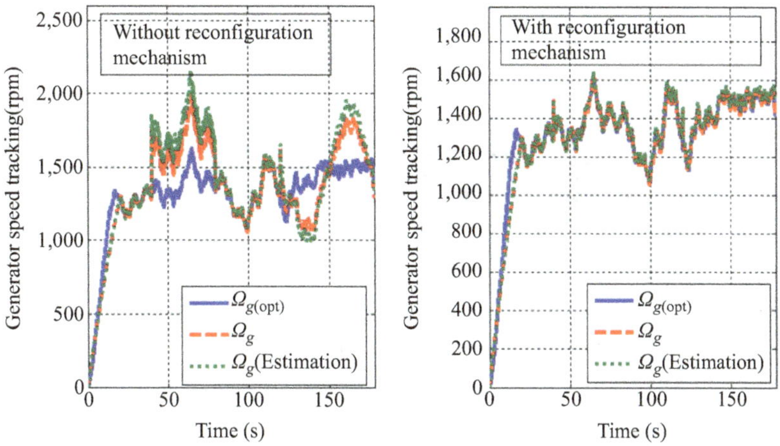

Figure 11.9 The trajectories of ω_r and its estimate without (left) and with (right) RFFTC strategy

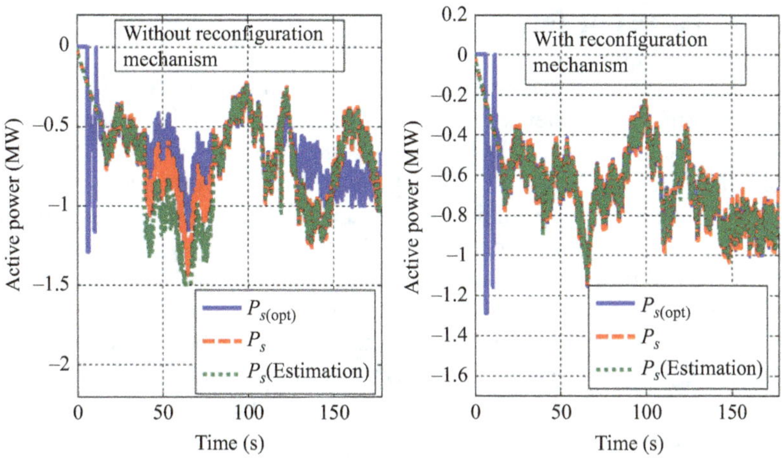

Figure 11.10 The trajectories of P_s and its estimate without (left) and with (right) RFFTC strategy

In summary, it has been shown that the proposed scheme is able to detect and isolate sensor faults, through a proper and feasible selection of the healthy observed variables. It can also compensate the actuator faults using the nonlinear unknown input observers. The simulation results demonstrate the effectiveness of the proposed control approach. The proposed control scheme can guarantee the stability of the closed-loop system and the convergence of the output tracking error.

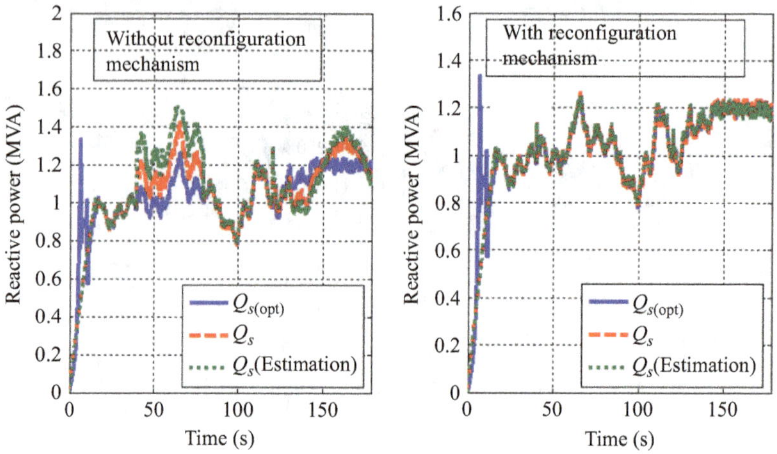

Figure 11.11 The trajectories of Q_s (dashed line) and its estimate (dotted line) without (left) and with (right) RFFTC strategy

11.3 RFSFTC of WES with DFIG subject to sensor and actuator faults

This chapter proposes a FDOS method using a nonlinear FPIEO with a RFSFTC algorithm for fuzzy TS systems subject to sensor faults, parametric uncertainties and time-varying actuator faults. FDOS provide residuals for detection and isolation of sensor faults which can affect a TS model. The TS fuzzy model is adopted for fuzzy modelling of the uncertain nonlinear system and establishing fuzzy state observers. The concept of PDC is employed to design RFSFTC and fuzzy observers from the TS fuzzy models. TS fuzzy systems are classified into three families based on the input matrices, and a RFSFTC synthesis procedure is given for each family. In each family, sufficient conditions are derived for robust stabilization, in the sense of Taylor series stability and Lyapunov method, for the TS fuzzy system with parametric uncertainties, sensor faults and actuator faults. The sufficient conditions are formulated in the format of LMEs. Detailed simulation results are given *to show the effectiveness of the proposed controller through a* WES model system with DFG [40]–[43] subject to parameter parametric uncertainties.

This section is organized as follows. Section 11.3.1 provides the proposed RFSFTC scheme, TS fuzzy model FDOS and FPIEO. Section 11.3.2 shows the stability and robustness conditions for the proposed algorithm followed by the calculation of state RFSFTC, FDOS and FPIEO gains. Section 11.3.3 shows simulation and results on WES model system with DFIG.

11.3.1 TS fuzzy plant model with actuator faults, sensor faults and parameter uncertainties

In this section, the TS fuzzy plant model subject to parameter uncertainties, sensor faults and actuator faults will be expressed as a weighted sum of a number of fuzzy systems. An augmented TS fuzzy plant model is formed by adding a fuzzy uncertainty regenerator. The inferred outputs of the fuzzy scheduler fault system for three families have been introduced in Chapter 11.3. Since this chapter considers the occurrence of sensor faults, and actuator faults in presence the parameter uncertainties, for the first family, the dynamics of fuzzy scheduler fault system for the first family is given by

$$\dot{x}(t) = \sum_{i=1}^{p}\sum_{l=1}^{s} \mu_i h_l[(A_i + \Delta\tilde{A}_l)\, x(t) + \alpha_i B\, u(t) + \bar{D}_i f_a(t)]$$

$$y(t) = \sum_{i=1}^{p} \mu_i(q(t))(I + F_s)C_i\, x(t) \tag{11.52}$$

In the same manner, we can induce the inferred output of the fuzzy scheduler for the other two cases.

11.3.2 Proposed RFSFTC algorithm based on FPIEO and FDOS

In this section, RFSFTC is developed for each member of the TS family to deal with a wide range of parameter uncertainties, sensor faults and actuator faults such that the closed loop system is stable. We use the same proposed scheme, illustrated in Figure 11.1, but we will change the RFFTC to RFSFTC.

11.3.2.1 Proposed RFSFTC controllers

Based on the analysis given in [47] to obtain the RFSFTC and based on the FPIEO (11.13) and FDOS (11.14), we can obtain the fuzzy scheduler fault tolerant controller. ***For the first family***, the inferred output of the modified RFSFTC will be as the following:

$$u(t) = \frac{\sum_{j=1}^{c}\sum_{l=1}^{s} \mu_j h_l[-G_{jl}\, \hat{x}_o(t) - D_j \hat{f}_a(t) + r(t)]}{\sum_{l=1}^{s}\sum_{j=1}^{c} \alpha_j \mu_j h_l} \tag{11.53}$$

In the same manner, we can induce the inferred output of the RFSFTC for the other two cases as follows:
For the second family, the modified RFSFTC becomes

$$\dot{u}(t) = Z_k\, u(t) + \sum_{j=1}^{c}\sum_{l=1}^{s} \mu_j h_l[-G_{jl}\, \hat{x}_o(t) - D_j \hat{f}_a(t) + r(t)] \tag{11.54}$$

For the last family, the RFSFTC becomes

$$u(t) = \sum_{j=1}^{c}\sum_{l=1}^{s} \mu_j h_l[-G_{jl}\, \hat{x}_o(t) - D_j \hat{f}_a(t) + r(t)] \tag{11.55}$$

11.3.2.2 Stability analysis for the proposed RFSFTC algorithm

In order to carry out the analysis, the closed-loop fuzzy system should be obtained first by establishing the conditions for the asymptotic convergence of the observers.

For the first family, based on (11.19), (11.22), (11.25) and (11.52), we can obtain the augmented fuzzy system.

$$\dot{X}(t) = \sum_{i=1}^{p}\sum_{j=1}^{p}\sum_{l=1}^{s}\mu_i\mu_j h_l[(H_{ijl} + \Delta\tilde{H}_{ijl})X(t) + Sr(t) + \Psi_a\phi(t)] \qquad (11.56)$$

with $X(t) = \begin{bmatrix} x(t) \\ e_1(t) \\ e_2(t) \\ \tilde{f}_a(t) \end{bmatrix}$, $\phi(t) = [\dot{f}_a(t)]$,

$$\Delta H_{ijl} = \begin{bmatrix} \Delta\tilde{A}_l & 0 & 0 & 0 \\ \Delta\tilde{A}_l - N_i F_s C_j & 0 & 0 & 0 \\ \Delta\tilde{A}_l - K_i F_s C_j & 0 & 0 & 0 \\ L_i F_s C_j & 0 & 0 & 0 \end{bmatrix},\ S = \begin{bmatrix} B \\ 0 \\ 0 \\ 0 \end{bmatrix},\ \Psi_a = \begin{bmatrix} 0 \\ 0 \\ 0 \\ I \end{bmatrix},$$

$$H_{ijl} = \begin{bmatrix} (A_i - BG_{jl}) & BG_{jl} & 0 & \bar{D}_j \\ 0 & (A_i - N_i G_{jl}) & 0 & \bar{D}_j \\ 0 & 0 & (A_i - K_i C_j) & \bar{D}_j \\ 0 & 0 & -L_i C_i & 0 \end{bmatrix}.$$

In the same manner, the augmented fuzzy system for the second and the third families can be deduced.

11.3.3 Derivation of the stability and robustness conditions

The main result for the global asymptotic stability of a TS fuzzy model with parameter uncertainties, sensor faults and actuator faults input is summarized by the following theorem.

Theorem 11.2. *The fuzzy control system as given by* (11.56) *is stable if the controller and the observer gains are set to* $G_j = M_{a11}^{-1}Y_j$ *and* $N_i = P_{a22}{}^{-1}\ O_i$ *and* $\bar{E}_i = P_2^{-1}X_i$ *with the matrices* X_i, M_{a11}, Y_j *and* O_i *satisfying the following LMEs.*

$$M_{a11}A_i^T + A_i M_{a11} - (B_i Y_{jl})^T - (B_i Y_{jl}) = -\sigma I \qquad (11.57)$$

$$A_i^T P_{a22} + P_{a22}A_i - (O_i C_j)^T - (O_i C_j) = -\sigma I \qquad (11.58)$$

$$H_{bi}^T P_2 + P_2 H_{bi} - (X_i \bar{C}_j)^T - (X_i \bar{C}_j) = -\sigma I \qquad (11.59)$$

where $P = \begin{bmatrix} P_1 & 0_{2x2} \\ 0_{2x2} & P_2 \end{bmatrix}$, $H_{bi} = \begin{bmatrix} A_i & \bar{D}_i \\ 0 & 0 \end{bmatrix}, \bar{E}_i = \begin{bmatrix} K_i \\ L_i \end{bmatrix}, \bar{C}_j = \begin{bmatrix} C_j \\ 0 \end{bmatrix}^T$,

$P_1 = \begin{bmatrix} P_{a11} & 0 \\ 0 & P_{a22} \end{bmatrix}$, $M_{a11} = P_{a11}^{-1}$, *σ is the robustness index.*

According to the analysis above, the procedure for finding the RFSFTC and the fuzzy observer is summarized as follows:

1. Obtain the mathematical model of the uncertain multivariable nonlinear plant.

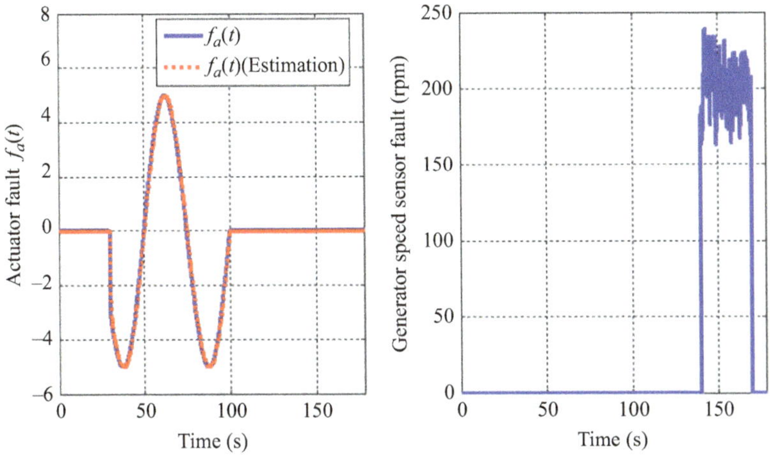

Figure 11.12 The actuator fault $f_a(t)$ and its estimate (left) and the proportional error on the generator speed sensor (right)

2. Obtain the TS fuzzy plant model (with parameter uncertainty information) of the plant obtained in previous step.
3. Determine the ranges of the parameter uncertainties, i.e., ΔA, so as to define the specified parameter space.
4. Model the parameter uncertainties with a fuzzy uncertainty.
5. Build the fuzzy scheduler using the same fuzzy rule antecedents of the fuzzy uncertainty regenerator.
6. Solve (11.57)–(11.59) to obtain X_i , Y_j , P_{a22} ,P_2, K_i, L_i, N_i, and $\bar{E}_i$ thus ($G_{jl} = Y_{jl}M_{a11}^{-1}$ and $N_i = P_{a22}{}^{-1}\ O_i$ and $\bar{E}_i = P_2^{-1}X_i$).
7. Construct the fuzzy observers (11.13), (11.14) and the RFSFTC (11.53)–(11.55).

11.3.4 WES with DFIG application and simulations and results

We use the random variation of wind speed as shown in Figure 11.5 also to test the proposed controller and the WES model system with DFIG system (11.43). The time varying actuator fault $f_a(t)$ (11.60) is shown in Figure 11.12 (left). For the testing purpose, it is required that at least one sensor fail every time. Faults are modeled as proportional signals to sensor outputs as shown in Figure 11.12 (right), and the parametric uncertainties R_s and R_r are considered within 40% of their nominal values. The actuator fault $f_a(t)$ is time varying and defined as follows:

$$f_a(t) = \begin{cases} 0 & t < 40\text{ s} \\ 4\sin(\pi t) & t \geq 40\text{ s} \end{cases} \tag{11.60}$$

Figures 11.13 and 11.14 show the rotational speed of the WT and generator (dashed line), respectively, and their estimates (dotted lines) in the presence

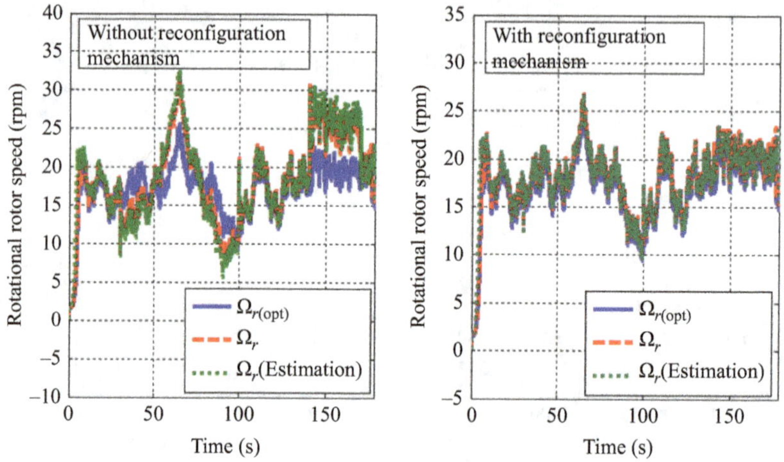

Figure 11.13 The trajectories of Ω_r and its estimate without (left) and with (right) RFSFTC strategy

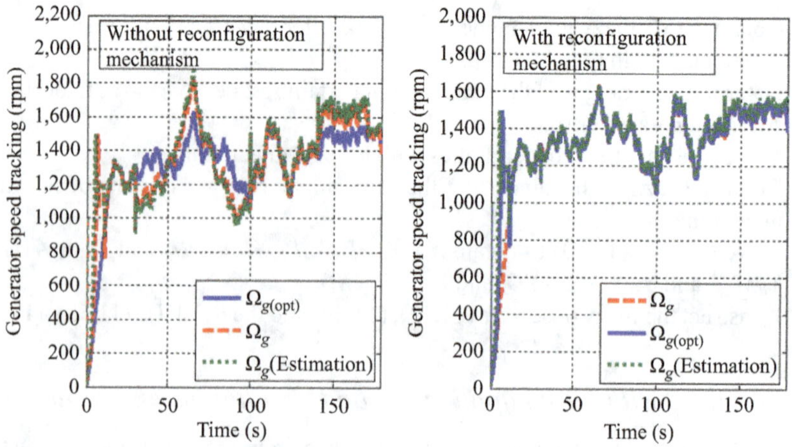

Figure 11.14 The trajectories of Ω_g and its estimate without (left) and with (right) RFSFTC strategy

of parametric uncertainties, sensor faults and actuator fault. A simulation without the RFSFTC scheme strategy is shown at left and one with the RFSFTC scheme strategy is shown at right. In order to obtain optimality, the $r(t) = \Omega_{gref} = \Omega_{opt} = \omega_s - n_b\lambda_{opt}V/R$ profile are chosen in such a way to follow the optimal speed profile and the optimal TSR (λ_{opt})(solid line). From the simulation results using the proposed control scheme, it can be seen that the outputs of the system are bounded, and good tracking performance can be obtained through the uncertain nonlinearities of the system, sensor faults and the actuator fault.

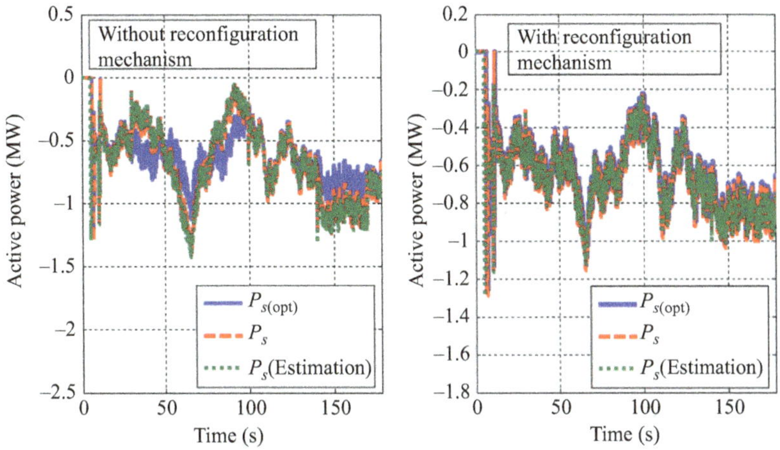

Figure 11.15 The trajectories of P_s and its estimate without (left) and with (right) RFSFTC strategy

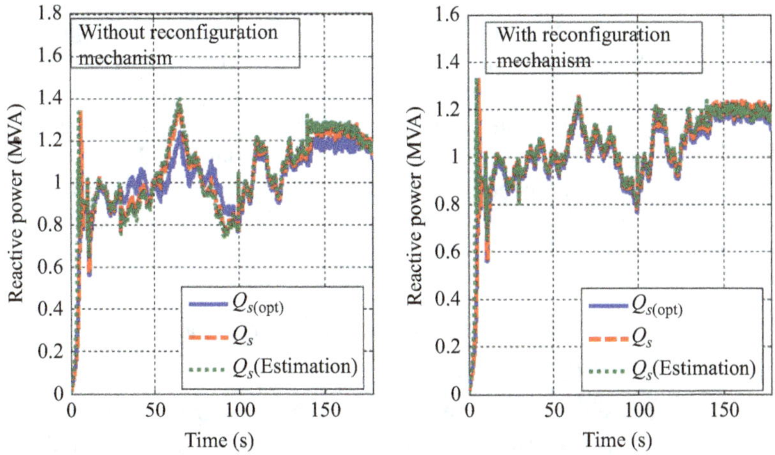

Figure 11.16 The trajectories of Q_s (dashed line) and its estimate (dotted line) without (left) and with (right) RFSFTC strategy

Figures 11.15 and 11.16 show the active power P_s (dashed line) and its estimate (dotted line) and the reactive power Q_s (dashed line) and its estimate (dotted line), respectively, without RFSFTC scheme strategy (left) and with RFSFTC scheme strategy (right).

From the simulation, it can be seen that without the reconfiguration mechanism, the WES lost performance just after the generator speed sensor became faulty, whereas for the same reference input and by using the RFSFTC scheme strategy proposed, the WES remains stable in the presence of sensor faults, parameter uncertainties and actuator fault which demonstrates the effectiveness of the proposed RFSFTC strategy.

In summary, it has been shown that the proposed scheme is able to detect and isolate sensor faults, through a proper and feasible selection of the healthy observed variables. It can also compensate for the actuator fault using the nonlinear FPIEO. The simulation results demonstrate the effectiveness of the proposed control approach. The proposed control scheme can guarantee the stability of the closed-loop system and the convergence of the output tracking error.

11.4 RDFFTC of HWDSS subject to actuator and sensor faults

This section extends the general ideas proposed in Section 11.3. A fuzzy FTC approach is proposed for WES with norm-bounded parameter uncertainties, sensor faults and actuator faults. The algorithm utilizes fuzzy systems based on TS fuzzy models to approximate uncertain nonlinear systems with sensor faults and actuator faults. Sufficient stabilization conditions of the fuzzy FTC systems are given, which are formulated in terms of linear matrix equalities (LMEs). The proposed algorithm combines the merits of (i) the states of the closed-loop system will follow those of a user-defined stable reference model despite the presence of bounded magnitude sensor faults, actuator faults and parameter uncertainties; (ii) the algorithm maximizes the power coefficient for a fixed pitch and reduces the voltage ripple.

This section is organized as follows: Section 11.4.1 describes the fuzzy system subject to sensor and actuator faults with parameter uncertainties. The proposed fault tolerant controller and the condition for stability are presented in Section 11.4.2. Section 11.4.3 presents TS fuzzy description for the WES and simulations of sensor and actuator faults and results analysis.

11.4.1 Fuzzy observer scheme for the uncertain system with sensor and actuator faults

The objective is to design a fuzzy observer for nonlinear system with parametric uncertainties, sensor fault and actuator faults. Consider initially the ith rule of the TS fuzzy model with parametric uncertainties, sensor and actuator faults [26,48] is given by

$$\begin{aligned} &\text{Plant Rule } i\text{: If } q_1(x) \text{ is } N_{i1} \text{ and } q_2(x) \text{ is } M_{i1} \\ &\quad \text{Then} \quad \dot{x}(t) = A_i x(t) + B_i u(t) + E_{ai} f_a(t) \\ &\qquad y(t) = C_i x(t) + E_{si} f_s(t), \quad i = 1, \ldots, 4. \end{aligned} \tag{11.61}$$

where $q(x(t)) = [q_1(x(t)), \ldots, q_\varepsilon(x(t))]$ are also measurable variables and do not depend on the faults [48]., i.e., the premise variables, $x(t) \in \kappa^{n\times 1}$ is the state vector, $U(t) \in \kappa^{m\times 1}$ is the control input vector, $y(t) \in \kappa^{g\times 1}$ is the output vector, $A_{fi} \in \kappa^{n\times n}$, $B_{fi} \in \kappa^{n\times m}$ and $C_{fi} \in \kappa^{g\times n}$ are system, input and output matrices, respectively, $\Delta A_{fi} \in \kappa^{n\times n}$ is the parametric uncertainties in the plant model, E_{ai} and E_{si} are actuator and sensor faults matrices, respectively, and $f_a(t)$ and $f_s(t)$ represents actuator and sensor faults, respectively, which is assumed to be bounded. The inferred system is given by

$$\dot{x}(t) = \sum_{i=1}^{p} \mu_i[(A_{fi} + \Delta A_{fi})\, x(t) + B_{fi} U(t) + E_{ai} f_a(t)],$$

$$y(t) = \sum_{i=1}^{p} \mu_i[C_{fi}\, x(t) + E_{si} f_s(t)] \tag{11.62}$$

Consider also the state $Z(t) \in \kappa^{r\times 1}$ that is a filtered version of the output $y(t)$ [49]. This state is given by

$$\dot{Z}(t) = \sum_{i=1}^{p} \mu_i[-A_{zi}\, z(t) + A_{zi} C_{fi} x(t) + A_{zi} E_{si} f_s(t)] \tag{11.63}$$

where $-A_{zi} \in \kappa^{rxr}$ is the stable matrix. From (11.62) and (11.63), one can obtain the augmented system:

$$\dot{X}(t) = \sum_{i=1}^{p} \mu_i[(A_i + \Delta A_i)\, X(t) + B_i U(t) + E_i f(t)],$$

$$Y(t) = \sum_{i=1}^{p} \mu_i C_i\, X(t) \tag{11.64}$$

where $X(t) = \begin{bmatrix} x(t) \\ Z(t) \end{bmatrix}$, $f(t) = \begin{bmatrix} f_a(t) \\ f_s(t) \end{bmatrix}$ $A_i = \begin{bmatrix} A_{fi} & 0 \\ A_{zi} C_{fi} & -A_{zi} \end{bmatrix}$, $\Delta A_i = \begin{bmatrix} \Delta A_{fi} & 0 \\ 0 & 0 \end{bmatrix}$, $B_i = \begin{bmatrix} B_{fi} \\ 0 \end{bmatrix}$, $E_i = \begin{bmatrix} E_{ai} & 0 \\ 0 & A_{zi} E_{si} \end{bmatrix}$, $C_i = [0 \quad I]$.

Based on (11.4), the overall fuzzy observer is represented as follows:

$$\dot{\hat{X}}(t) = \sum_{i=1}^{p} \mu_i[A_i \hat{X}(t) + B_i U(t) + E_i \hat{f}(t) + K_i(Y(t) - \hat{Y}(t))],$$

$$\dot{\hat{f}}(t) = \sum_{i=1}^{p} \mu_i L_i (Y - \hat{Y}) = \sum_{i=1}^{p} \mu_i L_i \tilde{Y},$$

$$\hat{Y}(t) = \sum_{i=1}^{p} \mu_i C_i \hat{X}(t) \tag{11.65}$$

where K_i is the proportional observer gain and L_i is its integral gain for the ith observer rule. $Y(t)$ and $\hat{Y}(t)$ are the final output of the fuzzy system and the fuzzy observer, respectively.

11.4.2 Proposed RDFFTC, reference model and stability analysis

In order to establish the conditions for the asymptotic convergence of the observers (11.65), consider the stable linear model without faults. The proposed FTC scheme in given in Figure 11.17.

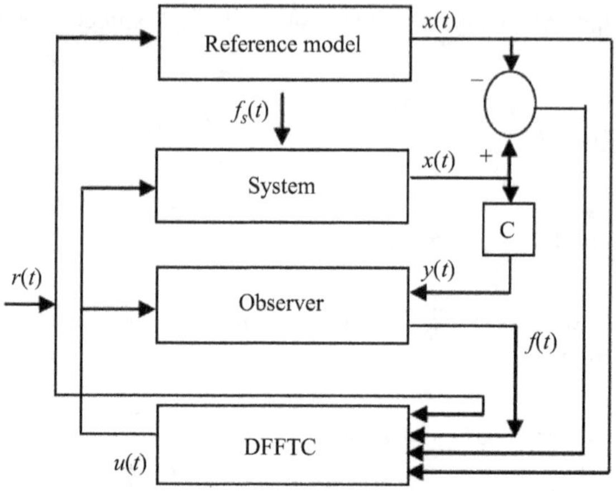

Figure 11.17 Block diagram of RDFFTC scheme

11.4.2.1 Proposed RDFFTC

On the basis of the analysis of [50], we construct the following fuzzy controller; the global output of the fuzzy fault tolerant controller is given by

$$U(t) = \sum_{i=1}^{p} \mu_i u_i(t) \tag{11.66}$$

From (11.63), (11.64) and (11.66), we have

$$\dot{X}(t) = \sum_{i=1}^{p} \mu_i (A_i + \Delta A_i) X(t) + BU(t) + Ef(t)$$

$$Y(t) = \sum_{i=1}^{p} \mu_i C_i X(t) \tag{11.67}$$

Note that B and E are known and defined as the following:

$$B = \sum_{i=1}^{p} \mu_i B_i, \quad E_s = \sum_{i=1}^{p} \mu_i E_{si}$$

11.4.2.2 Stability analyses of the proposed RDFFTC

The analysis procedures are based on [50], so the analysis results will be presented without proof. The main result for the global asymptotic stability of a TS fuzzy model with parameter uncertainties, sensor faults, and actuator faults are summarized by the following lemma and theorem.

Lemma 11.1. *The fuzzy control system as given by* (11.67) *is stable if* B_i *is nonsingular and the control laws of fuzzy controller of* (11.66) *are designed as*

$$u_i(t) = B_i^{-1} \{ [He_1(t) + \bar{A}\,\bar{x}(t) + \bar{B}r(t) \; - A_i\, x(t) - S\hat{f}(t)$$

$$
\left.\left.-\frac{e_1(t)\,\|e_1(t)\|\,\|P_1\|\,\|\Delta A_i\|_{\max}\,\|x(t)\|}{e_1(t)^T P_1 e_1(t)}
\quad \frac{e_1(t)\,\|x(t)\|\,\|P_1\|\,\|D\|_{\max}\,\|x(t)\|}{e_1(t)^T P_1 e_1(t)}
\quad -\frac{e_1(t)\,\|e_1(t)\|\,\|P_1\|\,\|E\|_{\max}\left\|\tilde{f}(t)\right\|}{e_1(t)^T P_1 e_1(t)}\right]\right\} \tag{11.68}
$$

$\|\cdot\|$ denotes the l_2 norm for vectors and l_2 induced norm for matrices, $\|\Delta A_i\| \leq \|\Delta A_i\|_{\max}$, $\|E\| \leq \|E\|_{\max}$, $\|D\| \leq \|D\|_{\max}$, $H \in \kappa^{n\times n}$ is a stable matrix to be designed and choosing S so that S=E and $D = B_{oi}^T B_{oi}$.

Theorem 11.3. *If there exists symmetric and positive definite matrices P_{11}, some matrices K_i and L_i, and matrices Z_{qi} and Q_i, such that the following LMEs are satisfied, then the TS fuzzy system* (11.67) *is asymptotically stabilizable via the TS fuzzy model-based output-feedback controller* (11.66) *and* (11.68).

$$A_i^T P_{11} + P_{11} A_i + (Z_{qi} C_i)^T + (Z_{qi} C_i) = -\sigma I \tag{11.69}$$

$$(Q_i C_i)^T + (Q_i C_i) = -\sigma I \tag{11.70}$$

According to the analysis above, the procedure for finding the proposed fuzzy FTC controller and the FPIEO observer is summarized as follows:

1. Obtain the mathematical model of the wind–diesel storage system (HWDSS) to be controlled.
2. Obtain the fuzzy plant model for the system stated in step (1) by means of a fuzzy modelling method.
3. Check if there exists B^{-1} by determining its rank.
4. Choose a stable reference model.
5. Solve LMEs (11.69) and (11.70) to obtain Q_i, Z_{qi}, P_{11}, K_i and L_i thus ($Z_{qi} = -P_{11}K_i$ and $Q_i = -P_{11}L_i$)
6. Construct fuzzy observer (11.65) according to the theorem and fuzzy controller (11.66) according to Lemma 11.1.

Figure 11.18 shows the robustness index σ for HWDSS based on (11.69) and (11.70).

11.4.3 HWDSS application and simulations and results

In this section, we present the description the hybrid HWDSS and its input and output relationship from the control point of view and TS fuzzy model for HWDSS description, and we study the simulations of sensor and actuator faults and results analysis.

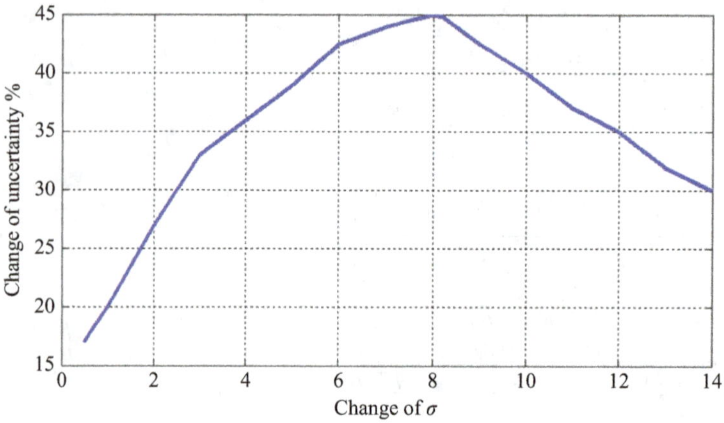

Figure 11.18 Change of uncertainty with σ

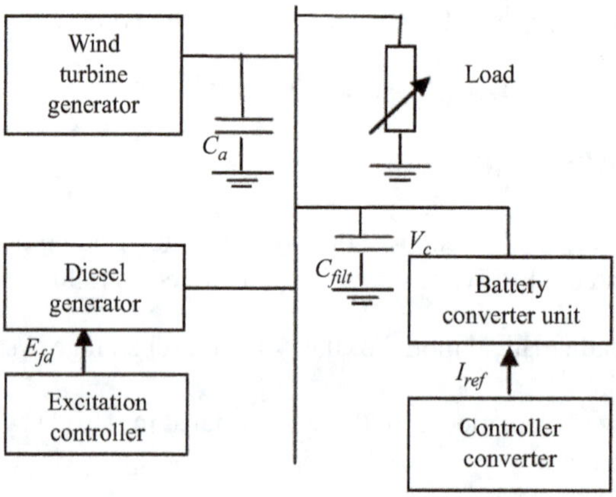

Figure 11.19 Structural diagram of hybrid wind-diesel storage system

11.4.3.1 HWDSS description

The HWDSS consists of a horizontal axis, three-bladed, stall-regulated WT with a rotor of 16.6 m diameter and equipped with an IG rated at 55 kW connected to an ac bus bar in parallel with a diesel-generator set consisting of a 50-kW turbocharged diesel engine (DE) driving a 65-kV-A brushless synchronous generator (SG) and an energy storage system. The two generators are connected to a common ac bus bar. The overall structure of the wind-battery system is shown in Figure 11.19 [50–52].

The dynamics of the nonlinear HWDSS can be characterized by the following equations [50–52]:

$$\dot{x} = A(x)x(t) + B(x)u(t), y = Cx(t) \tag{11.71}$$

Table 11.3 HWDSS parameters

Symbol	**Quantity**	**CGS EMU to SI**
	Rated power	55 [kW]
R	Blade radius	16.6 [m]
ρ	Air density	0.55 [kg/m^3]
	Rated line ac voltage	230 [V]
	AC rated current	138 [A]
	DC rated current	239 [A]
	Power rated of the IG	55 kW
	Power rated of the SG	55 kW
P_{load}	Rated load power	40 [kW]
J_s	The inertia of SG	1.11 [kg m^2]
J_t	The inertia of the IG	1.40 [kg m^2]
D_s	Torsional damping	0.557 [N m/rad]
r_a	Rotor resistance of SG	0.96 [Ω]
L_f	Stator *d-axis* inductance of SG	1.33 [mH]
L_d	Rotor inductance of SG	3.06 [mH]
L_{md}	*d*-Axis field mutual inductance	3 per unit
τ	The transient open circuit time constant	2.16 [ms]
	Nominal system frequency	50 Hz
	Frequency/angular speed	$2\pi 50$ rad/s

where $x(t) = [V_b \quad \omega_s]^T = [x_1(t) \quad x_2(t)]^T$,

$$u(t) = [E_{fd} \quad I_{ref}]^T = [u_1(t) \quad u_2(t)]^T$$

$$A(x) = \begin{bmatrix} 1 & 1 \\ 0 & 1 \end{bmatrix} \begin{bmatrix} \frac{L_f}{\tau L_{md}} q_1(x) & \frac{L_f}{\tau L_{md}} q_1(x)(L_d i_{sd-R_a i_{sq} q_1(x)}) \\ \frac{P_{ind}-P_{load}}{J_s} q_2(x) & -\frac{D_s}{J_s} \end{bmatrix},$$

$$C = \begin{bmatrix} 1 & 0 \\ 0 & 1 \end{bmatrix}$$

where $q_1(x(t)) = 1/\omega_s$ and $q_2(x(t)) = 1/V_b\omega_s$ are the nonlinear terms, V_c is the AC side voltage of the converter, E_{fd} is the SG field voltage, ω_s is the angular speed, which is proportional to frequency f_r, J_s and D_s are the inertia and frictional damping of SG, respectively, i_{sd} and i_{sq} are the direct and quadrature current components of SG, respectively, L_d and L_f are the stator d-axis and rotor inductances of SG, respectively, L_{md} is the d-axis field mutual inductance, τ is the transient open circuit time constant, r_a is the rotor resistance of SG, P_{ind} is the power of the IG, P_{load} is the power of the load, I_{ref} is the direct current set point and V_b is the bus voltage, C_a is the capacitor bank. Equation (11.71) indicates that the matrices $A(x)$ and $B(x)$ are not fixed but change as functions of state variables, thus making the model nonlinear. The selected system parameters for the detailed WT and machines are shown in Table 11.3 [50–52].

11.4.3.2 TS fuzzy HWDSS description

We first represent the system (11.71) by a TS fuzzy representation with the angular speed of SG (ω_s) and the bus voltage (V_b) as the measurable premise variables. Consequently, the HWDSS can be represented by a TS-fuzzy plant model having four rules. The ith rule can be written as follows ($i = 1, 2, 3, 4$):

Rule i: IF $q_1(t)$ is N_{ji} and $q_2(t)$ is M_{ji}

Then

$$\dot{x}(t) = (A_i + \Delta A_i)x(t) + B_i u(t) + E_{ai} f_a(t),$$
$$y(t) = C_i\, x(t) + E_{si} f_s(t) \qquad j = 1, 2;\; i = 1, 2, \ldots, 4 \quad (11.72)$$

Referring to (11.62), the fuzzy plant model given by:

$$\dot{x}(t) = \sum_{i=1}^{4} \mu_i[(A_i + \Delta A_i)\, x(t) + B_i U(t) + E_i f(t)],$$
$$y(t) = \sum_{i=1}^{4} \mu_i[C_i\, x(t) + E_{is} f_s(t)] \quad (11.73)$$

where $x(t)$ and $U(t)$ are the state vectors and the control input, respectively, ΔA_i represents the system parameters uncertainties but bounded; the elements of ΔA_i randomly achieve the values within 30% of their nominal values corresponding to A_i, which represent the change of parameter uncertainties and $\Delta B_i = 0$. The minimum and maximum values of $q_1(t)$ and $q_2(t)$ are given by $q_1(t) \in [q_{1min}\ q_{1max}] \in [1/240\ 1/220]$ and $q_2(t) \in [q_{2min}\ q_{2max}] \in [1/3\ 1/0.1]$, the membership functions ($N_i\ (q_1(t))$, $M_i(q_2(t))$) are related to the uncertain system parameters are shown in Figure 11.20 and given by

$$N_1(q_1(t)) = \frac{q_1(t) - q_{1\min}}{q_{1\max} - q_{1\min}}, \; N_2(q_1(t)) = 1 - N_1(q_1(t)) \quad (11.74)$$

$$M_1(q_2(t)) = \frac{q_2(t) - q_{2\min}}{q_{2\max} - q_{2\min}}, \; M_2(q_2(t)) = 1 - M_1(q_2(t)) \quad (11.75)$$

$$A_1 + \Delta A_1 = \begin{bmatrix} 1 & 1 \\ 0 & 1 \end{bmatrix} \begin{bmatrix} \frac{L_f + \Delta L_f}{\tau L_{md}} q_{1\max} & \frac{L_f + \Delta L_f}{\tau L_{md}} q_{1\max}((L_d + \Delta L_d) i_{sd} - r_a i_{sq} q_{1\max}) \\ \frac{P_{ind} - P_{load}}{J_s} q_{2\min} & -\frac{D_s + \Delta D_s}{J_s} \end{bmatrix},$$

$$A_2 + \Delta A_2 = \begin{bmatrix} 1 & 1 \\ 0 & 1 \end{bmatrix} \begin{bmatrix} \frac{L_f + \Delta L_f}{\tau L_{md}} q_{1\max} & \frac{L_f + \Delta L_f}{\tau L_{md}} q_{1\max}((L_d + \Delta L_d) i_{sd} - r_a i_{sq} q_{1\max}) \\ \frac{P_{ind} - P_{load}}{J_s} q_{2\max} & -\frac{D_s + \Delta D_s}{J_s} \end{bmatrix},$$

$$A_3 + \Delta A_3 = \begin{bmatrix} 1 & 1 \\ 0 & 1 \end{bmatrix} \begin{bmatrix} \frac{L_f + \Delta L_f}{\tau L_{md}} q_{1\min} & \frac{L_f + \Delta L_f}{\tau L_{md}} q_{1\min}((L_d + \Delta L_d) i_{sd} - r_a i_{sq} q_{1\min}) \\ \frac{P_{ind} - P_{load}}{J_s} q_{2\min} & -\frac{D_s + \Delta D_s}{J_s} \end{bmatrix},$$

$$A_4 + \Delta A_4 = \begin{bmatrix} 1 & 1 \\ 0 & 1 \end{bmatrix} \begin{bmatrix} \frac{L_f + \Delta L_f}{\tau L_{md}} q_{1\min} & \frac{L_f + \Delta L_f}{\tau L_{md}} q_{1\min}((L_d + \Delta L_d) i_{sd} - r_a i_{sq} q_{1\min}) \\ \frac{P_{ind} - P_{load}}{J_s} q_{2\max} & -\frac{D_s + \Delta D_s}{J_s} \end{bmatrix},$$

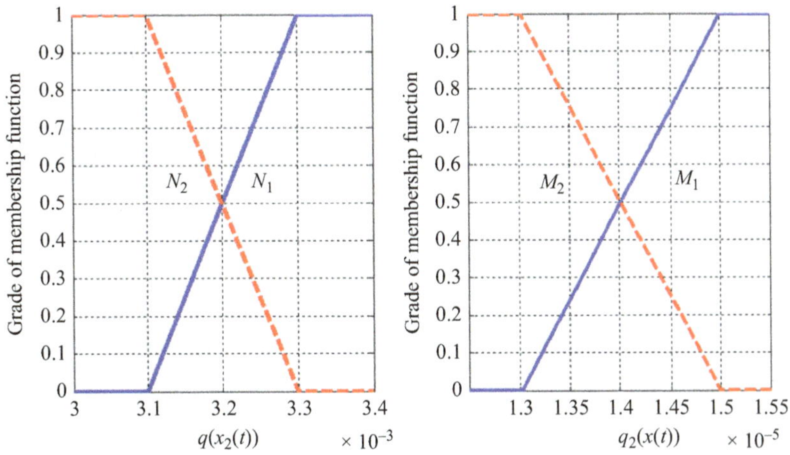

Figure 11.20 Membership functions of state variables

$$B_1 = B_2 = \begin{bmatrix} 1 & -\frac{V_c}{J_s} q_{1\,\max} \\ 0 & -\frac{V_c}{J_s} q_{1\,\max} \end{bmatrix}, \quad B_3 = B_4 = \begin{bmatrix} 1 & -\frac{V_c}{J_s} q_{1\,\min} \\ 0 & -\frac{V_c}{J_s} q_{1\,\min} \end{bmatrix},$$

$$E_{ai} = \begin{bmatrix} 1 & 1 \\ 0.1 & 0.1 \end{bmatrix}, \quad E_{si} = \begin{bmatrix} 10 & 1 \\ 0.1 & 0.01 \end{bmatrix}$$

To define the state Z, one chooses $A_{iz} = 20 \times I$, where I is the identity matrix. Consider $H = \begin{bmatrix} -4 & -4 \\ 0 & -1 \end{bmatrix}$ is a stable matrix. The stable reference model is chosen as follows:

$$\bar{A} = \begin{bmatrix} -4 & -4 \\ 0 & -4 \end{bmatrix}, \quad \bar{B} = \begin{bmatrix} 1 & 1 \\ 0 & 1 \end{bmatrix}, \quad \bar{C} = \begin{bmatrix} 1 & 0 \\ 0 & 1 \end{bmatrix}$$

Solved LMEs (11.69) and (11.70) give the computation of the matrices K_i and L_i. Based on Lemma 11.1, construct the fuzzy fault tolerant controller (11.66) and based on the theorem, construct the fuzzy observer (11.65).

11.4.3.3 Simulation studies

The proposed RDFFTC for the HWDSS (11.71) is tested, and we study the responses for the HWDSS subject to sensor faults, actuator faults and parameter uncertainties. The proposed controller is tested for same random profiles of wind speed signal as shown in Figure 11.21 to prove the effectiveness of the proposed algorithm. The actuator fault $f_a(t) = [f_{a1}\, f_{a2}]^T$ is modeled as follows:

$$f_{a1}(t) = \begin{cases} 0 & t < 13.34\text{ s} \\ 3\sin(\pi t) & t \geq 13.34\text{ s} \end{cases}, \quad f_{a2}(t) = \begin{cases} 0 & t \leq 13.34\text{ s} \\ \sin(\pi t) & t \geq 13.34\text{ s} \end{cases} \tag{11.76}$$

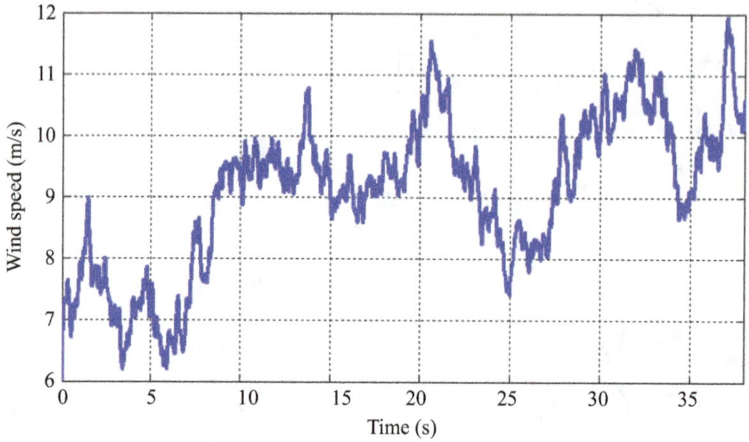

Figure 11.21 Wind speed profile

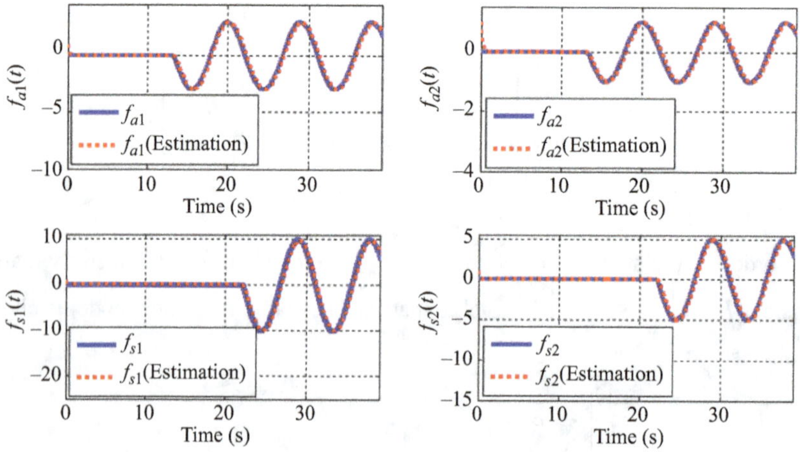

Figure 11.22 Actuator faults [$f_{a1}(t)$ and $f_{a2}(t)$] and their estimations (top) and sensor faults [$f_{s1}(t)$ and $f_{s2}(t)$] and their estimations (bottom)

where $f_{a1}(t)$ is the excitation field voltage actuator fault of the SG and the direct-current set point actuator fault of the converter is $f_{a2}(t)$. The sensor faults $f_s(t) = [f_{s1}\, f_{s2}]^T$ are modeled as follows:

$$f_{s1}(t) = \begin{cases} 0 & t < 22.22\,\text{s} \\ 10\sin(\pi t) & t \geq 22.22\,\text{s} \end{cases}, \quad f_{s2}(t) = \begin{cases} 0 & t < 22.22\,\text{s} \\ 5\sin(\pi t) & t \geq 22.22\,\text{s} \end{cases} \tag{11.77}$$

where f_{s1} is the bus voltage sensor and f_{s2} is the generator speed sensors. Figure 11.22 shows the actuator fault (top) and sensor faults (bottom), and their estimations are based on (11.76) and (11.77), respectively. The frequency and bus voltage of the system (dashed line), observer states (dotted line) and reference states (solid line) are

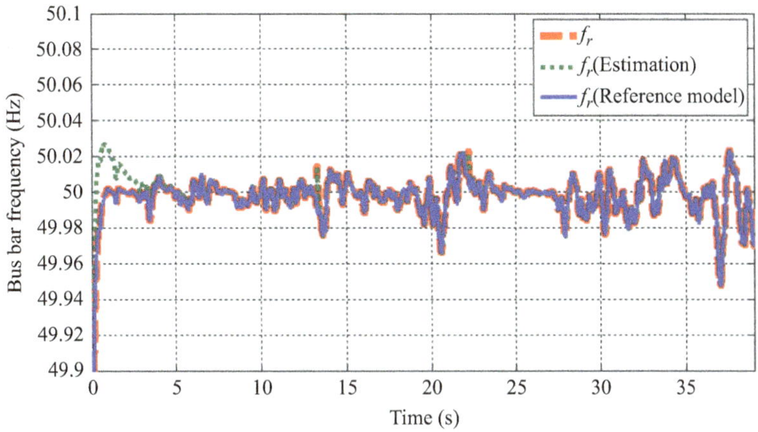

Figure 11.23 The frequency of the system (dashed line), observer (dotted line) and reference model (solid line)

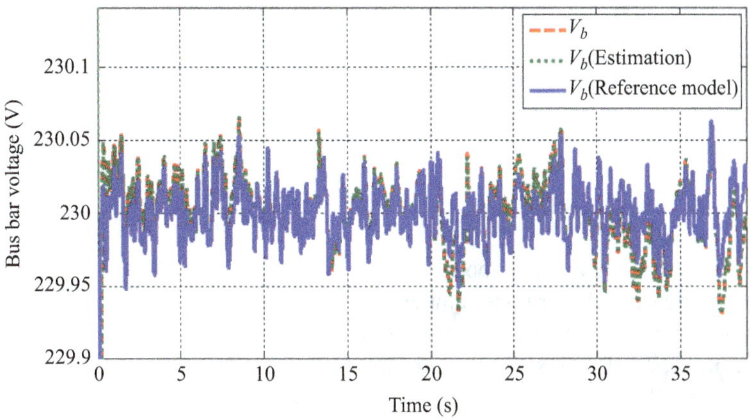

Figure 11.24 The bus voltage of the system (dashed line), observer (dotted line) and reference model (solid line)

shown in Figures 11.23 and 11.24, respectively. When the wind speed is varying, the produced power curve takes almost the wind speed curve as shown in Figure 11.25, but there is only spike when the fault is detected at 13.34 and 22.22 s, it is clear that a 35% increase is obtained in the maximum value compared with [53].

It can be seen from the simulation results there is a good tracking between the states of the nonlinear HWDSS subject to the norm-bounded parametric uncertainties, sensor faults and actuator faults and the reference model. Thus, the TS fuzzy model based controller through fuzzy observer is robust against norm-bounded parametric uncertainties, sensor faults and actuator faults. Comparing the results of the proposed

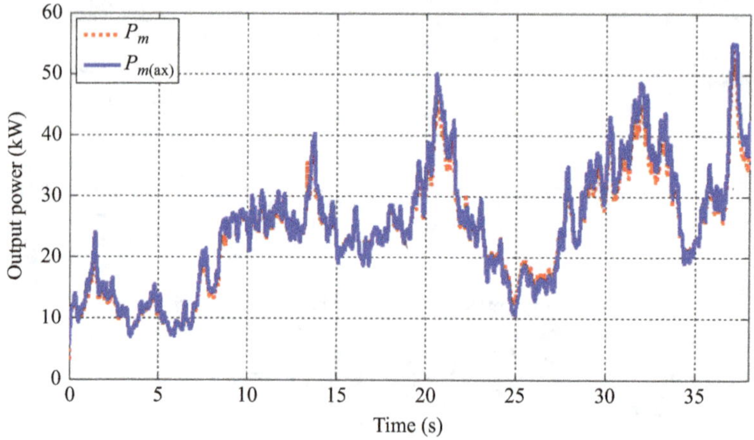

Figure 11.25 Produced power with the proposed control law

algorithm, with that given in the previous algorithms, we can be seen that the proposed controller has the following advantages: (i) It is stable over a wide range of uncertainty up to 30% compared with [53] and when the actuators and sensors become faulty at any time, (ii) the generated power is increased and bus voltage is nearly constant compared with [52].

11.5 Chapter conclusion

An augmented TS fuzzy plant model has been proposed to model the nonlinear plant subject to large parameter uncertainties, sensor faults and actuator faults. Based on this augmented TS fuzzy plant model, three different methods to design the fuzzy FTC have been proposed to tackle this nonlinear system. A design procedure of fuzzy fault tolerant controllers has been developed. The stability and robustness of the fuzzy FTC systems have been investigated based on the results of Chapters 11.2 and 11.3. An application example on stabilizing a WES with sensor faults, actuator faults and parameter uncertainties has been given to illustrate the design procedure and merits of the proposed fuzzy fault tolerant controller.

References

[1] H. O. Wang, K. Tanaka and M. F. Griffin, "An approach to fuzzy control of nonlinear systems: stability and the design issues," *IEEE Trans. Fuzzy Syst.*, vol. 4, no. 1, pp. 14–23, Feb. 1996.

[2] Y. Wang, D. Zhou, S. Joe Qin and H. Wang, "Active fault-tolerant control for a class of nonlinear systems with sensor faults," *Int. J. Control, Autom., Syst.*, vol. 6, no. 3, pp. 339–350, 2008.

[3] H. Wang and K. Tanaka, "An LMI-based stable fuzzy control of nonlinear systems and its application to control of chaos," *Proc. 5th IEEE International Conference on Fuzzy Systems*, pp. 1433–1438, 1996.
[4] D. Ichalal, B. Marx, D. Maquin and J. Ragot, "New fault tolerant control strategy for nonlinear systems with multiple model approach," *Control and Fault Tolerant Systems Conference*, pp. 606–611, 2010.
[5] Y. Zhang, S. Xu, Y. Zou and J. Lu, "Delay-dependent robust stabilization for uncertain discrete-time fuzzy Markovian jump systems with mode-dependent time delays," *Fuzzy Sets Syst.*, vol. 164, no. 1, pp. 66–81, 2011.
[6] B. Zhang, W. X. Zheng and S. Xu, "Passivity analysis and passive control of fuzzy systems with time-varying delays," *Fuzzy Sets Syst.*, vol. 174, no. 1, pp. 83–98, 2011.
[7] S. He and F. Liu, "Filtering-based robust fault detection of fuzzy jump systems," *Fuzzy Sets Syst.*, vol. 185, no. 1, pp. 95–110, 2011.
[8] I. Kar, P. Patchaikani and L. Behera, "Short communication on balancing a cart-pole system using T-S fuzzy model," *Fuzzy Sets Syst.*, vol. 207 , pp. 94–110, 2012.
[9] C. Hu, H. Jiang and Z. Teng, "General impulsive control of chaotic systems based on a TS fuzzy model," *Fuzzy Sets Syst.*, vol. 174, pp. 66–82, 2011.
[10] T. M. Guerra, M. Bernal, K. Guelton and S. Labiod, "Non-quadratic local stabilization for continuous-time Takagi–Sugeno models," *Fuzzy Sets Syst.*, vol. 201, pp. 40–54, 2012.
[11] O. M. Kwon, M. J. Park, S. M. Lee and J. H. Park, "Augmented Lyapunov–Krasovskii functional approaches to robust stability criteria for uncertain Takagi–Sugeno fuzzy systems with time-varying delays," *Fuzzy Sets Syst.*, vol. 201, pp. 1–19, 2012.
[12] Q. Gao, G. Feng, Y. Wang and J. Qiu, "Universal fuzzy controllers based on generalized T-S fuzzy models," *Fuzzy Sets Syst.*, vol. 201, pp. 55–70, 2012.
[13] K. Tanaka, T. Ikeda and H. O. Wang, "Robust stabilization of a class of uncertain nonlinear systems via fuzzy control: quadratic stability, H∞ control theory, and linear matrix inequalities," *IEEE Trans. Fuzzy Syst.*, vol. 4, no. 1, pp. 1–13, 1996.
[14] S. X. Guo, "Robust reliability as a measure of stability of controlled dynamic systems with bounded uncertain parameters," *J. Vib. Control*, vol. 16, no. 9, pp. 1351–1368, 2010.
[15] R. Patton, "Fault-tolerant control systems," *The 1997 situation, 3rd IFAC Symposium on Fault Detection, Supervision and Safety of Technical Processes*, pp. 1033–1054, 1997.
[16] H. Niemann and J. Stoustrup, "Passive fault tolerant control of a double inverted pendulum—a case study," *Control Eng. Pract.*, vol. 13, no. 8, pp. 1047–1059, 2005.
[17] M. M. Mufeed, J. Jiang and Z. Zhang, "Active Fault Tolerant Control Systems: Stochastic Analysis and Synthesis," Springer-Verlag, Berlin/Heidelberg, 2003.

[18] C. Ocampo-Martinez, J. De Dona and M. Seron, "Actuator fault-tolerant control based on set separation," *Int. J. Adapt Control Signal Process.*, vol. 24, no. 12, pp. 1070–1090, 2010.

[19] D. J. Leith and W. E. Leithead, "Survey of gain scheduling analysis design," *Int. J. Control*, vol. 73, no. 11, pp. 1001–1025, 1999.

[20] D. J. Stilwell and W. J. Rugh, "Interpolation of observer state feedback controllers for gain scheduling," *IEEE Trans. Autom. Control*, vol. 44, no. 6, pp. 1225–1229, 1997.

[21] B. Marx, D. Koenig and D. Georges, "A Robust fault tolerant control for descriptor systems," *IEEE Trans. Autom. Control,* vol. 49, no. 10, pp. 1869–1875, 2004.

[22] Y. Zhang and J. Jiang, "Bibliographical review on reconfigurable fault tolerant control systems," *Annu. Rev. Control*, vol. 32, no. 2, pp. 229–252, 2008.

[23] Z. Zuo, D. W.C. Ho and Y. Wang, "Fault tolerant control for singular systems with actuator saturation and nonlinear perturbation," *Automatica,* vol. 46, no. 3, pp. 569–576, 2010.

[24] Y. Shtessel, J. Buffington and S. Banda, "Tailless aircraft flight control using multiple time scale reconfigurable sliding modes," *IEEE Trans. Control Syst. Tech.*, vol. 10, no. 2, pp. 288–296, 2002.

[25] X. Yao, S. Liu, G. Shan, Z. Xing, C. Guo and C. Yi, "LQG controller for a variable speed pitch regulated wind turbine," *International Conference on Intelligent Human–Machine Systems and Cybernetics*, pp. 210–213, 2009.

[26] S. Tong and L. Han-Hiong, "Observer-based robust fuzzy control of nonlinear systems with parametric uncertainties," *Fuzzy Sets Syst.*, vol. 131, pp. 165–184, 2002.

[27] A. G. Abo-Khalil and L. Dong-Choon, "MPPT control of wind generation systems based on estimated wind speed using SVR," *IEEE Trans. Ind. Electron.*, vol. 55, no. 3, pp. 1489–1490, 2008.

[28] V. Aggarwal, R. K. Patidar and P. Patki, "A novel scheme for rapid tracking of maximum power point in wind energy generation systems," *IEEE Trans. Energy Convers.*, vol. 25, no. 1, pp. 228–236, 2010.

[29] H. Camblong, I. Martinez de Alegria, M. Rodriguez and G. Abad, "Experimental evaluation of wind turbines maximum power point tracking controllers," *Energy Convers. Manage.*, vol. 47, no. 18–19, pp. 2846–2858, 2006.

[30] R. Datta and V. T. Ranganathan, "A method of tracking the peak power points for a variable speed wind energy conversion system," *IEEE Trans. Energy Convers.*, vol. 18 no. 1, pp. 163–168, 2003.

[31] L. Whei-Min and H. Chih-Ming, "Intelligent approach to maximum power point tracking control strategy for variable-speed wind turbine generation system," *Energy*, vol. 35 no. 6, pp. 2440–2447, 2010.

[32] V. Galdi, A. Piccolo and P. Siano, "Exploiting maximum energy from variable speed wind power generation system by using an adaptive Takagi–Sugeno–Kang fuzzy model," *Energy Convers. Manage.*, vol. 50, pp. 413–420, 2009.

[33] S. Masoud Barakati, M. Kazerani and J. Dwight Aplevich, "Maximum power tracking control for a wind turbine system including a matrix converter," *IEEE Trans. Energy Convers.*, vol. 24, no. 3, pp. 705–713, 2009.

[34] E. Koutroulis and K. Kalaitzakis, "Design of a maximum power tracking system for wind-energy-conversion applications," *IEEE Trans. Ind. Electron.*, vol. 53, no. 2, pp. 486–494, 2006.

[35] A. Z. Mohamed, M. N. Eskander and F. A. Ghali, "Fuzzy logic control based maximum power tracking of a wind energy system," *Renewable Energy*, vol. 23, no. 2, pp. 235–245, 2001.

[36] K. Tanaka and M. Sugeno, "Stability analysis and design of fuzzy control systems," *Fuzzy Sets Syst.*, vol. 45, no. 2, pp. 135–156, 1992.

[37] W. Chen and M. Saif, "Design of a TS based fuzzy nonlinear unknown input observer with fault diagnosis applications," *American Control Conference*, pp. 2545–2550, 2007.

[38] S. A. Khan and M. I. Hossain, "Intelligent control based maximum power extraction strategy for wind energy conversion system," *24th Canadian Conference on Electrical and Computer Engineering (CCECE)*, pp. 1040–1043, 2011.

[39] E. Kamal, M. Koutb, A. A. Sobaih and B. Abozalam, "An intelligent maximum power extraction algorithm for hybrid wind–diesel-storage system," *Int. J. Electr. Power Energy Syst.*, vol. 32, pp. 170–177, 2010.

[40] F. Poitiers, T. Bouaouiche and M. Machmoum, "Advanced control of a doubly-fed induction generator for wind energy conversion," *Electr. Power Syst. Res.*, vol. 79, no. 7, pp. 1085–1096, 2009.

[41] R. Fadaeinedjad, M. Moallem and G. Moschopoulos, "Simulation of a wind turbine with doubly fed induction generator by FAST and Simulink," *IEEE Trans. Energy Convers.*, vol. 23, no. 2, pp. 690–700, 2008.

[42] M. Gálvez-Carrillo and M. Kinnaert, "Sensor fault detection and isolation in three-phase systems using a signal-based approach," *IET Control Theory Appl.*, vol. 4, no. 9, pp. 1838–1848, 2010.

[43] M. Gálvez-Carrillo and M. Kinnaert, "Sensor fault detection and isolation in doubly-fed induction generators accounting for parameter variations," *Renewable Energy*, vol. 36, no. 5, pp. 1447–1457, 2011.

[44] Q. Wang and L. Chang, "An intelligent maximum power extraction algorithm for inverter-based variable speed wind turbine systems," *IEEE Trans. Power Electron.*, vol. 19, no. 5, pp. 1242–1249, 2004.

[45] B. Boukhezzar and H. Siguerdidjane, "Nonlinear control of variable speed wind turbines for power regulation," *Proceedings of the 2005 IEEE Conference on Control Applications*, pp. 114–119, 2005.

[46] B. Boukhezzar and H. Siguerdidjane, "Comparison between linear and nonlinear control strategies for variable speed wind turbines," *Control Eng. Pract.*, vol. 18, pp. 1357–1368, 2010.

[47] E. Kamal, A. Aitouche and D. Abbes, "Robust fuzzy scheduler fault tolerant control of wind energy systems subject to sensor and actuator faults," *Int. J. Electr. Power Energy Syst.*, vol. 55, no. 1, pp. 402–419, 2014.

[48] J. Chen, R. Patton and H. Zhang, "Design of unknown input observers and robust fault detection filters," *Int. J. Control*, vol. 63, pp. 85–105, 1996.
[49] C. Edwards, "A comparison of sliding mode and unknown input observers for fault reconstruction," *Eur. J. Control*, vol. 12, no. 3, pp. 245–260, 2006.
[50] E. Kamal and A. Aitouche, "Fuzzy fault-tolerant control of wind–diesel hybrid systems subject to sensor faults," *IEEE Trans. Sustainable Energy*, vol. 4, no. 4, pp. 857–866, Oct. 2013.
[51] K. Hs, T. Niimura, J. Jatskevich, K. Hc and L. Ky, "Power quality control of hybrid wind power generation with battery storage using Fuzzy-LQR controller," *IEEE Power Engineering Society Meeting,* pp. 1722–1727, 2004.
[52] K. Hee-Sang, K. Y. Lee, M. Kang and H. Kim, "Power quality control of an autonomous wind–diesel power system based on hybrid intelligent controller," *Neural Networks*, vol. 21, pp. 1439–1446, 2008.
[53] R. B. Chedid, S. H. Karaki and C. El-Chamali, "Adaptive fuzzy control for wind diesel weak power systems," *IEEE Trans. Energy Convers.*, vol. 15, no. 1, pp. 71–78, 2000.

Index

www.ingramcontent.com/pod-product-compliance
Lightning Source LLC
LaVergne TN
LVHW011848200926
848484LV00002B/417